Linear Algebra

Linear Algebra is intended primarily as an undergraduate textbook but is written in such a way that it can also be a valuable resource for independent learning. The narrative of the book takes a matrix approach: the exposition is intertwined with matrices either as the main subject or as tools to explore the theory. Each chapter contains a description of its aims, a summary at the end of the chapter, exercises, and solutions. The reader is carefully guided through the theory and techniques presented which are outlined throughout in 'How to...' text boxes. Common mistakes and pitfalls are also pointed out as one goes along.

Features

- Written to be essentially self-contained
- Ideal as a primary textbook for an undergraduate course in linear algebra
- Applications of the general theory which are of interest to disciplines outside of mathematics, such as engineering

Linear Algebra

Lina Oliveira

CRC Press
Taylor & Francis Group
Boca Raton London New York

CRC Press is an imprint of the
Taylor & Francis Group, an **informa** business
A CHAPMAN & HALL BOOK

First edition published 2022
by CRC Press
6000 Broken Sound Parkway NW, Suite 300, Boca Raton, FL 33487-2742

and by CRC Press
4 Park Square, Milton Park, Abingdon, Oxon, OX14 4RN

CRC Press is an imprint of Taylor & Francis Group, LLC

Library of Congress Cataloging-in-Publication Data

Names: Oliveira, Lina, author.
Title: Linear algebra / Lina Oliveira.
Description: First edition. | Boca Raton : Chapman & Hall/CRC Press, 2022.
| Includes bibliographical references and index.
Identifiers: LCCN 2021061942 (print) | LCCN 2021061943 (ebook) | ISBN 9781032287812 (hardback) | ISBN 9780815373315 (paperback) | ISBN 9781351243452 (ebook)
Subjects: LCSH: Algebras, Linear.
Classification: LCC QA184.2 .O43 2022 (print) | LCC QA184.2 (ebook) | DDC 512/.5--dc23/eng20220415
LC record available at https://lccn.loc.gov/2021061942
LC ebook record available at https://lccn.loc.gov/2021061943

ISBN: 9781032287812 (hbk)
ISBN: 9780815373315 (pbk)
ISBN: 9781351243452 (ebk)

DOI: 10.1201/9781351243452

Typeset in CMR10 font
by KnowledgeWorks Global Ltd.

Publisher's note: This book has been prepared from camera-ready copy provided by the authors.

To my daughters.

Contents

Preface **xi**

Symbol Description **xiii**

Biography **xv**

1 Matrices **1**

1.1 Real and Complex Matrices 1
1.2 Matrix Calculus 17
1.3 Matrix Inverses 32
1.4 Elementary Matrices 37
1.4.1 LU and LDU factorisations 43
1.5 Exercises 48
1.6 At a Glance 53

2 Determinant **55**

2.1 Axiomatic Definition 55
2.2 Leibniz's Formula 66
2.3 Laplace's Formula 70
2.4 Exercises 76
2.5 At a Glance 79

3 Vector Spaces **81**

3.1 Vector Spaces 82
3.2 Linear Independence 87
3.3 Bases and Dimension 93
3.3.1 Matrix spaces and spaces of polynomials 99
3.3.2 Existence and construction of bases 101
3.4 Null Space, Row Space, and Column Space 107
3.4.1 $A\mathbf{x} = \mathbf{b}$ 115
3.5 Sum and Intersection of Subspaces 116
3.6 Change of Basis 119
3.7 Exercises 123
3.8 At a Glance 130

4 Eigenvalues and Eigenvectors **131**
4.1 Spectrum of a Matrix 131
4.2 Spectral Properties 134
4.3 Similarity and Diagonalisation 139
4.4 Jordan Canonical Form 149
4.4.1 Nilpotent matrices 149
4.4.2 Generalised eigenvectors 156
4.4.3 Jordan canonical form 160
4.5 Exercises 171
4.6 At a Glance 173

5 Linear Transformations **175**
5.1 Linear Transformations 176
5.2 Matrix Representations 179
5.3 Null Space and Image 185
5.3.1 Linear transformations $T : \mathbb{K}^n \to \mathbb{K}^k$ 185
5.3.2 Linear transformations $T : U \to V$ 187
5.4 Isomorphisms and Rank-nullity Theorem 189
5.5 Composition and Invertibility 191
5.6 Change of Basis 195
5.7 Spectrum and Diagonalisation 198
5.8 Exercises 201
5.9 At a Glance 203

6 Inner Product Spaces **205**
6.1 Real Inner Product Spaces 205
6.2 Complex Inner Product Spaces 214
6.3 Orthogonal Sets 218
6.3.1 Orthogonal complement 220
6.3.2 Orthogonal projections 228
6.3.3 Gram–Schmidt process 235
6.4 Orthogonal and Unitary Diagonalisation 238
6.5 Singular Value Decomposition 245
6.6 Affine Subspaces of $\mathbb{R}^n$ 249
6.7 Exercises 253
6.8 At a Glance 256

7 Special Matrices by Example **257**
7.1 Least Squares Solutions 257
7.2 Markov Chains 260
7.2.1 Google matrix and PageRank 265
7.3 Population Dynamics 266
7.4 Graphs 271
7.5 Differential Equations 275
7.6 Exercises 279

7.7 At a Glance 282

8 Appendix **285**
8.1 Uniqueness of Reduced Row Echelon Form 285
8.2 Uniqueness of Determinant 286
8.3 Direct Sum of Subspaces 287

9 Solutions **289**
9.1 Solutions to Chapter 1 289
9.2 Solutions to Chapter 2 294
9.3 Solutions to Chapter 3 294
9.4 Solutions to Chapter 4 299
9.5 Solutions to Chapter 5 300
9.6 Solutions to Chapter 6 301
9.7 Solutions to Chapter 7 303

Bibliography **307**

Index **309**

Preface

This is a first course in Linear Algebra. It grew out of several courses given at Instituto Superior Técnico, University of Lisbon, and is naturally shaped by the experience of teaching. But, perhaps more importantly, it also imparts the feedback of generations of students: the way they felt challenged and the challenges they set in turn.

This is a book on Linear Algebra which could also be easily described as a book about matrices. The opening chapter defines what a matrix is, establishes the basics, while the following chapters build on it to develop the theory and the closing chapter showcases special types of matrices present in concrete applications – for example, life sciences, statistics, or the internet.

The book aims at conciseness and simplicity, for it is intended as an undergraduate textbook which also allows for self-learning. In this trait, it summarises the theory throughout in 'How to...' text boxes and makes notes of common mistakes and pitfalls. Every aspect of the content is assessed in exercise and solution sections.

The narrative of the book is intertwined with matrices either as the main subject or tools to explore the theory. There is not a chapter where they are not present, be it in the forefront or the background. As it happens, each of the first five chapters is anchored on a particular number or numbers, in this order: the rank of a matrix, its determinant, the dimension of a vector space, the eigenvalues of a matrix, and the dimensions of the null space and the image of a linear transformation. The sixth chapter is about real and complex inner product spaces. Chapters 7 and 9 are applications of the theory and solutions to the exercises, respectively, whilst Chapter 8 is an appendix consisting of the proofs of some results relegated to a later stage, as not to impair the flow of the exposition.

Notwithstanding the simplicity goal of the presentation voiced above, the book ventures a few times into more advanced topics that, given the mostly self-contained nature of the book, call for more involved proofs. However, it is the reader's choice to avoid these topics at first (or even definitely), if so wished. This will not be an impediment to the understanding of the fundamentals of Linear Algebra.

At the end of each chapter, the respective contents are briefly highlighted in the very synthetic 'At a Glance' sections, which will mostly make sense for those who read the theory, solved the corresponding exercises, and, in the end, learned it well enough as to understand the meaningful *droplets* these sections consist of.

Finally, I wish to express my gratitude to all my students, without whom this particular book could not be.

Lisboa, December 2021

Symbol Description

$\bar{A}$	The conjugate of a matrix A
$\|A\|$	The determinant of A
$\mathrm{adj}\, A$	The adjugate matrix of A
$[A_{ij}]$	The submatrix of A obtained by deleting row i and column j
$\mathcal{B}_c$	The standard basis of $\mathrm{M}_n(\mathbb{K})$
$\mathbb{C}$	The complex numbers
cof A	The matrix of cofactors of A
$C(\mathbf{x}, \lambda)$	The set of vectors in a Jordan chain corresponding to the eigenvalue λ
$\det A$	The determinant of A
$D_i(\alpha)$	Elementary matrix
$E_{ij}(\alpha)$	Elementary matrix
$E(\lambda)$	The eigenspace corresponding to the eigenvalue λ
$\mathcal{E}_n$	The standard basis of $\mathbb{K}^n$
$G(\lambda)$	The generalised eigenspace corresponding to the eigenvalue λ
$I(T)$	The image of the linear transformation T
$\mathbb{K}$	The field of scalars
$M_{\mathcal{B}_2 \leftarrow \mathcal{B}_1}$	The change of basis matrix from basis $\mathcal{B}_1$ to basis $\mathcal{B}_2$
$\mathrm{M_{k,n}}(\mathbb{C})$	The set of $k \times n$ matrices over $\mathbb{C}$
$\mathrm{M_{k,n}}(\mathbb{K})$	The set of $k \times n$ matrices over the field $\mathbb{K}$
$\mathrm{M_{k,n}}(\mathbb{R})$	The set of $k \times n$ matrices over $\mathbb{R}$
$\mathrm{M_n}(\mathbb{C})$	The set of square complex matrices of order n
$\mathrm{M_n}(\mathbb{K})$	The set of square matrices of order n over the field $\mathbb{K}$
$\mathrm{M_n}(\mathbb{R})$	The set of square real matrices of order n
$N(T)$	The null space of the linear transformation T
$\mathrm{nul}(A)$	The nullity of matrix A
$\mathrm{nul}(T)$	The dimension of the null space of the linear transformation T
P_{ij}	Elementary matrix
$\mathbb{P}_n$	The space of real polynomials of degree less than or equal to n
$\mathbb{R}$	The real numbers
$\mathrm{rank}\,(A)$	The rank of matrix A
$\mathrm{rank}\,(T)$	The rank of the linear transformation T
$\sigma(T)$	The spectrum of the linear transformation T
$\mathrm{span}\, X$	The span of the set X
$U + V$	The sum of subspaces U and V
$U \oplus V$	The direct sum of subspaces U and V

Biography

Lina Oliveira has a DPhil in Mathematics from the University of Oxford and is a faculty member of the Department of Mathematics of Instituto Superior Técnico, University of Lisbon. She has taught several graduate and undergraduate courses at Instituto Superior Técnico where she regularly teaches Linear Algebra. Her research interests are in the areas of Functional Analysis, Operator Algebras, and Operator Theory. As of late, she has been responsible for the Linear Algebra course at the Portuguese Air Force Academy.

Chapter 1

Matrices

1.1	Real and Complex Matrices	1
1.2	Matrix Calculus	17
1.3	Matrix Inverses	32
1.4	Elementary Matrices	37
	1.4.1 LU and LDU factorisations	43
1.5	Exercises	48
1.6	At a Glance	53

This first chapter is about matrices, the link binding all the subjects presented in this book. After defining what a matrix is, we will move on into its properties by developing a toolkit to work with, namely, Gaussian and Gauss–Jordan eliminations, elementary matrices, and matrix calculus.

As said in the preface, almost every chapter has a particularly outstanding number associated. In this chapter, this number is the rank of a matrix: it will be used to classify systems of linear equations and decide if a matrix has an inverse, for example, and, mostly, the whole chapter will revolve around it.

1.1 Real and Complex Matrices

In what follows, $\mathbb{K} = \mathbb{R}, \mathbb{C}$, and the elements of $\mathbb{K}$ are called numbers or scalars.

A $k \times n$ **matrix** or a **matrix of size** $k \times n$ over $\mathbb{K}$ is an array

$$\begin{bmatrix} a_{11} & a_{12} & \dots & a_{1j} & \dots & a_{1n} \\ a_{21} & a_{22} & \dots & a_{2j} & \dots & a_{2n} \\ \vdots & \vdots & & \vdots & & \vdots \\ a_{i1} & a_{i2} & \dots & a_{ij} & \dots & a_{1n} \\ \vdots & \vdots & & \vdots & & \vdots \\ a_{k1} & a_{k2} & \dots & a_{kj} & \dots & a_{kn} \end{bmatrix} \tag{1.1}$$

of scalars in $\mathbb{K}$ having k **rows** and n **columns**. Each number a_{ij}, for all indices $i = 1, \dots, k$ and $j = 1, \dots, n$, is called an **entry** of the matrix. The indices

DOI: 10.1201/9781351243452-1

i and j correspond, respectively, to the number of the row and the number of the column where the **entry-*ij***, i.e., the scalar a_{ij} is located.

The rows of the matrix are numbered from 1 to k, starting from the top, and the columns of the matrix are numbered from 1 to n, starting from the left.

A matrix whose entries are real numbers is called a **real matrix** or a **matrix over** $\mathbb{R}$, and a matrix whose entries are complex numbers is called a **complex matrix** or a **matrix over** $\mathbb{C}$.

Example 1.1 *The entry-23 of*

$$\begin{bmatrix} 1 & -2 & 5 & 1 \\ 2 & -1 & 7 & 3 \end{bmatrix}$$

is the scalar which is located in row 2 and column 3, that is, $a_{23} = 7$. This matrix has two rows and four columns and therefore is a 2×4 matrix.

The sets of $k \times n$ matrices over $\mathbb{R}, \mathbb{C}$, and $\mathbb{K}$ are denoted, respectively, by $\mathrm{M}_{\mathrm{k,n}}(\mathbb{R})$, $\mathrm{M}_{\mathrm{k,n}}(\mathbb{C})$, and $\mathrm{M}_{\mathrm{k,n}}(\mathbb{K})$. When $k = n$, the notation is simplified to $\mathrm{M}_{\mathrm{n}}(\mathbb{R})$, $\mathrm{M}_{\mathrm{n}}(\mathbb{C})$, and $\mathrm{M}_{\mathrm{n}}(\mathbb{K})$, respectively.

In the matrix (1.1) above, the $i\,th$-row is

$$\mathbf{l}_i = \begin{bmatrix} a_{i1} & a_{i2} & \dots & a_{ij} & \dots & a_{1n} \end{bmatrix},$$

and the $j\,th$-column is

$$\mathbf{c}_j = \begin{bmatrix} a_{1j} \\ a_{2j} \\ \vdots \\ a_{ji} \\ \vdots \\ a_{kj} \end{bmatrix}.$$

Example 1.2 *In Example 1.1, the first row of the matrix is*

$$\mathbf{l}_1 = \begin{bmatrix} 1 & -2 & 5 & 1 \end{bmatrix},$$

and the second row is

$$\mathbf{l}_2 = \begin{bmatrix} 2 & -1 & 7 & 3 \end{bmatrix}.$$

Columns 1, 2, 3, and 4 of this matrix are, respectively,

$$\mathbf{c}_1 = \begin{bmatrix} 1 \\ 2 \end{bmatrix}, \quad \mathbf{c}_2 = \begin{bmatrix} -2 \\ -1 \end{bmatrix}, \quad \mathbf{c}_3 = \begin{bmatrix} 5 \\ 7 \end{bmatrix}, \quad \mathbf{c}_4 = \begin{bmatrix} 1 \\ 3 \end{bmatrix}.$$

The matrix (1.1) can also be presented as $[a_{ij}]_{\substack{i=1,\dots,k\\ j=1,\dots,n}}$, or simply as $[a_{ij}]$, whenever the matrix size is clear from the context.

The matrix is said to be

(a) a **rectangular matrix** if $k \neq n$;

(b) a **square matrix** if $k = n$, and in this case, the matrix is called a square matrix of **order** n (or k);

(c) a **column matrix** or **column vector** if $n = 1$;

(d) a **row matrix** or **row vector** if $k = 1$.

For reasons that will become apparent in Chapter 3, in most cases, column vectors will be referred to as vectors.

In what follows, a row of a matrix whose entries consist only of zeros will be called a **zero row**. A **zero column** is defined similarly.

Definition 1 *A matrix $A = [a_{ij}]$ is in* ***row echelon form*** *or is a* ***row echelon matrix*** *if the following two conditions hold.*

(i) There are no zero rows above non-zero rows.

(ii) If $\mathbf{l}_i$ and $\mathbf{l}_{i+1}$ are non-zero consecutive rows of A, then the first non-zero entry of row $\mathbf{l}_{i+1}$ is situated (in a column) to the right of (the column of) the first non-zero entry of $\mathbf{l}_i$.

The first non-zero entry in each row of a row echelon matrix is called a **pivot**.

Example 1.3 *Consider the matrices*

$$A = \begin{bmatrix} 1 & -2 & 3 & 0 & 9 \\ 0 & 0 & 4 & 0 & -2 \\ 0 & 0 & 0 & 6 & -7 \end{bmatrix}, \quad B = \begin{bmatrix} 0 & 1 & 0 & 4 & 5 \\ 0 & 0 & 0 & 8 & 2 \\ 0 & 0 & 0 & 7 & -1 \\ 0 & 0 & 0 & 0 & 0 \end{bmatrix}, \quad C = \begin{bmatrix} 1 & 0 & 0 & 4 \\ 0 & 0 & 0 & 0 \\ 0 & 1 & 0 & 8 \\ 0 & 0 & 1 & 1 \end{bmatrix}$$

Matrix A is in row echelon form and its pivots are $1, 4,$ and 6. The matrices B and C are not row echelon matrices because matrix B does not satisfy condition (ii) and matrix C does not satisfy condition (i) of Definition 1.

In the next section, we shall make use of matrices to solve systems of linear equations. A key tool to be used in the process is the concept of an elementary operation performed on the rows of a matrix, thereby obtaining a new matrix.

Definition 2 *There exist three kinds of* ***elementary row operations:***

(i) exchanging two rows, i.e., exchanging row $\mathbf{l}_i$ with row $\mathbf{l}_j$, with $i \neq j$;

(ii) replacing row $\mathbf{l}_i$ *by* $\alpha\mathbf{l}_i$*, where* α *is a non-zero scalar;*

(iii) replacing row $\mathbf{l}_i$ *by* $\mathbf{l}_i + \alpha\mathbf{l}_j$*, where* α *is a scalar and* $i \neq j$*;*

The scalars α in this definition lie in the same field as the entries in the matrix.

For simplicity, in what follows elementary row operations may also be called **elementary operations**.

Example 1.4 *The three types of elementary operations will be illustrated with the real matrix*

$$A = \begin{bmatrix} 3 & 1 & -1 & 2 \\ 1 & -1 & 2 & 2 \\ 1 & 1 & 1 & 3 \end{bmatrix}.$$

The first elementary operation to be performed is of type (i). We exchange rows 1 and 3 of matrix A, obtaining matrix B:

$$A = \begin{bmatrix} 1 & 1 & 1 & 3 \\ 1 & -1 & 2 & 2 \\ 3 & 1 & -1 & 2 \end{bmatrix} \xrightarrow[\mathbf{l}_1 \leftrightarrow \mathbf{l}_3]{} \begin{bmatrix} 3 & 1 & -1 & 2 \\ 1 & -1 & 2 & 2 \\ 1 & 1 & 1 & 3 \end{bmatrix} = B.$$

To illustrate an elementary operation of type (ii), now we replace the second row $\mathbf{l_2}$ *of matrix A by* $-3\mathbf{l_2}$*, obtaining matrix C:*

$$A = \begin{bmatrix} 1 & 1 & 1 & 3 \\ 1 & -1 & 2 & 2 \\ 3 & 1 & -1 & 2 \end{bmatrix} \xrightarrow[-3\mathbf{l}_2]{} \begin{bmatrix} 3 & 1 & -1 & 2 \\ -3 & 3 & -6 & -6 \\ 1 & 1 & 1 & 3 \end{bmatrix} = C.$$

Finally, row $\mathbf{l}_3$ *will be replaced by the new row* $\mathbf{l}_3 + 2\mathbf{l}_1$*, i.e.,*

$$A = \begin{bmatrix} 1 & 1 & 1 & 3 \\ 1 & -1 & 2 & 2 \\ 3 & 1 & -1 & 2 \end{bmatrix} \xrightarrow[\mathbf{l}_3 + 2\mathbf{l}_1]{} \begin{bmatrix} 1 & 1 & 1 & 3 \\ 1 & -1 & 2 & 2 \\ 5 & 3 & 1 & 8 \end{bmatrix} = M,$$

i.e., $\mathbf{l}_3$ *of M is*

$$\mathbf{l}_3 = \begin{bmatrix} a_{31} + 2a_{11} & a_{32} + 2a_{12} & a_{33} + 2a_{13} & a_{34} + 2a_{14} \end{bmatrix}.$$

Hence we have obtained three new matrices, B, C*, and* M*, by acting on the rows of A with the operations described above. Symbolically,*

$$A \xrightarrow[\mathbf{l}_1 \leftrightarrow \mathbf{l}_3]{} B, \qquad A \xrightarrow[-3\mathbf{l}_2]{} C, \qquad A \xrightarrow[\mathbf{l}_3 + 2\mathbf{l}_1]{} M.$$

A **linear equation** is an equation which can be presented in the form

$$a_1x_1 + a_2x_2 + \cdots + a_{n-1}x_{n-1} + a_nx_n = b, \tag{1.2}$$

where $a_1, a_2, \ldots, a_{n-1}, a_n, b$ are scalars and $x_1, x_2, \ldots, x_{n-1}, x_n$ are the **variables** or the **unknowns**. A **system of linear equations** is a conjunction of linear equations

$$\begin{cases} a_{11}x_1 + a_{12}x_2 + \cdots + a_{1n}x_n = b_1 \\ a_{21}x_1 + a_{22}x_2 + \cdots + a_{2n}x_n = b_2 \\ \vdots \\ a_{k1}x_1 + a_{k2}x_2 + \cdots + a_{kn}x_n = b_k \end{cases}.$$

Equation (1.2) is said to be a **homogeneous equation** if $b = 0$. Similarly, the system above is called a **homogeneous system** if $b_1 = b_2 = \cdots = b_k = 0$. Equations and systems are otherwise called **non-homogeneous**.

Solving a system of linear equations consists in determining the set of all n-tuples $(x_1, x_2, \ldots, x_n)$ of n scalars which satisfy all the equations in the system. This set is called the **solution set** or the **general solution** of the system.

The set consisting of all n-tuples $(x_1, x_2, \ldots, x_n)$ of real numbers (respectively, of complex numbers) will be denoted by $\mathbb{R}^n$ (respectively, $\mathbb{C}^n$). That is, for a positive integer n,

$$\mathbb{R}^n = \{(x_1, x_2, \ldots, x_n) \colon x_1, x_2, \ldots, x_n \in \mathbb{R}\},$$

$$\mathbb{C}^n = \{(x_1, x_2, \ldots, x_n) \colon x_1, x_2, \ldots, x_n \in \mathbb{C}\}.$$

Two systems of linear equations are said to be **equivalent** if they have the same solution set. Systems are classified according to the nature of their solution set. A system of linear equations is said to be:

(a) **consistent**, if its solution set is non-empty;

(b) **inconsistent**, if its solution set is the empty set, i.e., the system is not solvable.

We shall see in Proposition 1.3 that in case (a) the system either has a unique solution or has infinitely many solutions.

A homogeneous system of linear equations is always consistent.

In fact, such a system possesses, at least, the **trivial solution**, that is, the solution where all the variables take the value 0.

The system of linear equations above is associated with the following matrices

$$A = \begin{bmatrix} a_{11} & a_{12} & \dots & a_{1n} \\ a_{21} & a_{22} & \dots & a_{2n} \\ \vdots & \vdots & & \vdots \\ a_{k1} & a_{k22} & \dots & a_{kn} \end{bmatrix} \qquad \textbf{Coefficient matrix,} \tag{1.3}$$

$$\mathbf{b} = \begin{bmatrix} b_1 \\ b_2 \\ \vdots \\ b_k \end{bmatrix} \qquad \textbf{Column vector of independent terms,} \tag{1.4}$$

$$[A|\mathbf{b}] = \left[\begin{array}{cccc|c} a_{11} & a_{12} & \dots & a_{1n} & b_1 \\ a_{21} & a_{22} & \dots & a_{2n} & b_2 \\ \vdots & \vdots & & \vdots & \vdots \\ a_{k1} & a_{k22} & \dots & a_{kn} & b_k \end{array}\right] \qquad \textbf{Augmented matrix.} \tag{1.5}$$

The augmented matrix may also be presented without the vertical separation line, i.e.,

$$[A|\mathbf{b}] = \begin{bmatrix} a_{11} & a_{12} & \dots & a_{1n} & b_1 \\ a_{21} & a_{22} & \dots & a_{2n} & b_2 \\ \vdots & \vdots & & \vdots & \vdots \\ a_{k1} & a_{k22} & \dots & a_{kn} & b_k \end{bmatrix}.$$

Consider the following system of linear equations

$$\begin{cases} x + y + z = 3 \\ x - y + 2z = 2 \\ 2x + y - z = 2 \end{cases} . \tag{1.6}$$

The system's augmented matrix is

$$[A|\mathbf{b}] = \left[\begin{array}{ccc|c} 1 & 1 & 1 & 3 \\ 1 & -1 & 2 & 2 \\ 2 & 1 & -1 & 2 \end{array}\right].$$

If an elementary operation is performed on the augmented matrix $[A|\mathbf{b}]$, the new matrix thus obtained is the augmented matrix of an equivalent system of linear equations.

This fact will be used to 'simplify' the augmented matrix in order to obtain a system equivalent to the given one but easier to solve. The aim is to reduce the augmented matrix $[A|\mathbf{b}]$ to a row echelon matrix using elementary operations according to a method called Gaussian elimination.

The **Gaussian elimination** (GE) is a step-by-step method which is applied to a matrix and consists in:

1. placing all the zero rows below the non-zero rows, by exchanging the necessary rows;
2. choosing a row with a non-zero entry situated in a column as far left as possible in the matrix and making this row the first row, by possibly exchanging rows;
3. using the first row and elementary operations to place zeros in all the entries of the matrix in the rows below and in the same column as the column of the chosen entry;
4. repeating the preceding steps descending a row, i.e., considering only the rows below the first row;
5. repeating the steps above descending another row, i.e., considering only the rows below row 2;
6. keeping on repeating these steps as many times as necessary until obtaining a row echelon matrix.

In short, the Gaussian elimination is an specific sequence of elementary row operations whose final goal is to obtain a row echelon matrix.

The Gaussian elimination is now applied to the augmented matrix of system (1.6):

$$\left[\begin{array}{ccc|c} \textcircled{1} & 1 & 1 & 3 \\ 1 & -1 & 2 & 2 \\ 2 & 1 & -1 & 2 \end{array}\right] \xrightarrow[\mathrm{l}_3-2\mathrm{l}_1]{\mathrm{l}_2-\mathrm{l}_1} \left[\begin{array}{ccc|c} 1 & 1 & 1 & 3 \\ 0 & -2 & 1 & -1 \\ 0 & -1 & -3 & -4 \end{array}\right] \cdots$$

$$\cdots \xrightarrow[\mathrm{l}_2\leftrightarrow\mathrm{l}_3]{} \left[\begin{array}{ccc|c} 1 & 1 & 1 & 3 \\ 0 & \textcircled{-1} & -3 & -4 \\ 0 & -2 & 1 & -1 \end{array}\right] \xrightarrow[\mathrm{l}_3-2\mathrm{l}_2]{} \left[\begin{array}{ccc|c} 1 & 1 & 1 & 3 \\ 0 & -1 & -3 & -4 \\ 0 & 0 & 7 & 7 \end{array}\right].$$

The elementary operations appearing below the arrows are:

- $\mathrm{l}_2 - \mathrm{l}_1$ indicates that row 1 multiplied by -1 has been added to row 2, and $\mathrm{l}_3 - 2\mathrm{l}_1$ indicates that row 1 multiplied by -2 has been added to row 3;

- $l_2 \leftrightarrow l_3$ indicates that rows 2 and 3 have been exchanged;
- $l_3 - 2l_2$ indicates that row 2 multiplied by -2 has been added to row 3.

Notice that in the points above and throughout this book we use the following notation: *when indicating an elementary operation of type (iii), the first row to be written is the one to be modified.*

The matrix

$$\left[\begin{array}{ccc|c} 1 & 1 & 1 & 3 \\ 0 & -1 & -3 & -4 \\ 0 & 0 & 7 & 7 \end{array}\right]$$

is a row echelon matrix and the augmented matrix of the system of linear equations

$$\begin{cases} x + y + z = 3 \\ -y - 3z = -4 \\ 7z = 7 \end{cases} . \tag{1.7}$$

Moreover,

$$\begin{cases} x + y + z = 3 \\ x - y + 2z = 2 \\ 2x + y - z = 2 \end{cases} \iff \begin{cases} x + y + z = 3 \\ -y - 3z = -4 \\ 7z = 7 \end{cases} .$$

Beginning to solve the system by the simplest equation, i.e., the third equation, we have $z = 1$. Substituting z by its value in the second equation, we get

$$-y - 3 = -4,$$

that is, $y = 1$. Finally, using the first equation and the values of y and z already obtained, we have

$$x + 1 + 1 = 3$$

and, hence, $x = 1$. It immediately follows that the solution set $\mathcal{S}$ of system (1.7) or, equivalently, of system (1.6) is $\mathcal{S} = \{(1,1,1)\}$. Hence, we conclude that the system (1.6) is consistent and has a unique solution.

Observe that, once the row echelon augmented matrix is obtained, we begin solving the system from the bottom equation up, and we keep ascending in the system until the top equation is reached. This is called **back substitution**.

How to solve a system of linear equations

Summing up, recall that we solve a system of linear equations in three steps:

1. the system's augmented matrix $[A|\mathbf{b}]$ is obtained;
2. matrix $[A|\mathbf{b}]$ is reduced to a row echelon matrix R using Gaussian elimination or, symbolically,
$$[A|\mathbf{b}] \xrightarrow[\text{GE}]{} R;$$
3. the system corresponding to matrix R is solved (using back substitution).

In system (1.6), we have

$$[A|\mathbf{b}] = \begin{bmatrix} 1 & 1 & 1 & 3 \\ 1 & -1 & 2 & 2 \\ 2 & 1 & -1 & 2 \end{bmatrix} \xrightarrow[\text{GE}]{} \begin{bmatrix} 1 & 1 & 1 & 3 \\ 0 & -1 & -3 & -4 \\ 0 & 0 & 7 & 7 \end{bmatrix} = R.$$

Consider now the following system of linear equations

$$\begin{cases} x - y - 2z + w = 0 \\ 2x - 3y - 2z + 2w = 3 \\ -x + 2y - w = -3 \end{cases}. \tag{1.8}$$

Applying Gaussian elimination to the augmented matrix of system (1.8), we have

$$[A|\mathbf{b}] = \begin{bmatrix} 1 & -1 & -2 & 1 & 0 \\ 2 & -3 & -2 & 2 & 3 \\ -1 & 2 & 0 & -1 & -3 \end{bmatrix} \xrightarrow[\substack{l_2-2l_1 \\ l_3+l_1}]{} \begin{bmatrix} 1 & -1 & -2 & 1 & 0 \\ 0 & -1 & 2 & 0 & 3 \\ 0 & 1 & -2 & 0 & -3 \end{bmatrix}$$

$$\xrightarrow[l_3+l_2]{} \begin{bmatrix} 1 & -1 & -2 & 1 & 0 \\ 0 & -1 & 2 & 0 & 3 \\ 0 & 0 & 0 & 0 & 0 \end{bmatrix}.$$

Hence, the system (1.8) is equivalent to

$$\begin{cases} x - y - 2z + w = 0 \\ -y + 2z = 3 \end{cases} \iff \begin{cases} x = 4z - w + 3 \\ y = 2z - 3 \end{cases}. \tag{1.9}$$

It is obvious from (1.9) that the system (1.8) does not have a unique solution. In fact, fixing (any) values for z and w, the values of x and y are immediately determined. Hence, here the variables z, w can take any real values and, once fixed, determine the values of the remaining variables. We can say that z, w are the **independent variables** or **free variables**, whilst x, y are the **dependent variables**.

As before, we used back substitution in (1.9) and expressed the variables corresponding to the columns with pivots in terms of the variables corresponding to the columns without pivots in the coefficient matrix. This will be the rule throughout the book.

How to choose the dependent and the independent variables

In a consistent system:

- the dependent variables are those whose columns correspond to columns having pivots;
- the independent variables are the remaining ones, i.e., those whose columns correspond to columns without pivots.

A system of linear equations having a unique solution does not possess independent variables and, therefore, all its variables are dependent variables.

According to this rule, in (1.8) the dependent variables are x, y, and the independent variables are z, w. The solution set $\mathcal{S}$ of system (1.8) is

$$\mathcal{S} = \{(x, y, z, w) \in \mathbb{R}^4 : x = 4z + w - 3 \wedge y = 2z - 3\}$$

and, consequently, the system has infinitely many solutions.

For example, if we set $z = 2, w = 0$, then $(x, y, z, w) = (5, 1, 2, 0)$ is a **particular solution** of system (1.8). Another solution can be obtained setting, for example, $z = -1, w = 2$. In this case, we have that $(x, y, z, w) = (-5, -5, -1, 2)$ is the solution corresponding to $z = -1, w = 2$. Hence, whenever we fix particular values of z, w, we obtain a particular solution of system (1.8).

The way dependent and independent variables are chosen is decisive in how the solution set is obtained. That is, the number of pivots appearing after Gaussian elimination plays a crucial role in the system's solution.

Could it happen that, starting with the a given system and depending on the calculations, one might end up with a different number of pivots?

Fortunately, the answer is no, as will be shown in Proposition 1.2.

Consider now the system

$$\begin{cases} x+y-2z=1 \\ -x+y=0 \\ y-z=0 \end{cases} . \tag{1.10}$$

Using Gaussian elimination to solve (1.10), we have

$$[A|\mathbf{b}]=\begin{bmatrix} 1 & 1 & -2 & 1 \\ -1 & 1 & 0 & 0 \\ 0 & 1 & -1 & 0 \end{bmatrix} \xrightarrow[l_2+l_1]{} \begin{bmatrix} 1 & 1 & -2 & 1 \\ 0 & 2 & -2 & 1 \\ 0 & 1 & -1 & 0 \end{bmatrix} \xrightarrow[l_3-\frac{1}{2}l_2]{} \underbrace{\begin{bmatrix} 1 & 1 & -2 & 1 \\ 0 & 2 & -2 & 1 \\ 0 & 0 & 0 & -\frac{1}{2} \end{bmatrix}}_{R}.$$

Hence,

$$\begin{cases} x+y-2z=1 \\ -x+y=0 \\ y-z=0 \end{cases} \iff \begin{cases} x+y-2z=1 \\ 2y-2z=1 \\ 0=-\frac{1}{2} \end{cases} .$$

It follows that the system is inconsistent and its solution set is $\mathcal{S}=\emptyset$.

Observe that the key fact for this system to be inconsistent is that the matrix R has a pivot in a column corresponding to that of the independent terms in the augmented matrix (see the grey entry above).

Gaussian elimination has been used so far to reduce augmented matrices of systems of linear equations to row echelon matrices. However, this process can be used outside the setting of systems of linear equations. In fact, we can use Gaussian elimination to reduce any matrix to a row echelon matrix, which we shall need often to do.

Definition 3 *A $k \times n$ matrix A is said to be in* ***reduced row echelon form*** *or in* ***canonical row echelon form*** *if the following three conditions hold:*

(i) A is a row echelon matrix;

(ii) the pivots in A are all equal to 1;

(iii) in each column having a pivot, all remaining entries are equal to zero.

Example 1.5 *Let A and B be the matrices*

$$A=\begin{bmatrix} 1 & -2 & 3 & 9 \\ 0 & 0 & 4 & -2 \\ 0 & 0 & 0 & 6 \end{bmatrix}, \qquad B=\begin{bmatrix} 1 & 0 & -7 & 0 & 4 \\ 0 & 1 & 8 & 0 & 5 \\ 0 & 0 & 0 & 1 & 1 \end{bmatrix}.$$

Matrix A is a row echelon matrix but is not in reduced row echelon form because it does not satisfy conditions (ii) and (iii). Matrix B is a matrix in reduced row echelon form.

The Gauss–Jordan elimination is a method devised to reduce a given matrix to reduced row echelon form using elementary operations. It is a step-by-step method whose first part is Gaussian elimination.

The **Gauss–Jordan elimination method** (GJE) applied to a matrix A consists in

1) reducing A to a row echelon matrix using Gaussian elimination;

2) (a) using the right hand side most pivot (hence the bottom pivot) and the necessary elementary operations to put zeros in all entries in the same column as the pivot (hence the entries in the rows above the pivot);
   (b) repeating 2) (a) ascending one row, i.e., considering only the rows above that of the pivot in 2) (a);
   (c) repeating the process ascending one more row, i.e., considering only the rows above that of the pivot in 2) (b);
   (d) keeping on ascending in the matrix until reaching the first row (that is, until obtaining a matrix such that in each column having a pivot, the pivot is the only non-zero entry);

3) performing the necessary elementary operations in order to obtain a matrix where all pivots are equal to 1.

Example 1.6 *We apply now the Gauss–Jordan elimination to the matrix*

$$\begin{bmatrix} 1 & -2 & 0 & 2 \\ -\frac{1}{2} & 0 & 1 & 2 \\ -1 & -1 & 2 & 0 \\ 0 & -3 & 2 & 2 \end{bmatrix}$$

in order to obtain a matrix in reduced row echelon form. We have

$$\begin{bmatrix} 1 & -2 & 0 & 2 \\ -\frac{1}{2} & 0 & 1 & 2 \\ -1 & -1 & 2 & 0 \\ 0 & -3 & 2 & 2 \end{bmatrix} \xrightarrow[\substack{l_2+\frac{1}{2}l_1 \\ l_3+l_1}]{} \begin{bmatrix} 1 & -2 & 0 & 2 \\ 0 & -1 & 1 & 3 \\ 0 & -3 & 2 & 2 \\ 0 & -3 & 2 & 2 \end{bmatrix} \xrightarrow[\substack{l_3-3l_2 \\ l_4-3l_2}]{} \begin{bmatrix} 1 & -2 & 0 & 2 \\ 0 & -1 & 1 & 3 \\ 0 & 0 & -1 & -7 \\ 0 & 0 & -1 & -7 \end{bmatrix} \cdots$$

$$\cdots \xrightarrow[l_4-l_3]{} \begin{bmatrix} 1 & -2 & 0 & 2 \\ 0 & -1 & 1 & 3 \\ 0 & 0 & -1 & -7 \\ 0 & 0 & 0 & 0 \end{bmatrix} \xrightarrow[l_2+l_3]{} \begin{bmatrix} 1 & -2 & 0 & 2 \\ 0 & -1 & 0 & -4 \\ 0 & 0 & -1 & -7 \\ 0 & 0 & 0 & 0 \end{bmatrix} \cdots$$

$$\cdots \xrightarrow[l_1-2l_2]{} \begin{bmatrix} 1 & 0 & 0 & 10 \\ 0 & -1 & 0 & -4 \\ 0 & 0 & -1 & -7 \\ 0 & 0 & 0 & 0 \end{bmatrix} \xrightarrow[\substack{(-1)l_3 \\ (-1)l_2}]{} \begin{bmatrix} 1 & 0 & 0 & 10 \\ 0 & 1 & 0 & 4 \\ 0 & 0 & 1 & 7 \\ 0 & 0 & 0 & 0 \end{bmatrix}.$$

It should be noted that

- In the Gauss–Jordan elimination, 2) and 3) do not have necessarily to be performed in that order (i.e., first 2) and then 3)). It might be more convenient to make a pivot equal to 1 (or possibly all pivots) before completing 2), or even before point 2).
- In 2) and 3), it is not allowed to exchange rows.
- Gauss–Jordan elimination can also be used to solve systems of linear equations (see Example 1.8).

A common mistake. Halfway through the Gauss–Jordan elimination going from top to bottom when it was supposed to do a down-up elimination. Consequence: spoiling the matrix and wasting the work done so far. The first part of Gauss–Jordan elimination requires a up-down elimination which ends when a row echelon matrix is obtained. Then it must be followed by an down-up elimination to obtain a reduced row echelon form. *One must not mix these two eliminations.*

When reducing a matrix A to row echelon form using elementary operations, the final matrix may vary depending on the operations chosen and on the order under which they were applied to A. However, if the final matrix is in reduced row echelon form, then this matrix is always the same, depending neither on the operations nor on the sequence in which they were performed.

Proposition 1.1 *Let A be a $k \times n$ matrix over $\mathbb{K}$ and let R, R' be reduced row echelon matrices obtained from A through elementary row operations. Then $R = R'$.*

Proof *This proposition is proved in §8.1.*

Proposition 1.1 allows us to make the following definition.

Definition 4 *The* ***reduced row echelon form*** *or the* ***canonical row echelon form*** *of a matrix A is the matrix in reduced row echelon form obtained from A through elementary row operations.*

In Example 1.6, the matrix

$$\begin{bmatrix} 1 & 0 & 0 & 10 \\ 0 & 1 & 0 & 4 \\ 0 & 0 & 1 & 7 \\ 0 & 0 & 0 & 0 \end{bmatrix}$$

is <u>the</u> reduced row echelon from of

$$\begin{bmatrix} 1 & -2 & 0 & 2 \\ -\frac{1}{2} & 0 & 1 & 2 \\ -1 & -1 & 2 & 0 \\ 0 & -3 & 2 & 2 \end{bmatrix}.$$

As noted above, when reducing a matrix to a row echelon matrix through elementary operations, this row echelon matrix is not uniquely determined. However, the number of pivots is.

Proposition 1.2 *Let A be a $k \times n$ matrix over $\mathbb{K}$ and let R and R' be row echelon matrices obtained from A using elementary row operations. Then, the number of pivots of R and of R' coincide.*

Observe that in this proposition and, for that matter, also in Proposition 1.1, it is not required that R and R' be obtained using Gaussian elimination. R and R' are just row echelon matrices obtained from A applying elementary operations in no particular order.

Proof *Applying Gauss–Jordan elimination to the matrices A, R, and R', by Proposition 1.1, we obtain the same reduced row echelon matrix M. Notice that the Gauss–Jordan elimination forces both the number of pivots in R and the number of pivots in M to coincide. Similarly, the number of pivots in R' and M must also coincide. Hence, the result follows.*

We are now ready to make one of the crucial definitions in the book.

Definition 5 *The **rank** of a $k \times n$ matrix A is the number of pivots of any row echelon matrix obtained from A using elementary operations. The rank of A is denoted by* $\operatorname{rank}(A)$.

Example 1.7 *Let A be the matrix*

$$A = \begin{bmatrix} 1 & -1 & 1 & 0 \\ -1 & 1 & -1 & 0 \\ 3 & -3 & 6 & -3 \end{bmatrix}.$$

Using Gaussian elimination,

$$\begin{bmatrix} 1 & -1 & 1 & 0 \\ -1 & 1 & -1 & 0 \\ 3 & -3 & 6 & -3 \end{bmatrix} \xrightarrow[\substack{l_2+l_1 \\ l_3-3l_1}]{} \begin{bmatrix} 1 & -1 & 1 & 0 \\ 0 & 0 & 0 & 0 \\ 0 & 0 & 3 & -3 \end{bmatrix} \xrightarrow[l_2 \leftrightarrow l_3]{} \begin{bmatrix} 1 & -1 & 1 & 0 \\ 0 & 0 & 3 & -3 \\ 0 & 0 & 0 & 0 \end{bmatrix}.$$

Hence, since the last matrix is a row echelon matrix having two pivots, we conclude that the rank of matrix A is $\operatorname{rank}(A) = 2$.

> **How to calculate the rank of a matrix A**
>
> 1. Reduce the matrix A to a row echelon matrix R using Gaussian elimination;
>
> 2. the rank of A is the number of pivots in R, i.e.,
>
> $$\operatorname{rank}(A) = \operatorname{rank}(R) = \text{number of pivots of R}.$$

At this point, we have encountered already both consistent and inconsistent systems, and are in a position to obtain a general classification in terms of the ranks of the matrices associated with the systems.

Given a system with augmented matrix $[A|\mathbf{b}]$, either

$$\operatorname{rank}(A) = \operatorname{rank}([A|\mathbf{b}]) \qquad \text{or} \qquad \operatorname{rank}(A) \neq \operatorname{rank}([A|\mathbf{b}]).$$

In the latter case, the only possibility is $\operatorname{rank}(A) < \operatorname{rank}([A|\mathbf{b}])$, and it follows that the system is inconsistent. This means that a pivot appears in the column corresponding to the vector $\mathbf{b}$ of independent terms. This in turn corresponds to having an equation where the coefficients of all variables are zero whereas the independent term is non-zero.

In fact, recalling the inconsistent system (1.10), we saw that

$$\operatorname{rank}(A) < \operatorname{rank}(R) = \operatorname{rank}([A|\mathbf{b}]),$$

since an extra pivot appeared in the column corresponding to the independent terms.

On the other hand, if $\operatorname{rank}(A) = \operatorname{rank}([A|\mathbf{b}])$ then the columns of the pivots correspond to (some or all of) those in the matrix A. It follows that the system is consistent and either

$$\operatorname{rank}(A) = \text{number of columns of } A \tag{1.11}$$

or

$$\operatorname{rank}(A) < \text{number of columns of } A. \tag{1.12}$$

In case (1.11), the system has a unique solution. In case (1.12), the system has infinitely many solutions and, if A is a $k \times n$ matrix, then the number of independent variables is $n - \operatorname{rank}(A)$.

We summarise this discussion in the proposition below.

Proposition 1.3 *Let $[A|\mathbf{b}] \in M_{k,n+1}(\mathbb{K})$ be the augmented matrix of a system of linear equations. Then the following assertions hold.*

(i) If $\operatorname{rank}(A) < \operatorname{rank}([A|\mathbf{b}])$, then the system is inconsistent.

(ii) *If* $\text{rank}(A) = \text{rank}([A|\mathbf{b}])$, *then the system is consistent. In this case, the number of independent variables coincides with* $n - \text{rank}(A)$, *i.e.,*

$$\textit{number of independent variables} = n - \text{rank}(A).$$

Example 1.8 *Consider the system whose coefficient matrix* A *and vector* $\mathbf{b}$ *of independent terms are, respectively,*

$$A = \begin{bmatrix} 1 & 0 & 0 \\ 0 & 1-i & -2i \\ 0 & 1 & 1-i \end{bmatrix}, \qquad \mathbf{b} = \begin{bmatrix} 0 \\ 1-i \\ 1 \end{bmatrix}.$$

Hence, we see that this is the case of a system in three complex variables, say, z_1, z_2, z_3, *having three equations.*

Applying Gaussian elimination to the system's augmented matrix $[A|\mathbf{b}]$, *we have*

$$[A|\mathbf{b}] = \begin{bmatrix} 1 & 0 & 0 & 0 \\ 0 & 1-i & -2i & 1-i \\ 0 & 1 & 1-i & 1 \end{bmatrix} \xrightarrow{\mathbf{l}_2 \leftrightarrow \mathbf{l}_3} \begin{bmatrix} 1 & 0 & 0 & 0 \\ 0 & 1 & 1-i & 1 \\ 0 & 1-i & -2i & 1-i \end{bmatrix}$$

$$\xrightarrow{\mathbf{l}_3 - (1-i)\mathbf{l}_2} \underbrace{\begin{bmatrix} 1 & 0 & 0 & 0 \\ 0 & 1 & 1-i & 1 \\ 0 & 0 & 0 & 0 \end{bmatrix}}_{R}.$$

Since the pivots in the matrix R *are the grey entries, we conclude that* $\text{rank}(A) = \text{rank}([A|\mathbf{b}]) = 2$. *Hence, the system is consistent and*

$$\textit{number of independent variables} = 3 - \text{rank}(A) = 1.$$

Just by looking at the matrix R, *we know that there is one independent variable which, according to the rules previously established, is* z_3, *since* z_1, z_2 *correspond to the columns having pivots.*

Observe that R *is a reduced row echelon matrix and, therefore, the solution of the system can be* read *directly in* R. *Indeed, without back substitution, we have immediately that*

$$\begin{cases} z_1 = 0 \\ z_2 = (i-1)z_3 - 1 \end{cases},$$

from which follows that the solution set of the system is

$$\mathcal{S} = \{(0, (i-1)z_3 - 1, z_3) \colon z_3 \in \mathbb{C}\}.$$

How to solve a system of linear equations (continued)

- Use either Gaussian elimination or Gauss–Jordan elimination to reduce the system's augmented matrix to a row echelon matrix or a reduced row echelon matrix, respectively.
- If the system is consistent, the dependent variables are those corresponding to the columns having pivots, and the independent variables are those corresponding to the columns without pivots.
- The system's solution set $\mathcal{S}$ can be presented in several ways. We give below an example of some possibilities for presenting the solution set of system (1.8).

  (a) $\mathcal{S} = \{(x, y, z, w) \in \mathbb{R}^4 \colon x = 4z + w - 3 \wedge y = 2z - 3\}$

  (b) $\mathcal{S} = \{(4z + w - 3, 2z - 3, z, w) \colon z, w \in \mathbb{R}\}$

  (c) $\mathcal{S} = \{(x, y, z, w) \in \mathbb{R}^4 \colon x = 4t + s - 3, y = 2t - 3, z = t, w = s \quad (t, s \in \mathbb{R})\}$

  (d) $\mathcal{S} = \{(4t + s - 3, 2t - 3, t, s) \colon t, s \in \mathbb{R}\}$

1.2 Matrix Calculus

We have seen how matrices can be used as tools to solve systems of linear equations. But matrices stand alone in their own right and are not just useful in applications. In fact, in this section, we will pay close attention to the set $\mathrm{M}_{k,n}(\mathbb{K})$ of $k \times n$ matrices over $\mathbb{K}$ and will define three operations on this set. Namely, the addition of matrices, the multiplication of a matrix by a scalar and the multiplication of two matrices.

We start with the definition of addition of two matrices. It is worth noticing that one can only add matrices having the same size. Roughly speaking, the sum of matrices A and B is a matrix $A + B$ whose entry-ij $(A + B)_{ij}$ is the sum of the corresponding entries in A and B. More precisely,

Definition 6 *The **addition**, $+$, on $\mathrm{M}_{k,n}$ is the operation*

$$\begin{aligned} + : \mathrm{M}_{k,n}(\mathbb{K}) \times \mathrm{M}_{k,n}(\mathbb{K}) &\to \mathrm{M}_{k,n}(\mathbb{K}) \\ (A, B) &\mapsto A + B \end{aligned}$$

defined, for $A = [a_{ij}], B = [b_{ij}]$ and $i = 1, \dots, k, j = 1, \dots n$, by $(A + B)_{ij} = a_{ij} + b_{ij}$.

Example 1.9 *Let A and B be the 3×4 real matrices*

$$A = \begin{bmatrix} 1 & -2 & 3 & 7 \\ 0 & 0 & 4 & -2 \\ -3 & 0 & 0 & 6 \end{bmatrix} \qquad B = \begin{bmatrix} -1 & 0 & 4 & 5 \\ -2 & 3 & 11 & 2 \\ 0 & 6 & 7 & -1 \end{bmatrix}.$$

The sum $A + B$ of these two matrices is a real 3×4 matrix each entry of which is calculated by adding the homologous entries of A and B, i.e.,

$$A + B = \begin{bmatrix} 0 & -2 & 7 & 12 \\ -2 & 3 & 15 & 0 \\ -3 & 6 & 7 & 5 \end{bmatrix}.$$

Definition 7 *The* ***multiplication by a scalar****, μ, on $\mathrm{M}_{k,n}$ is the operation*

$$\mu : \mathbb{K} \times \mathrm{M}_{k,n}(\mathbb{K}) \to \mathrm{M}_{k,n}(\mathbb{K})$$
$$(\alpha, A) \mapsto \alpha A$$

defined, for $A = [a_{ij}]$, $\alpha \in \mathbb{K}$ and $i = 1, \dots, k, j = 1, \dots n$, by $(\alpha A)_{ij} = \alpha a_{ij}$.

Example 1.10 *Letting A be the matrix of Example 1.9, the matrix $2A$ is obtained by multiplying all the entries of A by the scalar 2, i.e.,*

$$2A = \begin{bmatrix} 2 & -4 & 6 & 14 \\ 0 & 0 & 8 & -4 \\ -6 & 0 & 0 & 12 \end{bmatrix}.$$

The next two propositions collect essential properties of these operations.

Proposition 1.4 *The following assertions hold, for all matrices A, B, C in $\mathrm{M}_{k,n}(\mathbb{K})$.*

(i) $A + B = B + A$, *i.e., the addition of matrices is* ***commutative****.*

(ii) $A + (B + C) = (A + B) + C$, *i.e., the addition of matrices is* ***associative****.*

(iii) There exists a unique ***additive identity****, 0, i.e.,*

$$A + 0 = A = 0 + A.$$

(iv) There exists a unique matrix $-A \in \mathrm{M}_{k,n}$, called the ***additive inverse*** *of A, such that*

$$A + (-A) = 0 = (-A) + A.$$

Proof *The commutativity (i) and associativity (ii) follow easily from the commutativity and associativity of the addition in $\mathbb{K}$, and it is obvious that the (unique) additive identity is the $k \times n$ zero matrix.*

It is immediate that, given $A = [a_{ij}]$, its additive inverse is $-A = [-a_{ij}]$.

Proposition 1.5 *The following assertions hold, for all matrices A, B in $\mathrm{M}_{k,n}(\mathbb{K})$ and $\alpha, \beta \in \mathbb{K}$.*

(i) $\alpha(A+B) = \alpha A + \alpha B$, *i.e., the multiplication by a scalar is distributive relative to the addition of matrices.*

(ii) $(\alpha\beta)A = \alpha(\beta A)$.

(iii) $(\alpha + \beta)A = \alpha A + \beta A$.

(iv) $1A = A$.

Proof *These assertions are an immediate consequence of the corresponding properties of the scalars in* $\mathbb{K}$.

We define now the multiplication of a matrix by a column vector.

Definition 8 *Let A be $k \times n$ over $\mathbb{K}$ and let $\mathbf{b}$ be the $n \times 1$ column vector*

$$\mathbf{b} = \begin{bmatrix} b_{11} \\ \cdot \\ \cdot \\ \cdot \\ b_{j1} \\ \cdot \\ \cdot \\ \cdot \\ b_{n1} \end{bmatrix}$$

*with entries in $\mathbb{K}$. The **product** $A\mathbf{b}$ of A and $\mathbf{b}$*

$$A\mathbf{b} = \begin{bmatrix} a_{11} & a_{12} & \dots & a_{1j} & \dots & a_{1n} \\ a_{21} & a_{22} & \dots & a_{2j} & \dots & a_{2n} \\ \vdots & \vdots & & \vdots & & \vdots \\ a_{i1} & a_{i2} & \dots & a_{ij} & \dots & a_{in} \\ \vdots & \vdots & & \vdots & & \vdots \\ a_{k1} & a_{k2} & \dots & a_{kj} & \dots & a_{kn} \end{bmatrix} \begin{bmatrix} b_{11} \\ b_{21} \\ \cdot \\ \cdot \\ b_{j1} \\ \cdot \\ \cdot \\ \cdot \\ b_{n1} \end{bmatrix} = \begin{bmatrix} c_{11} \\ \cdot \\ \cdot \\ \cdot \\ c_{i1} \\ \cdot \\ \cdot \\ \cdot \\ c_{k1} \end{bmatrix},$$

the $k \times 1$ column vector $A\mathbf{b}$ whose entry $(A\mathbf{b})_{i1}$ is defined, for all indices $i = 1, \dots, k$, by

$$(A\mathbf{b})_{i1} = a_{i1}b_{11} + a_{i2}b_{21} + \cdots + a_{ij}b_{j1} + \cdots + a_{in}b_{n1} = \sum_{j=1}^{n} a_{ij}b_{j1}. \quad (1.13)$$

It is possible to multiply A and $\mathbf{b}$ only if the number of columns of A coincides with the number of rows of $\mathbf{b}$.

Example 1.11 *Consider the matrices*

$$A = \begin{bmatrix} 2 & -1 & 5 \\ -4 & 6 & 3 \end{bmatrix} \qquad \mathbf{b} = \begin{bmatrix} 7 \\ -2 \\ 1 \end{bmatrix},$$

of size 2×3 and 3×1, respectively. We can calculate the product $A\mathbf{b}$, since the number of columns of A equals the number of rows of $\mathbf{b}$. Using Definition 8, $A\mathbf{b}$ is the 2×1 column vector

$$A\mathbf{b} = \begin{bmatrix} 2 & -1 & 5 \\ -4 & 6 & 3 \end{bmatrix} \begin{bmatrix} 7 \\ -2 \\ 1 \end{bmatrix} = \begin{bmatrix} 21 \\ -37 \end{bmatrix}.$$

Observe that, row 1 of $A\mathbf{b}$ is calculated using row 1 of A and the column vector $\mathbf{b}$ as shown below

$$2 \times 7 + (-1) \times (-2) + 5 \times 1 = 21.$$

Analogously, row 2 of $A\mathbf{b}$ is calculated using row 2 of A and the column vector $\mathbf{b}$, obtaining

$$-4 \times 7 + 6 \times (-2) + 3 \times 1 = -37.$$

To calculate row i of $A\mathbf{b}$, we need row i of matrix A and the column vector $\mathbf{b}$, as shown in (1.13).

Going back to the general definition of $A\mathbf{b}$, observe that

$$A\mathbf{b} = \begin{bmatrix} a_{11} & a_{12} & \dots & a_{1j} & \dots & a_{1n} \\ a_{21} & a_{22} & \dots & a_{2j} & \dots & a_{2n} \\ \vdots & \vdots & & \vdots & & \vdots \\ a_{i1} & a_{i2} & \dots & a_{ij} & \dots & a_{in} \\ \vdots & \vdots & & \vdots & & \vdots \\ a_{k1} & a_{k2} & \dots & a_{kj} & \dots & a_{kn} \end{bmatrix} \begin{bmatrix} b_{11} \\ b_{21} \\ \vdots \\ b_{j1} \\ \vdots \\ b_{n1} \end{bmatrix} = \begin{bmatrix} a_{11}b_{11} + a_{12}b_{21} + \cdots + a_{1n}b_{n1} \\ \vdots \\ a_{i1}b_{11} + a_{i2}b_{21} + \cdots + a_{in}b_{n1} \\ \vdots \\ a_{k1}b_{11} + a_{k2}b_{21} + \cdots + a_{kn}b_{n1} \end{bmatrix}.$$

Hence,

$$Ab = b_{11}\begin{bmatrix} a_{11} \\ \cdot \\ \cdot \\ \cdot \\ a_{i1} \\ \cdot \\ \cdot \\ \cdot \\ a_{k1} \end{bmatrix} + b_{21}\begin{bmatrix} a_{12} \\ \cdot \\ \cdot \\ \cdot \\ a_{i2} \\ \cdot \\ \cdot \\ \cdot \\ a_{k2} \end{bmatrix} + \cdots + b_{n1}\begin{bmatrix} a_{1n} \\ \cdot \\ \cdot \\ \cdot \\ a_{in} \\ \cdot \\ \cdot \\ \cdot \\ a_{kn} \end{bmatrix} = b_{11}\mathbf{c}_1 + b_{21}\mathbf{c}_2 + \cdots + b_{n1}\mathbf{c}_n, \tag{1.14}$$

where $\mathbf{c}_1, \mathbf{c}_2, \ldots, \mathbf{c}_n$ are the columns of A.

Definition 9 *Let* $A = \begin{bmatrix} \mathbf{c}_1 & \mathbf{c}_2 & \ldots & \mathbf{c}_n \end{bmatrix}$ *be a* $k \times n$ *matrix over* $\mathbb{K}$. *A* ***linear combination of the columns of*** A *is any (column) vector which can be expressed as*

$$\alpha_1\mathbf{c}_1 + \alpha_2\mathbf{c}_2 + \cdots + \alpha_n\mathbf{c}_n,$$

where $\alpha_1, \alpha_2, \ldots, \alpha_n$ *are scalars in* $\mathbb{K}$.

Hence, we have shown in (1.14) that $A\mathbf{b}$ is a linear combination of the columns of A. This fact is so important that it is worth making a note of it in the proposition below.

Proposition 1.6 *Let* $A = \begin{bmatrix} \mathbf{c}_1 & \mathbf{c}_2 & \ldots & \mathbf{c}_n \end{bmatrix}$ *be a* $k \times n$ *matrix over* $\mathbb{K}$ *and let*

$$\mathbf{b} = \begin{bmatrix} b_{11} \\ b_{21} \\ \cdot \\ \cdot \\ b_{j1} \\ \cdot \\ \cdot \\ \cdot \\ b_{n1} \end{bmatrix}$$

be an $n \times 1$ *column vector. Then the product* $A\mathbf{b}$ *is a linear combination of the columns of* A. *More precisely,*

$$A\mathbf{b} = b_{11}\mathbf{c}_1 + b_{21}\mathbf{c}_2 + \cdots + b_{n1}\mathbf{c}_n.$$

We are now ready to multiply two matrices.

Definition 10 *Given matrices A of size $k \times p$ and B of size $p \times n$ over $\mathbb{K}$, the product*

$$AB = [(AB)_{ij}]_{\substack{i=1,\dots,k\\ j=1,\dots,n}}$$

$$= \begin{bmatrix} a_{11} & a_{12} & \dots & a_{1l} & \dots & a_{1p} \\ a_{21} & a_{22} & \dots & a_{2l} & \dots & a_{2p} \\ \vdots & \vdots & & \vdots & & \vdots \\ a_{i1} & a_{i2} & \dots & a_{il} & \dots & a_{ip} \\ \vdots & \vdots & & \vdots & & \vdots \\ a_{k1} & a_{k2} & \dots & a_{kl} & \dots & a_{kp} \end{bmatrix} \begin{bmatrix} b_{11} & b_{12} & \dots & b_{1j} & \dots & b_{1n} \\ b_{21} & b_{22} & \dots & b_{2j} & \dots & b_{2n} \\ \vdots & \vdots & & \vdots & & \vdots \\ b_{l1} & b_{l2} & \dots & b_{lj} & \dots & b_{ln} \\ \vdots & \vdots & & \vdots & & \vdots \\ b_{p1} & b_{p2} & \dots & b_{pj} & \dots & b_{pn} \end{bmatrix},$$

is the $k \times n$ matrix such that, for all indices i, j, the entry-ij of AB is

$$(AB)_{ij} = a_{i1}b_{1j} + a_{i2}b_{2j} + \dots + a_{il}b_{lj} + \dots + a_{ip}b_{pj} = \sum_{l=1}^{p} a_{il}b_{lj}. \tag{1.15}$$

Hence, if $\mathbf{l}_i$ is the row i of A and $\mathbf{c}_j$ is the column j of B, then $(AB)_{ij} = \mathbf{l}_i\mathbf{c}_j$. By a slight abuse of notation, here we identify the 1×1 matrix $[(AB)_{ij}]$ with the scalar $(AB)_{ij}$.

It is possible to multiply two matrices A and B only if the number of columns of A coincides with the number of rows of B.

Example 1.12 *Let A and B be the matrices*

$$A = \begin{bmatrix} 2 & -1 & 5 \\ -4 & 6 & 3 \end{bmatrix} \qquad B = \begin{bmatrix} 7 & 0 & -1 \\ -2 & 1 & 0 \\ 1 & 0 & -3 \end{bmatrix},$$

of size 2×3 and 3×3, respectively. Observe that we can multiply these two matrices since the number of columns of A equals the number of rows of B. It follows that the product AB is the 2×3 matrix

$$AB = \begin{bmatrix} 2 & -1 & 5 \\ -4 & 6 & 3 \end{bmatrix} \begin{bmatrix} 7 & 0 & -1 \\ -2 & 1 & 0 \\ 1 & 0 & -3 \end{bmatrix} = \begin{bmatrix} 21 & -1 & -17 \\ -37 & 6 & -5 \end{bmatrix}.$$

The entry-11 of AB was calculated using row 1 of A and column 1 of B:

$$2 \times 7 + (-1) \times (-2) + 5 \times 1 = 21.$$

The entry-12 of AB was calculated using row 1 of A and column 2 of B:

$$2 \times 0 + (-1) \times 1 + 5 \times 0 = -1.$$

The entry-13 of AB was calculated using row 1 of A and column 3 of B:

$$2 \times (-1) + (-1) \times 0 + 5 \times (-3) = -17.$$

Analogously, row 2 of AB was calculated using row 2 of A and all columns of B.

It can easily be seen from (1.13) and (1.15) that

Entry-ij of AB is the product of row i of A and column j of B.

Exercise. Let A be a $k \times n$ matrix and let $\begin{bmatrix} b_{11} & b_{12} & \dots & b_{1n} \end{bmatrix}$ be a row vector. Show that

$$\begin{bmatrix} b_{11} & b_{12} & \dots & b_{1n} \end{bmatrix} A = b_{11}\mathbf{l}_1 + b_{12}\mathbf{l}_2 + \cdots + b_{1n}\mathbf{l}_n,$$

where $\mathbf{l}_1, \mathbf{l}_2, \dots, \mathbf{l}_n$ are the rows of A.

The next proposition describes the product of two matrices both by columns (cf. Proposition 1.7 (i)) and by rows (cf. Proposition 1.7 (ii)).

Proposition 1.7 *Let $A \in \mathrm{M}_{k,p}(\mathbb{K})$ and $B \in \mathrm{M}_{p,n}(\mathbb{K})$ be matrices such that $\mathbf{a_1}, \mathbf{a_2}, \dots, \mathbf{a_k}$ are the rows of A and $\mathbf{b_1}, \mathbf{b_2}, \dots, \mathbf{b_n}$ are the columns of B. Then,*

(i)

$$AB = \begin{bmatrix} A\mathbf{b_1} & | & A\mathbf{b_2} & | & \cdots & | & A\mathbf{b}_n \end{bmatrix};$$

(ii)

$$AB = \begin{bmatrix} \mathbf{a}_1 B \\ \mathbf{a}_2 B \\ \vdots \\ \mathbf{a}_k B \end{bmatrix}.$$

Proof *Assertion (i) is a consequence of* (1.15)*. The proof of (ii) is left as an exercise.*

The next proposition collects properties of matrix multiplication.

Proposition 1.8 *Let A, B, C be matrices over $\mathbb{K}$ of appropriate sizes and let $\alpha \in \mathbb{K}$. The following assertions hold.*

(i) $A(BC) = (AB)C$, i.e., the multiplication of matrices is an associative operation.

(ii) Matrix multiplication is distributive relative to matrix addition, i.e.,

$$(A+B)C = AC + BC$$
$$A(B+C) = AB + AC$$

(iii) $\alpha(AB) = (\alpha A)B = A(\alpha B)$

Proof *Assertions (ii) and (iii) are left as exercises. We prove now assertion (i).*

Let $A = [a_{ij}]$, $B = [b_{ij}]$, *and* $C = [c_{ij}]$ *be* $k\times m, m\times p$ *and* $p\times n$ *matrices, respectively. Since both* $A(BC) = [d_{ij}]$ *and* $(AB)C = [e_{ij}]$ *are* $k\times n$ *matrices, it only remains to show that, for all* $i = 1,\dots k, j = 1,\dots,n$, *the entries* d_{ij} *and* e_{ij} *coincide.*

The entry d_{ij} *is calculated by multiplying row* i *of* A *and column* j *of* BC *which, in turn, is the product of* B *and column* j *of* C. *That is,*

$$\begin{bmatrix} a_{i1} & a_{i2} & \dots & a_{im} \end{bmatrix} \left(B \begin{bmatrix} c_{1j} \\ c_{2j} \\ \vdots \\ c_{1p} \end{bmatrix} \right) = \begin{bmatrix} a_{i1} & a_{i2} & \dots & a_{im} \end{bmatrix} \begin{bmatrix} \sum_{s=1}^{p} b_{1s}c_{sj} \\ \sum_{s=1}^{p} b_{2s}c_{sj} \\ \vdots \\ \sum_{s=1}^{p} b_{ps}c_{sj} \end{bmatrix}.$$

Hence

$$d_{ij} = \sum_{r=1}^{m} \left(\sum_{s=1}^{p} a_{ir}b_{rs}c_{sj} \right).$$

On the other hand, e_{ij} *is the product of row* i *of* AB *and column* j *of* C, *i.e.,*

$$\left(\begin{bmatrix} a_{i1} & a_{i2} & \dots & a_{im} \end{bmatrix} B \right) \begin{bmatrix} c_{1j} \\ c_{2j} \\ \vdots \\ c_{1p} \end{bmatrix}$$

$$= \begin{bmatrix} \sum_{r=1}^{m} a_{ir}b_{r1} & \sum_{r=1}^{m} a_{ir}b_{r2} & \dots & \sum_{r=1}^{m} a_{ir}b_{rm} \end{bmatrix} \begin{bmatrix} c_{1j} \\ c_{2j} \\ \vdots \\ c_{pj} \end{bmatrix}.$$

It follows that

$$e_{ij} = \sum_{s=1}^{p} \left(\sum_{r=1}^{m} a_{ir}b_{rs}c_{sj} \right)$$

which, by the commutativity and associativity of scalar addition, yields finally

$$e_{ij} = \sum_{r=1}^{m} \left(\sum_{s=1}^{p} a_{ir}b_{rs}c_{sj} \right) = d_{ij}.$$

Unlike matrix addition, matrix multiplication is not a commutative operation.

For example,

$$\begin{bmatrix} 0 & 1 \\ 1 & 0 \end{bmatrix} \begin{bmatrix} 1 & 2 \\ 3 & 4 \end{bmatrix} = \begin{bmatrix} 3 & 4 \\ 1 & 2 \end{bmatrix},$$

whereas

$$\begin{bmatrix} 1 & 2 \\ 3 & 4 \end{bmatrix} \begin{bmatrix} 0 & 1 \\ 1 & 0 \end{bmatrix} = \begin{bmatrix} 2 & 1 \\ 4 & 3 \end{bmatrix}.$$

Matrix multiplication allows us to write systems of linear equations in a more compact manner. Consider a system having k linear equations, n variables $x_1, x_2, \ldots, x_n$, and let A be the coefficient matrix (hence, A is a $k \times n$ matrix). Let $\mathbf{b}$ be the column vector of independent terms and define the column vector of variables by

$$\mathbf{x} = \begin{bmatrix} x_1 \\ x_2 \\ \vdots \\ x_n \end{bmatrix}.$$

With this notation, this system can be presented as the **matrix equation**

$$A\mathbf{x} = \mathbf{b}. \tag{1.16}$$

Example 1.13 *As an example, we write the matrix equation associated with the system*

$$\begin{cases} x + y + z = 3 \\ x - y + 2z = 2. \end{cases}$$

In this case, we have the following matrix equation $A\mathbf{x} = \mathbf{b}$

$$\underbrace{\begin{bmatrix} 1 & 1 & 1 \\ 1 & -1 & 2 \end{bmatrix}}_{A} \underbrace{\begin{bmatrix} x \\ y \\ z \end{bmatrix}}_{\mathbf{x}} = \underbrace{\begin{bmatrix} 3 \\ 2 \end{bmatrix}}_{\mathbf{b}}.$$

Solving the system can now be reformulated as solving the matrix equation (1.16). As we saw, $A\mathbf{x}$ is a linear combination of the columns of A, hence:

Proposition 1.9 *Let A and $\mathbf{b}$ be, respectively, a $k \times n$ matrix and an $k \times 1$ vector over the field $\mathbb{K}$. The equation $A\mathbf{x} = \mathbf{b}$ is solvable if and only if $\mathbf{b}$ is a linear combination of the columns of A.*

If the system under consideration is homogeneous, that is, $\mathbf{b}$ is a $k \times 1$ zero vector, then the matrix equation corresponding to this system is

$$A\mathbf{x} = \mathbf{0}. \tag{1.17}$$

This homogeneous equation is always solvable since it has, at least, the trivial solution $\mathbf{x} = \mathbf{0}$.

Matrices can be partitioned into blocks and, as long as the sizes of the blocks and the partitions are compatible, it is possible to devise a block multiplication. For example, let

$$A = \left[\begin{array}{cc|cc} 1 & 2 & 3 & 4 \\ 1 & 2 & 3 & 4 \\ \hline 1 & 2 & 3 & 4 \end{array}\right] = \left[\begin{array}{c|c} A_{11} & A_{12} \\ \hline A_{21} & A_{22} \end{array}\right]$$

and

$$B = \left[\begin{array}{cc|cc|cc} 1 & 0 & -1 & 0 & 0 & 0 \\ 0 & 1 & 0 & -1 & 0 & 0 \\ \hline -1 & 0 & 0 & 0 & 1 & 0 \\ 0 & -1 & 0 & 0 & 0 & 1 \end{array}\right] = \left[\begin{array}{c|c|c} B_{11} & B_{12} & B_{13} \\ \hline B_{21} & B_{22} & B_{23} \end{array}\right].$$

Using this block partition of matrices A and B, we have

$$AB = \left[\begin{array}{c|c} A_{11} & A_{12} \\ \hline A_{21} & A_{22} \end{array}\right] \left[\begin{array}{c|c|c} B_{11} & B_{12} & B_{13} \\ \hline B_{21} & B_{22} & B_{23} \end{array}\right]$$

and it is not difficult to see that one can multiply the blocks as if they were numbers, i.e.,

$$AB = \left[\begin{array}{c|c|c} A_{11}B_{11} + A_{12}B_{21} & A_{11}B_{21} + A_{12}B_{22} & A_{11}B_{13} + A_{12}B_{23} \\ \hline A_{21}B_{11} + A_{22}B_{21} & A_{21}B_{12} + A_{22}B_{22} & A_{21}B_{13} + A_{22}B_{23} \end{array}\right].$$

Hence

$$AB = \left[\begin{array}{c|c|c} A_{11} - A_{12} & -A_{11} & A_{12} \\ \hline A_{21} - A_{22} & -A_{21} & A_{22} \end{array}\right] = \left[\begin{array}{cc|cc|cc} -2 & -2 & -1 & -2 & 3 & 4 \\ -2 & -2 & -1 & -2 & 3 & 4 \\ \hline -2 & -2 & -1 & -2 & 3 & 4 \end{array}\right].$$

Finally, we have

$$AB = \begin{bmatrix} -2 & -2 & -1 & -2 & 3 & 4 \\ -2 & -2 & -1 & -2 & 3 & 4 \\ -2 & -2 & -1 & -2 & 3 & 4 \end{bmatrix}.$$

Block multiplication may be rather convenient as it may both simplify considerably the calculations involved and make them more transparent. Proposition 1.7 (i) is an example of block multiplication (and so is (ii) of the same proposition).

If we now divide A into its columns and B into its rows, then AB coincides with the sum

$$AB = \begin{bmatrix} \mathbf{a}_1 & | & \mathbf{a}_2 & | & \dots & | & \mathbf{a}_p \end{bmatrix} \begin{bmatrix} \mathbf{b}_1 \\ \mathbf{b}_2 \\ \vdots \\ \mathbf{b}_p \end{bmatrix} = \mathbf{a}_1\mathbf{b}_1 + \mathbf{a}_2\mathbf{c}_2 + \dots + \mathbf{a}_p\mathbf{c}_p.$$

The following example is a concrete illustration of this way of calculating the product.

Example 1.14 *Consider the matrices*

$$A = \begin{bmatrix} 1 & 2 & -1 \\ 1 & 2 & -1 \end{bmatrix} \qquad B = \begin{bmatrix} 3 & 4 \\ 5 & 6 \\ 7 & 8 \end{bmatrix}.$$

Then

$$AB = \begin{bmatrix} 1 \\ 1 \end{bmatrix} \begin{bmatrix} 3 & 4 \end{bmatrix} + \begin{bmatrix} 2 \\ 2 \end{bmatrix} \begin{bmatrix} 5 & 6 \end{bmatrix} + \begin{bmatrix} -1 \\ -1 \end{bmatrix} \begin{bmatrix} 7 & 8 \end{bmatrix} = \begin{bmatrix} 6 & 8 \\ 6 & 8 \end{bmatrix}.$$

The next proposition summarises the above discussion and, with respect to Proposition 1.7, gives a third possible way of calculating the product of two matrices other than the definition (1.15).

Proposition 1.10 *Let* $A \in \mathrm{M}_{k,p}(\mathbb{K})$ *and* $B \in \mathrm{M}_{p,n}(\mathbb{K})$ *be matrices such that* $\mathbf{a_1}, \mathbf{a_2}, \dots, \mathbf{a_p}$ *are the columns of* A *and* $\mathbf{b_1}, \mathbf{b_2}, \dots, \mathbf{b_p}$ *are the rows of* B. *Then,*

$$AB = \begin{bmatrix} \mathbf{a}_1 & | & \mathbf{a}_2 & | & \dots & | & \mathbf{a}_p \end{bmatrix} \begin{bmatrix} \mathbf{b}_1 \\ \mathbf{b}_2 \\ \vdots \\ \mathbf{b}_p \end{bmatrix} = \mathbf{a}_1\mathbf{b}_1 + \mathbf{a}_2\mathbf{c}_2 + \dots + \mathbf{a}_p\mathbf{c}_p.$$

Proof *Exercise.*

We end this section outlining four ways in which the product of two matrices can be calculated.

How to multiply two matrices

Given a $k \times p$ matrix A and a $p \times n$ matrix B, the product AB is a $k \times n$ matrix which can be determined as follows.

- By entry: entry-ij of AB is the product of row i of A and column j of B (see (1.15)).
- By column: column j of AB coincides with a linear combination of the columns of A whose coefficients are the entries of column j of B (see Proposition 1.7 (i)).
- By row: row i of AB coincides with a 'linear combination' of the rows of B whose coefficients are the entries in row i of A (see Proposition 1.7 (ii)).
- By column-row: AB is a sum of all the products $a_i b_i$ of columns a_i of A and rows b_i of B (see Proposition 1.10).

When we restrict ourselves to considering square matrices of a fixed order, it is always possible to calculate the product of any two such matrices and obtain a matrix having the same size. In particular, we can multiply a matrix by itself as many times as desired.

Definition 11 *Let A be a square matrix of order k and let $n \in \mathbb{N}_0$ be a non-negative integer. The* **power** n **of** A *is defined recursively by*

$$A^0 = I_k, \qquad A^n = AA^{n-1} \qquad (n \geq 1)$$

In other words, when n is a positive integer, A^n is the product of n factors equal to A:

$$A^n = \underbrace{AA \cdots A}_{n}.$$

Proposition 1.11 *Let A be a square matrix and let $n, m \in \mathbb{N}_0$.*

(i) $A^{n+m} = A^n A^m$

(ii) $(A^n)^m = A^{nm}$

Proof *The assertions are an immediate consequence of the associativity of the multiplication of matrices.*

Definition 12 *Let A be a matrix over $\mathbb{K}$ of size $k \times n$. The* **transpose matrix of** A *is the $n \times k$ matrix defined by $(A^T)_{ij} = a_{ji}$.*

Example 1.15 *The transpose matrix A^T of the 4×3 matrix*

$$A = \begin{bmatrix} 1 & 3 & 2 \\ 1 & -1 & 2 \\ 1 & 1 & -1 \\ 9 & 8 & 7 \end{bmatrix}$$

is the 3×4 matrix

$$A^T = \begin{bmatrix} 1 & 1 & 1 & 9 \\ 3 & -1 & 1 & 8 \\ 2 & 2 & -1 & 7 \end{bmatrix}.$$

The transposition of a matrix A can be described in a nutshell as obtaining a new matrix whose rows are the columns of the initial matrix, i.e., row j of A^T is column j of A.

It is thus defined an operation $^T : \mathrm{M}_{k,n} \to \mathrm{M}_{n,k}$ which satisfies the properties listed in the next proposition.

Proposition 1.12 *Let A, B be matrices over $\mathbb{K}$ with appropriate sizes and let $\alpha \in \mathbb{K}$. Then,*

(i) $(A^T)^T = A$

(ii) $(A + B)^T = A^T + B^T$

(iii) $(\alpha A)^T = \alpha A^T$

(iv) $(AB)^T = B^T A^T$

It is worth noticing that properties (ii) and (iii) above can be seen in an informal manner as, respectively, "the transpose of a sum is the sum of the transposes" and "the transpose of a product is the product of the transposes in the reverse order".

A common mistake. An often overlooked fact is that, when calculating the transpose of the product of a certain number of matrices, one must reverse the order. More precisely, according to Proposition 1.12 (iv), if $A_1, A_2, \ldots, A_m$ are matrices over the same field having appropriate sizes, then

$$(A_1 A_2 \ldots A_m)^T = A_m^T \ldots A_2^T A_1^T.$$

Proof *We only prove (iv). Let A be a $k \times p$ and let B be a $p \times n$ matrix. Then $(AB)^T$ is an $n \times k$ matrix whose entry $((AB)^T)_{ij}$ coincides with $(AB)_{ji}$, by definition. Hence*

$$\begin{aligned} ((AB)^T)_{ij} = (AB)_{ji} &= \begin{bmatrix} a_{j1} & a_{j2} & \dots & a_{jp} \end{bmatrix} \begin{bmatrix} b_{1i} \\ b_{2i} \\ \vdots \\ b_{pi} \end{bmatrix} \\ &= \begin{bmatrix} b_{1i} & b_{2i} & \dots & b_{pi} \end{bmatrix} \begin{bmatrix} a_{j1} \\ a_{j2} \\ \vdots \\ a_{jp} \end{bmatrix} \\ &= (B^T A^T)_{ij}. \end{aligned}$$

Definition 13 *Let A be an $n \times n$ matrix over $\mathbb{K}$. The matrix A is said to be a **symmetric matrix** if $A = A^T$ or, equivalently, if $a_{ij} = a_{ji}$, for all indices $i, j = 1, \dots, n$. The matrix A is said to be an **anti-symmetric matrix** if $A = -A^T$ or, equivalently, if $a_{ij} = -a_{ji}$, for all indices $i, j = 1, \dots, n$.*

*Given a matrix $A = [a_{ij}]$ of order n, the **main diagonal** or, simply, the **diagonal** of A consists of the n entries a_{ii}, where $i = 1, \dots, n$. The **anti-diagonal** of A consists of the n entries a_{ij} such that $i + j = n + 1$.*

Example 1.16 *Consider the following matrices*

$$A = \begin{bmatrix} \mathbf{1} & 2 & -3 \\ 2 & \mathbf{4} & 5 \\ -3 & 5 & \mathbf{-6} \end{bmatrix} \qquad B = \begin{bmatrix} 0 & 2 & -3 \\ -2 & 0 & 5 \\ 3 & -5 & 0 \end{bmatrix}.$$

Matrix A is symmetric while B is an anti-symmetric matrix. The diagonal of A is in bold and the anti-diagonal of B is in grey.

The following consequences of Definition 13 are noteworthy.

- The diagonal of an anti-symmetric matrix is null, i.e., all its entries are equal to zero.
- The diagonal of a symmetric matrix is 'like a mirror' reflecting the entries on both sides of the diagonal.
- When finding the transpose of a square matrix, the diagonal entries remain unchanged.
- Any square matrix A can be decomposed into a sum of a symmetric matrix with an anti-symmetric matrix:

$$A = \tfrac{1}{2}(A + A^T) + \tfrac{1}{2}(A - A^T).$$

Exercise. Show that, given a square matrix A, the matrix $A + A^T$ is symmetric and the matrix $A - A^T$ is anti-symmetric.

Definition 14 *Let $A = [a_{ij}]$ be a square matrix of order n. The* ***trace*** *of A is the sum of all entries in the diagonal of A, i.e.,*

$$\operatorname{tr} A = \sum_{i=1}^{n} a_{ii}.$$

For example, considering the matrices A, B in Example 1.16, we have that $\operatorname{tr} A = -1$ and $\operatorname{tr} B = 0$ (indeed, the trace of an anti-symmetric matrix is always zero).

Proposition 1.13 *Let A, B be square matrices of order n over $\mathbb{K}$ and let α be a scalar. Then,*

(i) $\operatorname{tr}(A + B) = \operatorname{tr} A + \operatorname{tr} B$;

(ii) $\operatorname{tr}(\alpha A) = \alpha \operatorname{tr} A$;

(iii) $\operatorname{tr} A = \operatorname{tr} A^T$;

(iv) $\operatorname{tr}(AB) = \operatorname{tr}(BA)$.

Proof *Assertions (i)–(iii) are immediate.*
(iv) By Proposition 1.12 (iv) and (iii) of this proposition, we have

$$\operatorname{tr}(AB) = \operatorname{tr}((AB)^T) = \operatorname{tr}(B^T A^T).$$

It is now enough to show that $(AB)_{ii} = (B^T A^T)_{ii}$, for $i =, \dots, n$. But

$$(AB)_{ii} = \sum_{r=1}^{n} a_{ir} b_{ri} = \sum_{r=1}^{n} b_{ri} a_{ir} = (B^T A^T)_{ii},$$

as required.

Although matrix multiplication is not a commutative operation, it might happen that two particular matrices A, B commute, that is, $AB = BA$. For example, if $B = A$, or if A is any $n \times n$ matrix and B is the $n \times n$ zero matrix, it is clear that $AB = 0 = BA$.

In the next section, we will see a square matrix of order n which commutes with all square matrices of the same order. This is the identity matrix.

1.3 Matrix Inverses

Definition 15 *A square matrix $A = [a_{ij}]$ is said to be a* ***diagonal matrix*** *if $a_{ij} = 0$ whenever $i \neq j$. In other words, A is a diagonal matrix if all its off-diagonal entries are equal to 0.*

For example, the matrices

$$A = \begin{bmatrix} 1 & 0 \\ 0 & -3 \end{bmatrix} \quad B = \begin{bmatrix} 0 & 0 & 0 \\ 0 & 2 & 0 \\ 0 & 0 & 9 \end{bmatrix} \quad C = \begin{bmatrix} 0 & 0 & 0 \\ 0 & 0 & 0 \\ 0 & 0 & 0 \end{bmatrix} \quad D = \begin{bmatrix} 1 & 0 & 0 \\ 0 & 1 & 0 \\ 0 & 0 & 1 \end{bmatrix}$$

are all diagonal matrices.

The **identity matrix of order** n, I_n, is the $n \times n$ diagonal matrix whose diagonal entries are all equal to 1. The identity matrix will be named I whenever its size is clear from the context.

Matrix D above is the identity matrix of order 3, i.e., $D = I_3$.

Proposition 1.14 *Let A be an $n \times k$ matrix and let B be a $k \times n$ matrix over the same field, and let I_n be the identity matrix. Then,*

(i) $I_n A = A$;

(ii) $BI_n = B$.

Proof *Exercise.*

Multiplication of any two matrices in $\mathrm{M}_n(\mathbb{K})$ is always possible, and the resulting product is again a square matrix of order n. Moreover, in $\mathrm{M}_n(\mathbb{K})$, by Proposition 1.14, I_n is the multiplicative identity, that is, for all $A \in \mathrm{M}_n(\mathbb{K})$,

$$AI = A = IA. \tag{1.18}$$

Observe that I is the unique $n \times n$ matrix satisfying (1.18). Indeed, if one supposes that J is an $n \times n$ matrix such that, for all $A \in \mathrm{M}_n(\mathbb{K})$, $AJ = A = JA$, then $JI = I$ and also, by (1.18), $JI = J$. Hence $J = I$.

Definition 16 *Let A be a square matrix of order n over the field $\mathbb{K}$. A matrix B over $\mathbb{K}$ is called an* ***inverse*** *of A if*

$$AB = I = BA. \tag{1.19}$$

Observe that, by (1.19), if such a matrix B exists then B must be a square matrix of order n.

Lemma 1.1 *Let $A \in \mathrm{M}_n(\mathbb{K})$ be a matrix. If there exists a matrix B satisfying Definition 16, then B is unique.*

Proof *Suppose that $B, C \in \mathrm{M}_n(\mathbb{K})$ are inverses of A. Then*

$$B(AC) = BI = B \qquad \text{and} \qquad (BA)C = IC = C.$$

Since matrix multiplication is associative, we have $B(AC) = (BA)C$. It follows that $B = C$.

This lemma allows us to define the (unique) **inverse matrix** A^{-1} of A, in case it exists. A matrix is said to be **invertible** or **non-singular** if it has an inverse and is called **singular**, otherwise.

Example 1.17 *It is not difficult to find the inverses of the matrices below.*

(a) Observing that the product of the identity matrix (of any order) with itself is again the identity matrix, we have

$$\begin{bmatrix} 1 & 0 \\ 0 & 1 \end{bmatrix}^{-1} = \begin{bmatrix} 1 & 0 \\ 0 & 1 \end{bmatrix}.$$

This shows that the identity matrix of order 2 is invertible and is its own inverse. Clearly, the same applies to the identity matrix of any order.

(b) Consider now the matrix $A = 2I$. Since

$$(2I)\frac{1}{2}I = I = \frac{1}{2}I(2I),$$

we see that

$$A^{-1} = \begin{bmatrix} 2 & 0 \\ 0 & 2 \end{bmatrix}^{-1} = \left(2\begin{bmatrix} 1 & 0 \\ 0 & 1 \end{bmatrix}\right)^{-1} = \begin{bmatrix} \frac{1}{2} & 0 \\ 0 & \frac{1}{2} \end{bmatrix}.$$

Proposition 1.15 *Any square matrix having a zero row or a zero column is not invertible.*

Proof *Suppose that A is an $n \times n$ matrix whose row $\mathbf{l}_i = \mathbf{0}$. Then, row i of the product AB of the matrix A and any matrix B is a zero row. Consequently, it is impossible to have AB equal to the identity matrix and, therefore, A is singular.*

Let now the column j of A be null. It follows that the column j of BA, where B is any $n \times n$ matrix, is also a zero column. Hence, it is not possible for BA to be the identity from which follows that A is not invertible.

Proposition 1.16 *Let A be an $n \times n$ invertible matrix over $\mathbb{K}$ and let $\mathbf{b}$ be an $n \times 1$ column vector over the same field $\mathbb{K}$. Then the system of linear equations $A\mathbf{x} = \mathbf{b}$ has the unique solution $\mathbf{x} = A^{-1}\mathbf{b}$.*

Proof *Let A^{-1} be the inverse of A and consider the system $A\mathbf{x} = \mathbf{b}$. Then, multiplying both members of this equality on the left by A^{-1}, it follows that*

$$\begin{aligned} A^{-1}(A\mathbf{x}) = A^{-1}\mathbf{b} \quad &\Leftrightarrow (A^{-1}A)\mathbf{x} = A^{-1}\mathbf{b} \\ &\Leftrightarrow I\mathbf{x} = A^{-1}\mathbf{b} \\ &\Leftrightarrow \mathbf{x} = A^{-1}\mathbf{b}. \end{aligned}$$

Hence, the system is consistent and has the unique solution $\mathbf{x} = A^{-1}\mathbf{b}$.

In Example 1.17, we found the inverses of two exceptionally simple matrices. We need however to devise a method to obtain the inverse of any matrix, should it exist. That is precisely what we will do next, using a concrete matrix as a model.

Consider the matrix

$$A = \begin{bmatrix} 1 & -2 \\ -1 & 1 \end{bmatrix}.$$

We aim to find whether this matrix is invertible and, in the affirmative situation, find its inverse. That is, we need to find, if possible, a matrix

$$B = \begin{bmatrix} x_1 & x_2 \\ y_1 & y_2 \end{bmatrix}$$

such that $AB = I$ and $BA = I$.

Starting with equation $AB = I$, we have

$$A \begin{bmatrix} x_1 & x_2 \\ y_1 & y_2 \end{bmatrix} = \begin{bmatrix} 1 & 0 \\ 0 & 1 \end{bmatrix}.$$

Hence, we must solve the systems

$$A \begin{bmatrix} x_1 \\ y_1 \end{bmatrix} = \begin{bmatrix} 1 \\ 0 \end{bmatrix} \qquad A \begin{bmatrix} x_2 \\ y_2 \end{bmatrix} = \begin{bmatrix} 0 \\ 1 \end{bmatrix}.$$

Observing that both systems have the same matrix of coefficients, we will solve them simultaneously using the Gauss–Jordan elimination method. Thus,

$$\left[\begin{array}{cc|cc} 1 & -2 & 1 & 0 \\ -1 & 1 & 0 & 1 \end{array}\right] \xrightarrow[l_2+l_1]{} \left[\begin{array}{cc|cc} 1 & -2 & 1 & 0 \\ 0 & -1 & 1 & 1 \end{array}\right] \xrightarrow[l_1-2l_2]{} \left[\begin{array}{cc|cc} 1 & 0 & -1 & -2 \\ 0 & -1 & 1 & 1 \end{array}\right] \cdots$$

$$\cdots \xrightarrow[-l_2]{} \left[\begin{array}{cc|cc} 1 & 0 & -1 & -2 \\ 0 & 1 & -1 & -1 \end{array}\right].$$

Recall that the grey column corresponds to the system with variables x_2, y_2, whereas the adjacent column corresponds to the system with variables x_1, y_1.

Keeping this in mind, it follows that the unique matrix B satisfying equation $AB = I$ is

$$B = \begin{bmatrix} -1 & -2 \\ -1 & -1 \end{bmatrix}.$$

At this point, we know that, if A is invertible, then B has to be its inverse. Hence, to end our search for the inverse of A, it only remains to show that $BA = I$. We only have to calculate the product BA and see that it equals I. Hence,

$$A^{-1} = \begin{bmatrix} -1 & -2 \\ -1 & -1 \end{bmatrix}.$$

The calculations above can be summed up as follows. In order to find the inverse of

$$A = \begin{bmatrix} 1 & -2 \\ -1 & 1 \end{bmatrix},$$

1. we solved the systems of linear equations $AB = I$ using Gauss–Jordan elimination:

$$[A|I] = \left[\begin{array}{cc|cc} 1 & -2 & 1 & 0 \\ -1 & 1 & 0 & 1 \end{array}\right] \xrightarrow[\text{GJE}]{} \left[\begin{array}{cc|cc} 1 & 0 & -1 & -2 \\ 0 & 1 & -1 & -1 \end{array}\right] = [I|B];$$

2. we verified that $BA = I$;
3. we concluded that $A^{-1} = B$.

Point 2. above can be avoided, as is shown in the proposition below.

Proposition 1.17 *Let A, B be square matrices of order n over $\mathbb{K}$ and let I be the identity matrix of order n. Then, $AB = I$ if and only if $BA = I$.*

Proof *This result will be proved further on in the book (cf. Proposition 1.20),*

Finally, the general procedure to obtain the inverse of a square matrix A, should it exist, is summarised in the box below.

How to calculate the inverse of an invertible matrix A

Let A be an $n \times n$ matrix and let I be the $n \times n$ identity matrix.

The matrix $[A|I]$ is reduced to the matrix $[I|A^{-1}]$, using Gauss–Jordan elimination. Symbolically,

$$[A|I] \xrightarrow[\text{GJE}]{} [I|A^{-1}]$$

The next proposition lists some relevant properties of the inverse of a matrix.

Proposition 1.18 *Let A, B be order k invertible matrices over $\mathbb{K}$, let $\alpha \in \mathbb{K}$ be a non-zero scalar and let $n \in \mathbb{N}_0$. Then A^{-1}, AB, A^n, αA, and A^T are invertible matrices such that*

(i) $(A^{-1})^{-1} = A$;

(ii) $(AB)^{-1} = B^{-1}A^{-1}$;

(iii) $(A^n)^{-1} = (A^{-1})^n$;

(iv) $(\alpha A)^{-1} = \frac{1}{\alpha}A^{-1}$;

(v) $(A^T)^{-1} = (A^{-1})^T$.

Informally, (v) can be seen as expressing the fact that "the inverse of the transpose is the transpose of the inverse".

A common mistake. As observed in (ii) above, calculating the inverse of the product of invertible matrices requires reversing the order of multiplication. That is, if $A_1, A_2, \dots, A_m$ are $n \times n$ invertible matrices over the same field, then

$$(A_1 A_2 \dots A_m)^{-1} = A_m^{-1} \dots A_2^{-1} A_1^{-1}.$$

Proof *We prove only (ii), the remaining assertions are left as an exercise.*

Directly evaluating $(B^{-1}A^{-1})(AB)$ and keeping in mind that matrix multiplication is an associative operation , we have

$$(B^{-1}A^{-1})(AB) = B^{-1}(A^{-1}A)B = B^{-1}IB = B^{-1}B = I.$$

It now follows from Proposition 1.17 that $B^{-1}A^{-1}$ is the inverse matrix of AB.

We have defined already the non-negative powers of a square matrix (see Definition 11). We extend now the definition to negative powers of invertible matrices by means of Proposition 1.18 (iii). In fact, this proposition allows for unequivocally defining the integer powers of an invertible matrix A.

Definition 17 *Let A be an invertible matrix of order k and let $n \in \mathbb{N}$ be a positive integer. The* ***power*** $-n$ ***of*** *A is defined by*

$$A^{-n} = (A^n)^{-1} = (A^{-1})^n.$$

Proposition 1.19 *Let A be an invertible matrix and let $r, s \in \mathbb{Z}$.*

(i) $A^{r+s} = A^r A^s$

(ii) $(A^r)^s = A^{rs}$

Proof *Exercise.*

1.4 Elementary Matrices

This section is devoted to the elementary matrices and to their outstanding role amongst the invertible matrices. We shall see that they are generators of the invertible matrices inasmuch as any such matrix is a product of elementary matrices. In broad strokes, an elementary matrix of order n is a matrix obtained from the identity through a single elementary operation (hence the name). Since there are three types of elementary operations, we get three types of elementary matrices.

Definition 18 *A square matrix of order n is said to be an* **elementary matrix** *if it coincides with one of the matrices P_{ij}, $E_{ij}(\alpha)$, $D_i(\alpha)$ below.*

- *P_{ij} (with $i < j$): the matrix that is obtained from the identity matrix (of order n) by exchanging rows i and j;*
- *$E_{ij}(\alpha)$ (with $i \neq j$) : the matrix that is obtained from the identity matrix by adding to row i row j multiplied by $\alpha \in \mathbb{K}$;*
- *$D_i(\alpha)$ (with $\alpha \neq 0$): the matrix that is obtained from the identity matrix multiplying row i by α.*

The three types of elementary are illustrated below. The rows and columns i are coloured light grey and the rows and columns j are coloured dark grey.

$$P_{ij} = \begin{bmatrix} 1 & 0 & \cdots & 0 & \cdots & 0 & \cdots & 0 \\ 0 & 1 & \cdots & 0 & \cdots & 0 & \cdots & 0 \\ \vdots & \vdots & \ddots & \vdots & & \vdots & & \vdots \\ 0 & 0 & \cdots & 0 & \cdots & 1 & \cdots & 0 \\ \vdots & \vdots & & \vdots & \ddots & \vdots & & \vdots \\ 0 & 0 & \cdots & 1 & \cdots & 0 & \cdots & 0 \\ \vdots & \vdots & & \vdots & & \vdots & \ddots & \vdots \\ 0 & 0 & \cdots & 0 & \cdots & 0 & \cdots & 1 \end{bmatrix}$$

$$E_{ij}(\alpha) = \begin{bmatrix} 1 & 0 & \cdots & 0 & \cdots & 0 & \cdots & 0 \\ 0 & 1 & \cdots & 0 & \cdots & 0 & \cdots & 0 \\ \vdots & \vdots & \ddots & \vdots & & \vdots & & \vdots \\ 0 & 0 & \cdots & 1 & \cdots & \alpha & \cdots & 0 \\ \vdots & \vdots & & \vdots & \ddots & \vdots & & \vdots \\ 0 & 0 & \cdots & 0 & \cdots & 1 & \cdots & 0 \\ \vdots & \vdots & & \vdots & & \vdots & \ddots & \vdots \\ 0 & 0 & \cdots & 0 & \cdots & 0 & \cdots & 1 \end{bmatrix}$$

$$D_i(\alpha) = \begin{bmatrix} 1 & 0 & \cdots & 0 & \cdots & 0 & \cdots & 0 \\ 0 & 1 & \cdots & 0 & \cdots & 0 & \cdots & 0 \\ \vdots & \vdots & \ddots & \vdots & & \vdots & & \vdots \\ 0 & 0 & \cdots & \alpha & \cdots & 0 & \cdots & 0 \\ \vdots & \vdots & & \vdots & \ddots & \vdots & & \vdots \\ 0 & 0 & \cdots & 0 & \cdots & 1 & \cdots & 0 \\ \vdots & \vdots & & \vdots & & \vdots & \ddots & \vdots \\ 0 & 0 & \cdots & 0 & \cdots & 0 & \cdots & 1 \end{bmatrix}$$

Given an $n \times p$ matrix A over $\mathbb{K}$, we describe next how these elementary matrices act on A when multiplied on the left. Hence all elementary matrices considered must be of order n, obviously.

- $A' = P_{ij}A$: the matrix A' is obtained by exchanging rows i and j of A, i.e., in the pre-established notation,

$$A \xrightarrow[\mathbf{l}_i \leftrightarrow \mathbf{l}_j]{} A' = P_{ij}A$$

- $A' = E_{ij}(\alpha)A$: the matrix A' is obtained from A adding to row i row j multiplied by α, i.e.,

$$A \xrightarrow[\mathbf{l}_i + \alpha\mathbf{l}_j]{} A' = E_{ij}(\alpha)A$$

- $A' = D_i(\alpha)A$: the matrix A' is obtained multiplying row i of A by α, i.e.,

$$A \xrightarrow[\alpha\mathbf{l}_i]{} A' = D_i(\alpha)A$$

It is a simple exercise to see that the results above are true. It is however desirable that one convinces oneself that the results do hold.

Having reached this point, we see that performing an elementary operation on a matrix A amounts to multiplying A on the left by the appropriate elementary matrix. Summing it all up in a sort of

Dictionary

Elementary operation	Elementary matrix
$A \xrightarrow[\mathbf{l}_i \leftrightarrow \mathbf{l}_j]{} A'$	$P_{ij}A = A'$
$A \xrightarrow[\mathbf{l}_i + \alpha \mathbf{l}_j]{} A'$	$E_{ij}(\alpha)A = A'$
$A \xrightarrow[\alpha \mathbf{l}_i]{} A'$	$D_i(\alpha)A = A'$

Example 1.18 *We examine now under this new light of elementary matrices the calculations we made to find the inverse matrix of*

$$A = \begin{bmatrix} 1 & -2 \\ -1 & 1 \end{bmatrix}.$$

The elementary operations that were performed on matrix $[A|I]$ *correspond to the following sequential multiplications by elementary matrices:*

Elementary operations	*Multiplication by elementary matrices*
$\mathbf{l}_2 + \mathbf{l}_1$	$\left[E_{21}(1)A \mid E_{21}(1)I\right]$
$\mathbf{l}_1 + (-2)\mathbf{l}_2$	$\left[E_{12}(-2)E_{21}(1)A \mid E_{12}(-2)E_{21}(1)I\right]$
$(-1)\mathbf{l}_2$	$\left[D_2(-1)E_{12}(-2)E_{21}(1)A \mid D_2(-1)E_{12}(-2)E_{21}(1)I\right]$

Hence

$$D_2(-1)E_{12}(-2)E_{21}(1)A = I \qquad D_2(-1)E_{12}(-2)E_{21}(1)I = A^{-1}$$

(cf. §1.3). It follows that A^{-1} *is a product of elementary matrices, i.e.,*

$$A^{-1} = D_2(-1)E_{12}(-2)E_{21}(1).$$

A common mistake. One should keep in mind that, in the Gaussian elimination or in the Gauss-Jordan elimination, the elementary matrices are always sequentially multiplied on the left.

All elementary matrices are invertible and their inverses are also elementary matrices. It is left as an exercise to see that

Inverses of elementary matrices

$$(P_{ij})^{-1} = P_{ij}$$

$$(E_{ij}(\alpha))^{-1} = E_{ij}(-\alpha)$$

$$(D_i(\alpha))^{-1} = D_i\left(\frac{1}{\alpha}\right)$$

Next we prove Proposition 1.17 whose statement we recall. Before, however, we make a note of a simple but useful fact to be applied in the proof of this proposition.

The reduced row echelon form of an invertible matrix is the identity matrix.

(Why?)

Proposition 1.20 *Let A, B be $n\times n$ matrices over $\mathbb{K}$ and let I be the identity matrix of order n. Then $AB = I$ if and only if $BA = I$.*

Proof *Suppose initially that $AB = I$, and let $E_1, E_2, \ldots, E_k$ be elementary matrices such that $E_1E_2\ldots E_kA$ is the reduced row echelon form of A. The reduced row echelon form of A cannot have any zero row. In fact, should it have a zero row, so would the matrix*

$$E_1E_2\ldots E_kAB = E_1E_2\ldots E_kI.$$

But this is impossible since $E_1E_2\ldots E_kI$ is a product of invertible matrices and, by Proposition 1.18 (ii), is itself invertible. It follows that

$$\underbrace{E_1E_2\ldots E_kA}_{I}B = E_1E_2\ldots E_k.$$

That is, B is a product of invertible matrices and, hence, is invertible. Multiplying on the right both members of the equality $AB = I$ by B^{-1}, we have

$$(AB)B^{-1} = IB^{-1} \quad\Longleftrightarrow\quad A = B^{-1}.$$

Hence, by the definition of inverse matrix, we have $AB = BA = I$.

Exchanging the roles of the matrices A and B in the above reasoning, we can show similarly that $BA = I$ implies $AB = I$.

Proposition 1.21 *Let A be a square matrix of order n over $\mathbb{K}$. Then A is invertible if and only if* $\operatorname{rank}(A) = n$.

Proof *We must prove the implications*

$$\operatorname{rank}(A) = n \Rightarrow A \textit{ is invertible}$$

and

$$A \textit{ is invertible} \Rightarrow \operatorname{rank}(A) = n.$$

We begin by proving the first implication. Suppose that A is a square matrix of order n having rank n. Then, by Proposition 1.3 (ii), the systems of linear equations

$$A\begin{bmatrix} x_{11} \\ x_{21} \\ \vdots \\ x_{n1} \end{bmatrix} = \begin{bmatrix} 1 \\ 0 \\ \vdots \\ 0 \end{bmatrix} \qquad A\begin{bmatrix} x_{12} \\ x_{22} \\ \vdots \\ x_{n2} \end{bmatrix} = \begin{bmatrix} 0 \\ 1 \\ \vdots \\ 0 \end{bmatrix} \qquad \dots \qquad A\begin{bmatrix} x_{1n} \\ x_{2n} \\ \vdots \\ x_{nn} \end{bmatrix} = \begin{bmatrix} 0 \\ 0 \\ \vdots \\ 1 \end{bmatrix}$$

are consistent and each one of them has a unique solution. Hence, there exists an $n \times n$ matrix B such that $AB = I$. By Proposition 1.20, it follows that A is invertible.

As to the second implication, we shall prove equivalently that

$$\operatorname{rank}(A) \neq n \Rightarrow A \textit{ is not invertible.}$$

If $\operatorname{rank}(A) < n$, then, by Proposition 1.3 (ii), each of the systems above is either inconsistent or consistent with dependent variables. If some system is inconsistent, then A is not invertible. If on the other hand the remaining situation occurs, i.e., all systems are consistent, we would have infinitely many solutions, contradicting the uniqueness of the inverse matrix (see Lemma 1.1).

The following theorem gives necessary and sufficient conditions for a square matrix to be invertible. This is the first of four sets of conditions that will appear in this book.

Theorem 1.1 (Necessary and sufficient conditions of invertibility (I)) *Let A be a square matrix of order n over $\mathbb{K}$. The following assertions are equivalent.*

(i) *A is invertible.*

(ii) $\operatorname{rank}(A) = n$.

(iii) *A is a product of elementary matrices.*

(iv) *A can be transformed in the identity matrix by elementary operations.*

(v) *The reduced row echelon form of A is the identity matrix.*

(vi) *The homogeneous system of linear equations $A\mathbf{x} = \mathbf{0}$ admits only the trivial solution.*

(vii) Given a column vector $\mathbf{b}$ *of size* $n \times 1$*, the system of linear equations* $A\mathbf{x} = \mathbf{b}$ *is consistent and has a unique solution.*

Before the proof of this theorem, observe that (iii) is a very striking assertion. It says, in other words, that

Any invertible matrix, and hence all invertible matrices, can be expressed as products of elementary matrices. In this sense, elementary matrices are 'generators' of the invertible matrices.

Proof *We shall show that*

$$(i) \Rightarrow (ii) \Rightarrow (iii) \Rightarrow (iv) \Rightarrow (v) \Rightarrow (vi) \Rightarrow (vii) \Rightarrow (i).$$

The equivalence between (i) and (ii) has been proved already (cf. Proposition 1.21).

(ii) $\Rightarrow$ *(iii) Let* R *be the reduced row echelon form of* A*. Then there exist elementary matrices* $E_1, E_2, \dots, E_k$ *such that* $E_1 E_2 \dots E_k A = R$*.*

Since by definition $\text{rank}\,(A) = \text{rank}\,(R)$*, we have* $\text{rank}\,(R) = n$*. It follows that* $R = I$ *and* $E_1 E_2 \dots E_k A = I$*. Multiplying on the left both members of this equality sequentially by* ${E_1}^{-1}, {E_2}^{-1}, \dots, {E_k}^{-1}$*,*

$$A = {E_k}^{-1} \cdots {E_2}^{-1} {E_1}^{-1}.$$

Hence we see that A *is a product of elementary matrices.*

(iii) $\Rightarrow$ *(iv) Since* A *is a product of elementary matrices, hence invertible matrices,* A *is itself invertible (cf. Proposition 1.18). It now follows from Proposition 1.21 that* $\text{rank}\,(A) = n$*. Hence the reduced row echelon form of* A *is the identity matrix.*

(iv) $\Rightarrow$ *(v) This implication is obvious. (It is an equivalence, in fact.)*

(v) $\Rightarrow$ *(vi) Suppose that the reduced row echelon of* A *is the identity matrix. Then there exist elementary matrices* $E_1, E_2, \dots, E_k$ *such that*

$$E_1 E_2 \dots E_k A = I.$$

Multiplying both members of the equation $A\mathbf{x} = \mathbf{0}$ *by* $E_1 E_2 \dots E_k$*,*

$$\underbrace{E_1 E_2 \dots E_k A}_{I}\,\mathbf{x} = \mathbf{0} \qquad \Longleftrightarrow \qquad \mathbf{x} = \mathbf{0}.$$

(vi) $\Rightarrow$ *(vii) We show firstly that, for each column vector* $\mathbf{b}$*, the system* $A\mathbf{x} = \mathbf{b}$ *is consistent.*

Suppose that, on the contrary, there exists $\mathbf{b}$ *such that* $A\mathbf{x} = \mathbf{b}$ *is inconsistent. Hence, by Proposition 1.3, we must have* $\text{rank}\,(A) < n$ *and, consequently, the homogeneous system* $A\mathbf{x} = \mathbf{0}$ *has infinitely many solutions which contradicts the initial assumption.*

We see next that the system $A\mathbf{x} = \mathbf{b}$ *has a unique solution. Suppose that* $\mathbf{x}_1, \mathbf{x}_2$ *are solutions of* $A\mathbf{x} = \mathbf{b}$. *Then*

$$A\mathbf{x}_1 = A\mathbf{x}_2 \qquad \Longleftrightarrow \qquad A(\mathbf{x}_1 - \mathbf{x}_2) = \mathbf{0}.$$

Since we assumed that the homogeneous system only admits the trivial solution, then

$$\mathbf{x}_1 - \mathbf{x}_2 = \mathbf{0} \qquad \Longleftrightarrow \qquad \mathbf{x}_1 = \mathbf{x}_2.$$

(vii) $\Rightarrow$ *(i) We want to show that* A *is invertible, that is, we want to show that there exists a matrix* B *such that* $AB = I$ *(cf. Proposition 1.20). In other words, we must show that the* n *systems below are all consistent* [1]*:*

$$A\mathbf{x} = \begin{bmatrix} 1 \\ 0 \\ \vdots \\ 0 \\ 0 \end{bmatrix} \qquad A\mathbf{x} = \begin{bmatrix} 0 \\ 1 \\ \vdots \\ 0 \\ 0 \end{bmatrix} \qquad \dots \qquad A\mathbf{x} = \begin{bmatrix} 0 \\ 0 \\ \vdots \\ 0 \\ 1 \end{bmatrix}$$

But this is exactly what assertion (vii) guarantees, since whichever vector $\mathbf{b}$ *might be considered the system* $A\mathbf{x} = \mathbf{b}$ *is consistent (and has a unique solution). Hence* A *is invertible, as required.*

1.4.1 LU and LDU factorisations

A square matrix $A = [a_{ij}]$ of order n is said to be **upper triangular** if, for all $i, j = 1, \dots, n$ with $j < i$, then $a_{ij} = 0$. Hence in an upper triangular matrix A all entries below the diagonal are equal to zero, i.e.,

$$A = \begin{bmatrix} a_{11} & a_{12} & \cdots & \cdots & a_{1n} \\ 0 & a_{22} & \cdots & \cdots & a_{2n} \\ 0 & 0 & a_{33} & \cdots & a_{3n} \\ \vdots & \vdots & & \ddots & \vdots \\ 0 & 0 & \cdots & 0 & a_{nn} \end{bmatrix}.$$

Similarly, a matrix $A = [a_{ij}]$ of order n is said to be **lower triangular** if, for all $i, j = 1, \dots, n$ with $i < j$, then $a_{ij} = 0$. Hence in a lower triangular matrix all entries above the diagonal are equal to zero. For example, any elementary matrix $E_{ij}(\alpha)$ can be upper triangular or lower triangular and each elementary matrix $D_i(\alpha)$ is both upper and lower triangular.

Proposition 1.22 *A product of two* $n \times n$ *upper triangular matrices (respectively, lower triangular matrices) is an* $n \times n$ *upper triangular matrix (respectively, a lower triangular matrix).*

---

[1]Notice that, if these n systems are simultaneously consistent, then each one of them has a unique solution, given the uniqueness of the inverse matrix (cf. Lemma 1.1).

Morevorer, if an $n \times n$ upper (respectively, lower) triangular matrix A is invertible, then its inverse A^{-1} is also upper (respectively, lower) triangular.

Proof *We prove the proposition only in the case of upper triangular matrices. The proof for lower triangular matrices is similar (or use transposition).*

Suppose that A, B are $n \times n$ upper triangular matrices. By Proposition 1.6 (ii), for each $j = 1, \dots, n$, column j of AB is a linear combination of the columns of A whose coefficients are the entries in column j of B.

Since B is also upper triangular, it follows that column j of AB is a linear combination of the first j columns of A. To see this, observe that in column j of B, all entries in rows $j+1, \dots, n$ coincide with zero.

Since each of the first j columns of A has zeros below row j, it follows that the same occurs in column j of AB.

Suppose now that A is an invertible upper triangular matrix. By Theorem 1.1 (v), the reduced row echelon form of A is the identity. Observe that this immediately implies that all diagonal entries of A are non-zero.

Since A is already upper triangular, to reduce it to the identity matrix, one uses Gauss–Jordan elimination where the Gaussian elimination is not required.

It follows that the elementary operations needed correspond to multiplications by upper triangular elementary matrices $E_1, E_2, \dots E_k$. More precisely,

$$E_1 E_2 \dots E_k A = I, \qquad A^{-1} = E_k^{-1} E_2^{-1} \dots E_1^{-1}.$$

Since $E_1^{-1}, E_2^{-1}, \dots E_k^{-1}$ are upper triangular, it follows from the first part of this proposition, that A^{-1} is also an upper triangular matrix.

We have seen that applying Gaussian elimination to a given matrix A amounts to successively multiplying A on the left by elementary matrices. Since this process involves 'descending' in the matrix A, whenever matrices of type $E_{ij}(\alpha)$ are used, they must be lower triangular.

Notice that, if A is an $n \times n$ square matrix, Gaussian elimination transforms A in an upper triangular matrix U.

Suppose also that no exchange of rows is used in obtaining U. Then, there exists an integer k such that

$$E_k E_2 \dots E_1 A = U,$$

where $E_1, E_2, \dots, E_k$ are all lower triangular elementary matrices. Hence

$$A = E_1^{-1} \dots E_2^{-1} E_k^{-1} U.$$

Since each inverse matrix is also a lower triangular elementary matrix, it follows, by Proposition 1.22, that there exists a lower triangular matrix L such that

$$A = LU.$$

Notice that if a matrix A (be it square or not) can be reduced to row echelon form without row exchange, then this reduction can be done using only the type (iii) operations of Definition 2.

Example 1.19 *Let A be the matrix*

$$A = \begin{bmatrix} 1 & 1 & -1 \\ 2 & 0 & 3 \\ 0 & 2 & -2 \end{bmatrix}$$

for which we want to find a LU factorisation. Gaussian elimination gives

$$E_{32}(1)E_{21}(-2)A = \underbrace{\begin{bmatrix} 1 & 1 & -1 \\ 0 & -2 & 5 \\ 0 & 0 & 3 \end{bmatrix}}_{U},$$

from which follows that

$$A = \underbrace{E_{21}(2)E_{32}(-1)}_{L}U,$$

where

$$L = \begin{bmatrix} 1 & 0 & 0 \\ 2 & 1 & 0 \\ 0 & -1 & 1 \end{bmatrix}.$$

We can now summarise this discussion in the following result.

Proposition 1.23 *Let A be an $n \times n$ matrix which can be reduced to a row echelon matrix without row exchange. Then, there exist an upper triangular matrix U and a lower triangular matrix L, whose entries $l_{ii} = 1$, for all $i = 1, \ldots, n$, such that $A = LU$. Moreover, if A is invertible then the matrices L and U are unique.*

Proof *It only remains to show the uniqueness part of the proposition.*

Let $A = L_1U_1$ and $A = L_2U_2$ be factorisations of A. Since A, L_1, L_2 are invertible matrices, then U_1, U_2 must also be invertible. If follows that $L_2^{-1}L_1 = U_2U_1^{-1}$. Hence, $L_2^{-1}L_1$ and $U_2U_1^{-1}$ must be diagonal matrices. However, for all $i = 1, \ldots, n$,

$$(L_1)_{ii} = 1 = (L_2)_{ii} = (L_2^{-1})_{ii},$$

from which follows that $L_2^{-1}L_1 = I = U_2U_1^{-1}$. Consequently, it must be the case that $L_2 = L_1$ and $U_2 = U_1$, concluding the proof.

How to obtain a LU factorisation of a square matrix

Let A be a square matrix which can be reduced to a row echelon form without exchanging rows.

1. Reduce A to a row echelon form U using Gaussian elimination. That is, find (lower triangular) elementary matrices $E_1, E_2, \dots, E_k$ of type $E_{ij}(\alpha)$ such that
$$E_k E_2 \dots E_1 A = U,$$
where U is an upper triangular matrix.
2. Construct a lower triangular matrix $L = [l_{ij}]$ such that, for all $i = 1, \dots, n$,
   (a) $l_{ii} = 1$;
   (b) for $j < i$, the entry $l_{ij} = -\alpha$, whenever $E_{ij}(\alpha)$ is used in the Gaussian elimination in 1., and $l_{ij} = 0$ otherwise.
3. Then $A = LU$.

Suppose that A is a square matrix whose reduction to row echelon form does not require row exchange. Similarly to Example 1.19, when looking for a LU factorisation of A, the resulting matrix L has all its diagonal entries equal to 1, since we can choose to use only elementary matrices of type $E_{ij}(\alpha)$. Notice also that each entry-ij below the diagonal of L is either the additive inverse of the multiplier α, if a matrix $E_{ij}(\alpha)$ was used in the Gaussian elimination, or is otherwise zero.

As is clear in Example 1.19, the diagonal entries of matrix U are not necessarily equal to 1. We can nevertheless obtain a factorisation of A where U has all its non-zero diagonal entries equal to 1 by means of a suitable diagonal matrix D.

Example 1.20 *Considering matrix U of Example1.19, we have*

$$U = \begin{bmatrix} 1 & 1 & -1 \\ 0 & -2 & 5 \\ 0 & 0 & 3 \end{bmatrix} = \underbrace{\begin{bmatrix} 1 & 0 & 0 \\ 0 & -2 & 0 \\ 0 & 0 & 3 \end{bmatrix}}_{D} \underbrace{\begin{bmatrix} 1 & 1 & -1 \\ 0 & 1 & -5/2 \\ 0 & 0 & 1 \end{bmatrix}}_{U'},$$

yielding

$$A = LDU' = \underbrace{\begin{bmatrix} 1 & 0 & 0 \\ 2 & 1 & 0 \\ 0 & -1 & 1 \end{bmatrix}}_{L} \underbrace{\begin{bmatrix} 1 & 0 & 0 \\ 0 & -2 & 0 \\ 0 & 0 & 3 \end{bmatrix}}_{D} \underbrace{\begin{bmatrix} 1 & 1 & -1 \\ 0 & 1 & -5/2 \\ 0 & 0 & 1 \end{bmatrix}}_{U'}.$$

Corollary 1.1 *Let A be an invertible $n \times n$ matrix which can be reduced to a row echelon matrix without row exchange. Then, there exist uniquely a lower triangular matrix $L = [l_{ij}]$, an upper triangular matrix $U = [u_{ij}]$, and a diagonal matriz D such that $A = LDU$ with $l_{ii} = 1 = u_{ii}$, for all $i = 1, \ldots, n$.*

Proof *By Proposition 1.23, there exists a unique factorisation $A = LU$ with U upper triangular and L lower triangular with $l_{ii} = 1$. Let D be the diagonal matrix whose entries $d_{ii} = u_{ii}$, for all $i = 1, \ldots, n$. Then*

$$A = LU = LDU',$$

$u'_{ii} = 1$, *for all $i = 1, \ldots, n$, and the uniqueness of L and U implies that of D and U'.*

How to obtain the LDU factorisation of an invertible matrix

Let A be an invertible matrix which can be reduced to a row echelon form without exchanging rows.

1. Calculate the unique factorisation $A = LU$, as described before.
2. Then $A = LDU'$ where D is the diagonal matrix with entries $d_{ii} = u_{ii}$, for all $i = 1, \ldots, n$, and

$$U' = D_1(u_{11}^{-1})D_2(u_{22}^{-1}) \ldots D_n(u_{nn}^{-1})U.$$

The LU factorisation can be used to solve systems of linear equations in two simpler steps. Suppose that we want to solve the system $A\mathbf{x} = \mathbf{b}$. If one has a LU factorisation of A, then

$$A\mathbf{x} = L\underbrace{U\mathbf{x}}_{\mathbf{y}} = \mathbf{b}.$$

One can solve firstly $L\mathbf{y} = \mathbf{b}$, by **forward substitution**, and then $U\mathbf{x} = \mathbf{y}$, by back substitution, as shown in the example below.

Example 1.21 *Consider the matrix A of Example 1.19 and its LU factorisation. We shall make use of $A = LU$ to solve the system*

$$A\mathbf{x} = \begin{bmatrix} 1 & 1 & -1 \\ 2 & 0 & 3 \\ 0 & 2 & -2 \end{bmatrix} \begin{bmatrix} x_1 \\ x_2 \\ x_3 \end{bmatrix} = \begin{bmatrix} 1 \\ 0 \\ -1 \end{bmatrix}$$

in two steps. In the first step, we solve the system

$$L\mathbf{y} = \begin{bmatrix} 1 & 0 & 0 \\ 2 & 1 & 0 \\ 0 & -1 & 1 \end{bmatrix} \begin{bmatrix} y_1 \\ y_2 \\ y_3 \end{bmatrix} = \begin{bmatrix} 1 \\ 0 \\ -1 \end{bmatrix}.$$

Using forward substitution, we have $y_1 = 1$,

$$y_2 = -2y_1 = -2$$

and

$$y_3 = y_2 - 1 = -3.$$

In the second step, we have

$$U\mathbf{x} = \begin{bmatrix} 1 & 1 & -1 \\ 0 & -2 & 5 \\ 0 & 0 & 3 \end{bmatrix} \begin{bmatrix} x_1 \\ x_2 \\ x_3 \end{bmatrix} = \begin{bmatrix} 1 \\ -2 \\ -3 \end{bmatrix}.$$

Using back substitution, it follows that $x_3 = -1$,

$$x_2 = -\frac{1}{2}(-5x_3 - 2) = -\frac{3}{2}$$

and

$$x_1 = -x_2 + x_3 - 1 = \frac{3}{2} - 1 - 1 = -\frac{1}{2}.$$

1.5 Exercises

EX 1.5.1. Find which of the equations are linear.

(a) $\sqrt{12}x_1 + x_2 - 8^{-\frac{1}{7}}x_3 = 10$ (b) $-x_1 + x_1x_2 + 2x_3 = 0$

(c) $v - \pi = \sqrt{e}u + e^{\frac{2}{3}}z - 2\pi w$ (d) $y^{\frac{1}{7}} - 6x + z = 5^{\frac{1}{4}}$

EX 1.5.2. Use Gaussian elimination or Gauss–Jordan elimination to solve the homogeneous systems of linear equations below.

(a) $\begin{cases} 2x_1 + 3x_2 = 0 \\ -2x_1 - 2x_2 - 6x_3 = 0 \\ -x_2 - x_3 = 0 \end{cases}$ (b) $\begin{cases} -10x_1 + 2x_2 - 2x_3 + 2x_4 = 0 \\ 2x_1 + 2x_2 + 2x_3 + 2x_4 = 0 \end{cases}$

(c) $\begin{cases} 2ix + 2iy + 4iz = 0 \\ w - y - 3z = 0 \\ 2w + 3x + y + z = 0 \\ -2iw + ix + 3iy - 2iz = 0 \end{cases}$

EX 1.5.3. Without any calculations, find which of the homogeneous systems might have non-trivial solutions. (It is allowed to write the coefficient matrix of the sytems.)

(a) $\begin{cases} x_1 - 3x_2 + 5x_3 - x_4 = 0 \\ 3x_1 + x_2 - 10x_3 + x_4 = 0 \\ x_1 + x_2 + x_3 - 11x_4 = 0 \end{cases}$ (b) $\begin{cases} a_{11}x_1 + a_{12}x_2 + a_{13}x_3 = 0 \\ a_{21}x_1 + a_{22}x_2 + a_{23}x_3 = 0 \\ a_{31}x_1 + a_{32}x_2 + a_{33}x_3 = 0 \\ a_{41}x_1 + a_{42}x_2 + a_{43}x_3 = 0 \end{cases}$

(c) $\begin{cases} -2v + 3u = 0 \\ 6u - 4v = 0 \end{cases}$ (d) $\begin{cases} 20x + 3y - z = 0 \\ -10y - 8z = 0 \\ 10y + 4z = 0 \end{cases}$

EX 1.5.4. Use Gaussian elimination or Gauss–Jordan elimination to solve the non-homogeneous systems of linear equations below.

(a) $\begin{cases} x + 2y - 3z = -1 \\ -3x - 3y - 6z = -24 \\ \frac{3}{2}x - \frac{7}{2}y + z = 5 \end{cases}$ (b) $\begin{cases} 8x_1 + x_2 + 4x_3 = -1 \\ -2x_1 + 5x_2 + 2x_3 = 1 \\ x_1 + x_2 + x_3 = 0 \end{cases}$

(c) $\begin{cases} -v + \frac{3}{2}w = \frac{1}{2} \\ 2u + 2v + w = \frac{5}{3} \\ 6u + 12v - 6w = -4 \end{cases}$ (d) $\begin{cases} 2w + 4x - 2y = 8 \\ -3x + 3y = -9 \\ -2u - 4v - w - 7x = -7 \\ 2w + 6x - 4y = 14 \end{cases}$

EX 1.5.5. For each of the sets listed, find a system of linear equations whose solution set is that set.

(a) $\{(1, 2, 3)\}$
(b) $\{(1, 2, t) : t \in \mathbb{R}\}$
(c) $\{(y, -3y, y) : y \in \mathbb{R}\}$
(d) $\{(x, 2x - z - w, z) : x, z, w \in \mathbb{R}\}$

EX 1.5.6. Which of the 3×3 matrices are row echelon matrices? Which matrices are in reduced row echelon form? What is the rank of each matrix?

(a) $\begin{bmatrix} 3 & 0 & 0 \\ 0 & 3 & 0 \\ 0 & 0 & 3 \end{bmatrix}$ (b) $\begin{bmatrix} 2 & 0 & 0 \\ 0 & 2 & 0 \\ 0 & 0 & 0 \end{bmatrix}$ (c) $\begin{bmatrix} 0 & 1+i & 0 \\ 0 & 0 & 12i \\ 0 & 0 & 0 \end{bmatrix}$

(d) $\begin{bmatrix} 1 & 0 & 0 \\ 0 & 0 & -1 \\ 0 & 0 & 0 \end{bmatrix}$ (e) $\begin{bmatrix} 0 & 5 & 0 \\ 3 & 0 & 0 \\ 0 & 0 & 0 \end{bmatrix}$ (f) $\begin{bmatrix} 1 & 6 & 0 \\ 0 & 1 & 0 \\ 0 & 0 & 0 \end{bmatrix}$

(g) $\begin{bmatrix} 1+i & 0 & 0 \\ 0 & 0 & 0 \\ 0 & 0 & 1-i \end{bmatrix}$ (h) $\begin{bmatrix} 0 & 0 & 0 \\ 0 & 0 & 0 \\ 0 & 0 & 0 \end{bmatrix}$ (i) $\begin{bmatrix} 0 & 2 & 0 \\ 0 & 1 & 0 \\ 0 & 0 & 0 \end{bmatrix}$

(j) $\begin{bmatrix} 2 & 1 & 0 \\ 0 & -2+i & 0 \\ 0 & 1+5i & 1+i \end{bmatrix}$ (k) $\begin{bmatrix} 20 & -10 & 0 \\ 0 & 0 & -60 \\ 0 & 0 & 2 \end{bmatrix}$ (l) $\begin{bmatrix} 2 & 10 & 0 \\ 0 & -1 & 20 \\ 0 & 0 & 0 \end{bmatrix}$

EX 1.5.7. Find the reduced row echelon form and the rank of the matrix

$$A = \begin{bmatrix} 1 & 1 & 1 & 2 & 0 \\ 2 & 1 & 2 & 2 & 1 \\ 1 & 1 & 2 & 1 & 0 \end{bmatrix}.$$

EX 1.5.8. Let

$$A_\alpha = \begin{bmatrix} 1 & 1 & 0 \\ 0 & \alpha & \alpha \\ -1 & 0 & \alpha^2 \end{bmatrix} \quad \text{and consider the system} \quad A_\alpha \begin{bmatrix} x \\ y \\ z \end{bmatrix} = \begin{bmatrix} 0 \\ 0 \\ \beta \end{bmatrix},$$

where $\alpha, \beta \in \mathbb{R}$. Determine the rank of A_α in terms of α. Classify the systems for all $\alpha, \beta \in \mathbb{R}$ and, whenever the systems are consistent, indicate the number of independent variables.

Solve the systems when $\alpha = 2, \beta \in \mathbb{R}$.

EX 1.5.9. Discuss the solutions of the homogeneous system $A_\alpha \mathbf{x} = 0$ in terms of α, where

$$A_\alpha = \begin{bmatrix} 0 & 1 & 1 & 1 \\ 1 & 1 & -1 & 1 \\ 4 & 4 & -\alpha^2 & \alpha^2 \\ 2 & 2 & -2 & \alpha \end{bmatrix}.$$

EX 1.5.10. If possible, calculate $A + B$, $B + C$, $2A$, AB, BA, CB, $\operatorname{tr} B$, and $\operatorname{tr} C$ for

$$A = \begin{bmatrix} 1 & -2 \\ 4 & 1 \\ \sqrt{2} & 3 \end{bmatrix} \qquad B = \begin{bmatrix} 1 & \sqrt{3} & 0 \\ 2 & -1 & 1 \\ \pi & 2 & -1 \end{bmatrix} \qquad C = \begin{bmatrix} 6 & 0 & 0 \\ 0 & -4 & 0 \\ 0 & 0 & 10 \end{bmatrix}$$

EX 1.5.11. Let A and B be the matrices

$$A = \begin{bmatrix} 1 & 0 & 1 \\ 2 & -7 & 1 \\ -1 & -1 & 1 \end{bmatrix} \qquad B = \begin{bmatrix} -1 & 0 & -1 \\ 1 & -1 & 1 \\ 1 & 1 & -1 \end{bmatrix}.$$

a) Find the entry-(23) and the second column of AB.

b) Calculate $(A - B)^T$ and its trace.

EX 1.5.12. Let A, B, C, and D be matrices such that

A is a 1×3 matrix $\qquad$ B is a 3×1 matrix

C is a 1×3 matrix $\qquad$ D is a 3×3 matrix

Choose the correct assertion.

A) The size of $(A + C)A$ is 1×3.

B) The trace of A might be zero.

C) $(5CB)^T = 5C^T B^T$.

D) $D(CBA + C)^T$ is a 3×1 matrix.

EX 1.5.13. Find the 3×3 anti-symmetric matrix $A = [a_{ij}]$ such that, for all $j < i$,

$$a_{ij} = i - j.$$

EX 1.5.14. Find a rank one symmetric matrix A such that

$$\begin{bmatrix} 2 \\ 4 \\ 8 \end{bmatrix}$$

is a column of A.

EX 1.5.15. Find an expression for A^n, where $A = \begin{bmatrix} 0 & -i \\ i & 0 \end{bmatrix}$.

EX 1.5.16. Fill in the entries of the matrix

$$A = \begin{bmatrix} -1 & 2 & \dots \\ 0 & 1 & \dots \\ 1 & \dots & \dots \end{bmatrix}$$

such that $\operatorname{rank}(A) = 2$ and $(-1, -1, 1)$ is a solution of $A\mathbf{x} = 0$.

EX 1.5.17. Use Gauss–Jordan elimination to find the inverse, if possible, of the following matrices.

$$\text{(a)} \begin{bmatrix} 1 & 2 \\ 4 & 7 \end{bmatrix} \qquad \text{(b)} \begin{bmatrix} -3 & 4 \\ 6 & 5 \end{bmatrix} \qquad \text{(c)} \begin{bmatrix} 6 & -3 \\ -4 & 2 \end{bmatrix}$$

$$\text{(d)} \begin{bmatrix} 3 & 1 & 2 \\ 4 & 0 & 5 \\ -1 & 3 & -4 \end{bmatrix} \qquad \text{(e)} \begin{bmatrix} -1 & 0 & -4 \\ 2 & 0 & 1 \\ -4 & 0 & -9 \end{bmatrix} \qquad \text{(f)} \begin{bmatrix} 2 & 2 & 2 \\ 6 & 7 & 7 \\ 6 & 6 & 7 \end{bmatrix}$$

$$\text{(g)} \begin{bmatrix} 1 & 1 & 1 & 1 \\ 0 & 3 & 3 & 3 \\ 0 & 0 & 5 & 5 \\ 0 & 0 & 0 & 7 \end{bmatrix} \qquad \text{(h)} \begin{bmatrix} 0 & 0 & 0 & 0 \\ 4 & 0 & \frac{2}{5} & -9 \\ -8 & 17 & 2 & \frac{1}{3} \\ -1 & 13 & 4 & 2 \end{bmatrix}$$

EX 1.5.18. For

$$A = \begin{bmatrix} 1 & 2 \\ 0 & 1 \end{bmatrix},$$

find A^3, A^{-3}, $A^2 - 2A + I$, and $(A - I)^2$. Solve the equation

$$A^{-1}X(A + I)^2 = A + A^T.$$

EX 1.5.19. Find the elementary operation and the corresponding elementary matrix needed to obtain the identity matrix from each of the following matrices.

$$\text{(a)} \begin{bmatrix} 1 & 0 \\ 5 & 1 \end{bmatrix} \quad \text{(b)} \begin{bmatrix} 1 & 0 & 0 \\ 0 & 1 & 0 \\ 0 & 0 & -3 \end{bmatrix} \quad \text{(c)} \begin{bmatrix} 1 & 0 & 0 & 0 \\ 0 & 0 & 0 & 1 \\ 0 & 0 & 1 & 0 \\ 0 & 1 & 0 & 0 \end{bmatrix} \quad \text{(d)} \begin{bmatrix} 1 & 0 & 0 & 0 \\ 0 & 1 & 0 & 0 \\ 0 & -\frac{1}{2} & 1 & 0 \\ 0 & 0 & 0 & 1 \end{bmatrix}$$

EX 1.5.20. Find elementary matrices E_1 and E_2 such that the matrix

$$A = \begin{bmatrix} 1 & 0 & 0 & 0 \\ 0 & -5 & 0 & -1 \\ 0 & 0 & 1 & 0 \\ 0 & 0 & 0 & 1 \end{bmatrix}$$

is such that $A = E_1E_2$. Write A^{-1} as a matrix product, using E_1 and E_2.

EX 1.5.21. Let A be a 3×3 real matrix such that

$$A = E_1E_2R,$$

where R is a rank-2 row echelon matrix, and

$$E_1 = D_3(-1) \qquad E_2 = E_{21}(3).$$

Find all the correct assertions in the following list.

I. $A = E_2R$.

II. A ia an invertible matrix.

III. A has a single zero row.

IV. The system of linear equations
$A\mathbf{x} = \begin{bmatrix}0\\0\\1\end{bmatrix}$ might be consistent.

A) II e III B) I e II e III C) I e IV D) I e III

EX 1.5.22. Show that the only matrices which commute with all $n \times n$ matrices are the $n \times n$ **scalar matrices**, that is, matrices which are multiples of the identity matrix.

EX 1.5.23. Find the LU and LDU decompositions of

$$A = \begin{bmatrix} 4 & -8 & 20 \\ -20 & 45 & -105 \\ -12 & 44 & -79 \end{bmatrix}.$$

Use the LU decomposition to solve the system $A\mathbf{x} = \mathbf{b}$ with $\boldsymbol{b} = (12, -65, -56)$.

1.6 At a Glance

In a nutshell, this chapter is about the introduction of an object, the matrix, and of the development of a toolkit to effectively extract knowledge about the object, which will be put to use in the following chapters.

Matrices are crucial in the book, and several fundamental notions related with matrices were established here and will be relied upon in the remainder of the book, notably, the rank and the inverse of a matrix. To be determined, both rank and inverse lean on two methods known as Gaussian elimination and Gauss–Jordan elimination. These methods aim at finding, respectively, a row echelon form and the (unique) reduced row echelon form of a matrix. Gaussian and Gauss–Jordan eliminations will be used extensively throughout the book and matrices are mostly what this book is about.

We know now how to operate with matrices (addition, multiplication by a scalar, and multiplication) and to do elementary operations by means of elementary matrices. In fact, invertible matrices are exactly the products of elementary matrices.

In this chapter, matrices were applied to the solution of systems of linear equations via Gaussian elimination or Gauss–Jordan elimination. These eliminations led to the LU and the LDU factorisations of matrices involving the products of diagonal, lower triangular, and upper triangular matrices. Matrices of all these types will play a decisive role in what follows.

Chapter 2

Determinant

2.1	Axiomatic Definition	55
2.2	Leibniz's Formula	66
2.3	Laplace's Formula	70
2.4	Exercises	76
2.5	At a Glance	79

In the previous chapter, we learned how to calculate the rank of a given matrix. This is a remarkable number whose importance we have only begun to unveil. It is indeed outstanding that we can use it to classify systems of linear equations and to determine the invertibility of matrices (see Proposition 1.3 and Theorem 1.1).

Still in this trait, the present chapter is entirely devoted to another striking number: the determinant of a square matrix. Here we shall see how to use it to decide about the invertibility of a matrix, calculate matrix inverses and solve systems of linear equations (Cramer's rule). The influential role of the determinant does not stop here however since it will be absolutely essential in Chapter 4 and, as a consequence, will be in the background of the following chapters.

Chapter 2 presents two approaches to calculate the determinant, either axiomatically or through formulas (Leibniz's and Laplace's). The axiomatic approach seems at first glance to be the most detached from 'reality'. In fact, it is quite the opposite as it boils down to using Gaussian elimination to calculate the determinant and, in this way, being the most time-saving process.

2.1 Axiomatic Definition

As said above, we wish to assign a number, the determinant, to each square matrix of a given size. Hence, if we fix the size of the matrices, say $n \times n$ matrices, and A is an $n \times n$ matrix over $\mathbb{K}$, we shall define the determinant, $\det A$, of the matrix A as a scalar in $\mathbb{K}$. In other words, we have a function defined from $\mathrm{M}_n(\mathbb{K})$ to $\mathbb{K}$ assigning to each matrix A the number $\det A$.

A word of advice: if seeing a function defined using axioms rather than formulas causes some anxiety, then the reader is advised to go firstly to §2.2,

DOI: 10.1201/9781351243452-2

where this function is defined in a traditional way, and then come back to §2.1. In the end, however, it will be apparent that the easiest way of calculating the determinant is that of the present section.

Definition 19 *The* ***determinant function***

$$\begin{aligned} \det : \mathrm{M}_n(\mathbb{K}) &\to \mathbb{K} \\ A &\mapsto \det A \end{aligned}$$

is the only function satisfying the following axioms:

(Ax1) $\det I = 1$;

(Ax2) $\det(P_{ij}A) = -\det A$ *(with* $i \neq j$, $i, j = 1, \ldots, n$*);*

(Ax3) Given $\alpha \in \mathbb{K}$ *and* $i \in \{1, \ldots, n\}$,

$$\det \begin{bmatrix} \vdots \\ \alpha \mathbf{l}_i \\ \vdots \end{bmatrix} = \alpha \det \begin{bmatrix} \vdots \\ \mathbf{l}_i \\ \vdots \end{bmatrix}$$

$$\det \begin{bmatrix} \mathbf{l}_1 \\ \vdots \\ \mathbf{l}_{i-1} \\ \mathbf{l}_i + \mathbf{l}'_i \\ \mathbf{l}_{i+1} \\ \vdots \\ \mathbf{l}_n \end{bmatrix} = \det \begin{bmatrix} \mathbf{l}_1 \\ \vdots \\ \mathbf{l}_{i-1} \\ \mathbf{l}_i \\ \mathbf{l}_{i+1} \\ \vdots \\ \mathbf{l}_n \end{bmatrix} + \det \begin{bmatrix} \mathbf{l}_1 \\ \vdots \\ \mathbf{l}_{i-1} \\ \mathbf{l}'_i \\ \mathbf{l}_{i+1} \\ \vdots \\ \mathbf{l}_n \end{bmatrix},$$

where $\mathbf{l}_i, \mathbf{l}'_i$ *are matrix rows. The number* $\det A$ *is called the* ***determinant*** *of the matrix* A.

Notice that, in (Ax3), α may take the value 0.

Axiom (Ax1) is clear enough, it says that the determinant of the $n \times n$ identity matrix is 1. Axiom (Ax2) tells you that if two rows of a matrix A are exchanged then the determinant changes sign or, equivalently, $\det A$ is multiplied by -1. Axiom (Ax3) establishes that if one has two matrices

$$A = \begin{bmatrix} \mathbf{l}_1 \\ \vdots \\ \mathbf{l}_{i-1} \\ \mathbf{l}_i \\ \mathbf{l}_{i+1} \\ \vdots \\ \mathbf{l}_n \end{bmatrix}, \qquad B = \det \begin{bmatrix} \mathbf{l}_1 \\ \vdots \\ \mathbf{l}_{i-1} \\ \mathbf{l}'_i \\ \mathbf{l}_{i+1} \\ \vdots \\ \mathbf{l}_n \end{bmatrix} \tag{2.1}$$

whose rows coincide except possibly for row i, then one can construct a new matrix

$$C = \det \begin{bmatrix} \mathbf{l}_1 \\ \vdots \\ \mathbf{l}_{i-1} \\ \mathbf{l}_i + \mathbf{l}'_i \\ \mathbf{l}_{i+1} \\ \vdots \\ \mathbf{l}_n \end{bmatrix} \tag{2.2}$$

whose rows coincide with those of A and B, except possibly for row i which is the sum of the corresponding rows of A and B, one has that the determinant $\det C$ of the matrix C satisfies

$$\det C = \det A + \det B. \tag{2.3}$$

It is not obvious that these axioms define a function or even that this function is uniquely defined. In fact, this is the case: *there exists a unique function* satisfying (Ax1)–(Ax3) in Definition 19. The existence and uniqueness of the determinant function will be shown in due course (see Sections 2.2 and 8.2), but first we want to make clear that we can calculate the determinant of any given matrix just by abiding to the rules (Ax1)–(Ax3) above.

Notation. In what follows, the determinant of A may be denoted by $\det A$ or $|A|$.

Example 2.1 *Firstly consider the* 1×1 *matrix* $A = [\alpha]$, *where* $\alpha \in \mathbb{K}$. *Then, by (Ax3),*

$$\det[\alpha] = \alpha \det[1].$$

It now follows from (Ax1) that

$$\det[\alpha] = \alpha \det[1] = \alpha 1 = \alpha.$$

Example 2.2 *Now let* A *be a diagonal matrix. Repeatedly using (Ax3),*

$$\det A = \det \begin{bmatrix} a_{11} & 0 & 0 & \cdots & 0 \\ 0 & a_{22} & 0 & \cdots & 0 \\ 0 & 0 & a_{33} & \cdots & 0 \\ \vdots & \vdots & \vdots & \ddots & \vdots \\ 0 & 0 & 0 & \cdots & a_{nn} \end{bmatrix} = a_{11} \det \begin{bmatrix} 1 & 0 & 0 & \cdots & 0 \\ 0 & a_{22} & 0 & \cdots & 0 \\ 0 & 0 & a_{33} & \cdots & 0 \\ \vdots & \vdots & \vdots & \ddots & \vdots \\ 0 & 0 & 0 & \cdots & a_{nn} \end{bmatrix} =$$

$$= a_{11}a_{22} \det \begin{bmatrix} 1 & 0 & 0 & \cdots & 0 \\ 0 & 1 & 0 & \cdots & 0 \\ 0 & 0 & a_{33} & \cdots & 0 \\ \vdots & \vdots & \vdots & \ddots & \vdots \\ 0 & 0 & 0 & \cdots & a_{nn} \end{bmatrix} = a_{11}a_{22}a_{33} \ldots a_{nn} \det \begin{bmatrix} 1 & 0 & 0 & \cdots & 0 \\ 0 & 1 & 0 & \cdots & 0 \\ 0 & 0 & 1 & \cdots & 0 \\ \vdots & \vdots & \vdots & \ddots & \vdots \\ 0 & 0 & 0 & \cdots & 1 \end{bmatrix}.$$

Hence, applying now (Ax1),

$$\det \begin{bmatrix} a_{11} & 0 & 0 & \cdots & 0 \\ 0 & a_{22} & 0 & \cdots & 0 \\ 0 & 0 & a_{33} & \cdots & 0 \\ \vdots & \vdots & \vdots & \ddots & \vdots \\ 0 & 0 & 0 & \cdots & a_{nn} \end{bmatrix} = a_{11}a_{22}a_{33}\dots a_{nn} \det I$$

$$= a_{11}a_{22}a_{33}\dots a_{nn}1$$

Hence

$$\det A = a_{11}a_{22}a_{33}\dots a_{nn}.$$

Summarising both examples,

The determinant of a diagonal matrix is the product of its diagonal entries.

Proposition 2.1 *Let A be a square matrix over $\mathbb{K}$. The following hold.*

(i) If A has two equal rows then $|A| = 0$.

(ii) If A has a zero row then $|A| = 0$.

(iii) The determinant remains unchanged if one replaces row $\mathbf{l}_i$ of A by $\mathbf{l}_i + \alpha\mathbf{l}_j$, where $i \neq j, \alpha \in \mathbb{K}$.

Notice that Proposition 2.1 (iii) refers to an elementary operation and asserts that for this particular operation the determinant does not change. Hence now we know that, whenever doing Gaussian elimination, this type of operation leaves unchanged the determinant of the initial matrix.

Proof *(i) If A is a matrix such that its rows i and j (with $i \neq j$) are equal, then $A = P_{ij}A$ and, consequently,*

$$\det A = \det(P_{ij}A).$$

On the other hand, by Axiom (Ax2), $\det(P_{ij}A) = -\det A$. *It then follows that*

$$\det A = \det(P_{ij}A) = -\det A.$$

Hence

$$\det A = -\det A \Leftrightarrow 2\det A = 0 \Leftrightarrow \det A = 0.$$

(ii) Let $\mathbf{l}_i$ be a zero row of matrix A and let A' be the matrix obtained from A by multiplying row $\mathbf{l}_i$ by $\alpha = 0$. By Axiom (Ax3), we have

$$\det A = \det A' = \alpha \det A = 0 \det A = 0.$$

(iii) Consider the matrix

$$A = \begin{bmatrix} \mathbf{l}_1 \\ \vdots \\ \mathbf{l}_i \\ \vdots \\ \mathbf{l}_j \\ \vdots \\ \mathbf{l}_n \end{bmatrix}.$$

Using Axiom (Ax3), we have

$$\det \begin{bmatrix} \mathbf{l}_1 \\ \vdots \\ \mathbf{l}_{i-1} \\ \mathbf{l}_i + \alpha \mathbf{l}_j \\ \mathbf{l}_{i+1} \\ \vdots \\ \mathbf{l}_j \\ \vdots \\ \mathbf{l}_n \end{bmatrix} = \det \begin{bmatrix} \mathbf{l}_1 \\ \vdots \\ \mathbf{l}_{i-1} \\ \mathbf{l}_i \\ \mathbf{l}_{i+1} \\ \vdots \\ \mathbf{l}_j \\ \vdots \\ \mathbf{l}_n \end{bmatrix} + \det \begin{bmatrix} \mathbf{l}_1 \\ \vdots \\ \mathbf{l}_{i-1} \\ \alpha \mathbf{l}_j \\ \mathbf{l}_{i+1} \\ \vdots \\ \mathbf{l}_j \\ \vdots \\ \mathbf{l}_n \end{bmatrix}$$

$$= \det \begin{bmatrix} \mathbf{l}_1 \\ \vdots \\ \mathbf{l}_{i-1} \\ \mathbf{l}_i \\ \mathbf{l}_{i+1} \\ \vdots \\ \mathbf{l}_j \\ \vdots \\ \mathbf{l}_n \end{bmatrix} + \alpha \det \begin{bmatrix} \mathbf{l}_1 \\ \vdots \\ \mathbf{l}_{i-1} \\ \mathbf{l}_j \\ \mathbf{l}_{i+1} \\ \vdots \\ \mathbf{l}_j \\ \vdots \\ \mathbf{l}_n \end{bmatrix}.$$

Observing that the last matrix has two equal rows, assertion (i) of this proposition yields

$$\det \begin{bmatrix} \mathbf{l}_1 \\ \vdots \\ \mathbf{l}_{i-1} \\ \mathbf{l}_i + \alpha \mathbf{l}_j \\ \mathbf{l}_{i+1} \\ \vdots \\ \mathbf{l}_j \\ \vdots \\ \mathbf{l}_n \end{bmatrix} = \det \begin{bmatrix} \mathbf{l}_1 \\ \vdots \\ \mathbf{l}_{i-1} \\ \mathbf{l}_i \\ \mathbf{l}_{i+1} \\ \vdots \\ \mathbf{l}_j \\ \vdots \\ \mathbf{l}_n \end{bmatrix} + 0 = \det A,$$

which ends the proof.

If A is an upper triangular matrix, two possibilities can occur: either (1) all diagonal entries of A are non-zero or (2) some diagonal entry of A is equal to zero.

In the first case, the Gauss-Jordan elimination process applied to A will yield a diagonal matrix, this being done using elementary operations of type

(iii) only (see Definition 2). Hence

$$A = \begin{bmatrix} a_{11} & a_{12} & \cdots & a_{1n} \\ 0 & a_{22} & \cdots & a_{2n} \\ \vdots & \vdots & \ddots & \vdots \\ 0 & 0 & \cdots & a_{nn} \end{bmatrix} \xrightarrow[\text{GJE}]{} \begin{bmatrix} a_{11} & 0 & \cdots & 0 \\ 0 & a_{22} & \cdots & 0 \\ \vdots & \vdots & \ddots & \vdots \\ 0 & 0 & \cdots & a_{nn} \end{bmatrix}$$

Observe that here we stopped the Gauss–Jordan elimination before changing all pivots to 1. By Proposition 2.1 (iii), we have now

$$\det A = \begin{bmatrix} a_{11} & a_{12} & \cdots & a_{1n} \\ 0 & a_{22} & \cdots & a_{2n} \\ \vdots & \vdots & \ddots & \vdots \\ 0 & 0 & \cdots & a_{nn} \end{bmatrix} = \det \begin{bmatrix} a_{11} & 0 & \cdots & 0 \\ 0 & a_{22} & \cdots & 0 \\ \vdots & \vdots & \ddots & \vdots \\ 0 & 0 & \cdots & a_{nn} \end{bmatrix} = a_{11}a_{22}\cdots a_{nn}.$$

In case (2), let a_{kk} be the first from below zero entry in the diagonal. Using only elementary operations of type (iii) in Definition 2 and the (non-zero) entries $a_{nn}, \dots, a_{k+1,k+1}$, we can change row k into a zero row (see the grey rows in the matrices below).

$$A = \begin{bmatrix} a_{11} & a_{12} & \cdots & a_{1k} & a_{1,k+1} & \cdots & a_{1n} \\ 0 & a_{22} & \cdots & a_{2k} & a_{2,k+1} & \cdots & a_{2n} \\ \vdots & \vdots & \ddots & \vdots & \vdots & & \vdots \\ 0 & 0 & \cdots & 0 & a_{k,k+1} & \cdots & a_{kn} \\ 0 & 0 & \cdots & 0 & a_{k+1,k+1} & \cdots & a_{k+1,n} \\ \vdots & \vdots & & \vdots & \vdots & \ddots & \vdots \\ 0 & 0 & \cdots & 0 & 0 & \cdots & a_{nn} \end{bmatrix} \xrightarrow[\text{GJE}]{}$$

$$A = \begin{bmatrix} a_{11} & a_{12} & \cdots & a_{1k} & 0 & \cdots & a_{1n} \\ 0 & a_{22} & \cdots & a_{2k} & 0 & \cdots & a_{2n} \\ \vdots & \vdots & \ddots & \vdots & \vdots & & \vdots \\ 0 & 0 & \cdots & 0 & 0 & \cdots & 0 \\ 0 & 0 & \cdots & 0 & a_{k+1,k+1} & \cdots & 0 \\ \vdots & \vdots & & \vdots & \vdots & \ddots & \vdots \\ 0 & 0 & \cdots & 0 & 0 & \cdots & a_{nn} \end{bmatrix}$$

Observing that, by Proposition 2.1 (iii), the elementary operations used do not change the determinant,

$$\det A = \det \begin{bmatrix} a_{11} & a_{12} & \cdots & a_{1k} & 0 & \cdots & a_{1n} \\ 0 & a_{22} & \cdots & a_{2k} & 0 & \cdots & a_{2n} \\ \vdots & \vdots & \ddots & \vdots & \vdots & & \vdots \\ 0 & 0 & \cdots & 0 & 0 & \cdots & 0 \\ 0 & 0 & \cdots & 0 & a_{k+1,k+1} & \cdots & 0 \\ \vdots & \vdots & & \vdots & \vdots & \ddots & \vdots \\ 0 & 0 & \cdots & 0 & 0 & \cdots & a_{nn} \end{bmatrix} = 0.$$

Hence, the analysis done of (1) and (2) above yields

The determinant of an upper triangular matrix is the product of its diagonal entries:

$$\det \begin{bmatrix} a_{11} & a_{12} & \cdots & a_{1n} \\ 0 & a_{22} & \cdots & a_{2n} \\ \vdots & & \ddots & \vdots \\ 0 & \cdots & 0 & a_{nn} \end{bmatrix} = a_{11}a_{22}\cdots a_{nn}.$$

The axioms of the determinant function together with Proposition 2.1 describe completely how the elementary operations change the determinant. On the other hand, any square matrix can be reduced to an upper triangular matrix by Gaussian elimination. These two observations combined, make it clear that we have now all the knowledge required to calculate the determinant of a matrix.

Example 2.3 *Let A be the matrix*

$$\begin{bmatrix} 3 & -3 & -3 \\ 0 & 1 & -1 \\ -1 & 0 & 0 \end{bmatrix}.$$

The determinant of A is

$$|A| = \begin{vmatrix} 3 & -3 & -3 \\ 0 & 1 & -1 \\ -1 & 0 & 0 \end{vmatrix} = 3 \underbrace{\begin{vmatrix} 1 & -1 & -1 \\ 0 & 1 & -1 \\ -1 & 0 & 0 \end{vmatrix}}_{|B|} = 3 \begin{vmatrix} 1 & -1 & -1 \\ 0 & 1 & -1 \\ 0 & -1 & -1 \end{vmatrix} =$$

$$= 3 \begin{vmatrix} 1 & -1 & -1 \\ 0 & 1 & -1 \\ 0 & 0 & -2 \end{vmatrix} = 3(-2) = -6.$$

Notice that matrix A is obtained from matrix B by multiplying row 1 by 3 (see (Ax2)).

How to calculate the determinant of a square matrix using Gaussian elimination

Let A be a square matrix. To obtain the value of $\det A$, we proceed as follows.

1. Reduce A to a row echelon matrix, i.e., an upper triangular matrix, A' using Gaussian elimination and calculate the determinant of A'.
2. Calculate $\det A$, keeping track of how the elementary operations in 1. force $\det A$ to differ from $\det A'$.

Example 2.4 *In this example we obtain the formula of the determinant of a 2×2 matrix. Let A be the matrix over $\mathbb{K}$,*

$$A = \begin{bmatrix} a & b \\ c & d \end{bmatrix}.$$

We shall calculate $\det A$ *by considering separately the cases $a \neq 0$ and $a = 0$.*

- $a \neq 0$

  By Gaussian elimination,

$$\begin{bmatrix} a & b \\ c & d \end{bmatrix} \xrightarrow[l_2 - \frac{c}{a} l_1]{} \begin{bmatrix} a & b \\ 0 & d - \frac{c}{a} b \end{bmatrix}.$$

  Hence

$$|A| = \begin{vmatrix} a & b \\ c & d \end{vmatrix} = \begin{vmatrix} a & b \\ 0 & d - \frac{c}{a} b \end{vmatrix} = ad - bc.$$

- $a = 0$

$$A = \begin{bmatrix} 0 & b \\ c & d \end{bmatrix} \xrightarrow[L_1 \leftrightarrow L_2]{} \begin{bmatrix} c & d \\ 0 & b \end{bmatrix},$$

  We have

$$|A| = \begin{vmatrix} 0 & b \\ c & d \end{vmatrix} = - \begin{vmatrix} c & d \\ 0 & b \end{vmatrix} = -cb = ad - bc.$$

In summary,

$$|A| = \begin{vmatrix} a & b \\ c & d \end{vmatrix} = ad - bc.$$

It is now clear what the determinants of elementary matrices will be:

- $\det P_{ij} = -1$ (by (Ax2) in Definition 19);
- $\det E_{ij}(\alpha) = 1$ (by Proposition 2.1 (iii));
- $\det D_i(\alpha) = \alpha$ (by (Ax3) in Definition 19).

Proposition 2.2 *Let A be a $n \times n$ matrix and let E be a $n \times n$ elementary matrix. Then*

$$|EA| = |E||A|.$$

Proof *Let A be a square matrix. Recalling how A is changed when it is multiplied on the left by the three types of elementary matrices, we have*

$$|P_{ij}A| = -|A| = |P_{ij}||A|,$$

$$|E_{ij}(\alpha)A| = |A| = |E_{ij}(\alpha)||A|,$$

and

$$|D(\alpha)A| = \alpha|A| = |D(\alpha)||A|,$$

as required.

We are now able to charaterise the invertibility of a matrix in terms of its determinant. This leads to the necessary and sufficient condition of invertibility in the next theorem.

Theorem 2.1 (Necessary and sufficient condition of invertibility (II)) *Let A be a square matrix of order n. The following assertions are equivalent.*

(i) *A is invertible.*

(ii) $|A| \neq 0$.

Proof *(i) $\Rightarrow$ (ii) Suppose that A invertible. Then, by Theorem 1.1, there exist elementary matrices $E_1, \ldots, E_k$ such that*

$$A = E_1 E_2 \cdots E_k.$$

Hence, by Proposition 2.2,

$$|A| = |E_1||E_2| \cdots |E_k|$$

and, since the determinant of each of the elementary matrices is non-zero, it finally follows that $|A| \neq 0$.

(ii) $\Rightarrow$ (i) We shall prove equivalently that

$$A \text{ not invertible} \Rightarrow |A| = 0.$$

Suppose then that A is singular. By Theorem 1.1, we know that $\operatorname{rank} A < n$ *and that, consequently, the reduced row echelon form R of matrix A has, at least, one zero row.*

It is also the case that there exist elementary matrices $E_1, \dots, E_m$ such that

$$E_1 E_2 \cdots E_m A = R.$$

By Propositions 2.1, 2.2, we have

$$|E_1||E_2| \cdots |E_m||A| = |R| = 0.$$

Since the determinants of the elementary matrices are all different from zero, it follows that $|A| = 0$.

Proposition 2.3 *Let A and B be square matrices of order n. Then*

$$|AB| = |A||B|.$$

Proof *Suppose initially that A is non-singular. Hence, by Theorem 1.1, there exist elementary matrices $E_1, \dots, E_m$ such that $A = E_1 E_2 \cdots E_m$. It follows, by Proposition 2.2, that*

$$\begin{aligned} |AB| &= |E_1 E_2 \cdots E_m B| \\ &= |E_1||E_2 \cdots E_m B| \\ &= |E_1||E_2| \cdots |E_m||B| \\ &= |E_1 E_2 \cdots E_m||B| \\ &= |A||B|. \end{aligned}$$

If A is singular, then its reduced row echelon form R has a zero row (see Theorem 1.1). That is, there exist elementary matrices $E_1, \dots, E_r$ such that

$$|E_1 E_2 \cdots E_r AB| = |RB| = 0,$$

since RB has a zero row (see Proposition 2.1 (ii)). By Proposition 2.2,

$$|E_1 E_2 \cdots E_r AB| = \underbrace{|E_1||E_2| \cdots |E_r|}_{\neq 0} |AB| = 0.$$

Since $|A| = 0$ (see Theorem 2.1), we have

$$0 = |AB| = |A||B|,$$

as required.

An easy consequence of this proposition is the following result.

Corollary 2.1 *Let A be an invertible $n \times n$ matrix. Then*

$$|A^{-1}| = |A|^{-1}.$$

Proof *By Proposition 2.3,*

$$|AA^{-1}| = |A||A^{-1}|,$$

from which follows that

$$1 = |I| = |AA^{-1}| = |A||A^{-1}| \iff |A^{-1}| = \frac{1}{|A|}.$$

Lemma 2.1 *Let E be an elementary matrix of order n. Then $|E^T| = |E|$.*

Proof *If E is an elementary matrix of the form P_{ij} or $D_i(\alpha)$, the assertion is immediate, since these matrices are symmetric. On the other hand, if the matrix E is of the form $E_{ij}(\alpha)$, then*

$$E_{ij}(\alpha)^T = E_{ji}(\alpha)$$

and, therefore,

$$|E_{ij}(\alpha)^T| = |E_{ji}(\alpha)| = 1 = |E_{ij}(\alpha)|.$$

Proposition 2.4 *Let A be a square matrix of order n. Then*

$$|A^T| = |A|.$$

Proof *If A is non-singular, then, by Theorem 1.1, there exist elementary matrices $E_1, \ldots, E_m$ such that $A = E_1E_2 \cdots E_m$. It follows, by Proposition 2.3 and Lemma 2.1, that*

$$\begin{aligned}
|A^T| &= |(E_1E_2 \cdots E_m)^T| \\
&= |E_m^T \cdots E_2^T E_1^T| \\
&= |E_m^T| \cdots |E_2^T||E_1^T| \\
&= |E_m| \cdots |E_2||E_1| \\
&= |E_1||E_2| \cdots |E_m| \\
&= |E_1E_2 \cdots E_m| \\
&= |A|.
\end{aligned}$$

If A is singular, then matrix A^T is also singular (see Proposition 1.12 (i) and Proposition 1.18 (v)). Hence, by Theorem 2.1, we have

$$|A^T| = 0 = |A|.$$

Observe that any lower triangular matrix can be obtained as the transposed matrix of an upper triangular matrix and that transposition does not change the determinant, as seen in Proposition 2.4. This leads immediately to the next proposition.

Proposition 2.5 *Let $A = [a_{ij}]$ be an $n \times n$ upper triangular (respectively, lower triangular) matrix. Then,*

$$|A| = \prod_{i=1}^{n} a_{ii}.$$

An immediate corollary of Propositions 2.1, 2.5 is the following 'column'-version of Proposition 2.1 (i),(ii).

Corollary 2.2 *Let A be a square matrix. The following assertions hold.*

(i) If A has two equal columns then $|A| = 0$.

(ii) If A has a zero column then $|A| = 0$.

2.2 Leibniz's Formula

We already know how to calculate the determinant of any $n \times n$ matrix A, albeit not having an explicit expression to do so. As seen before, basically we just have to know the determinant of the identity matrix I (which is equal to 1) and make a careful use of Gaussian elimination. That is, we reduce A to an upper triangular matrix U, whose determinant is the product of its diagonal entries, and keep a record of how Gaussian elimination changes the determinant of U to get the determinant of A.

We have defined the determinant function axiomatically and it might seem that we could have stopped there. However, there are two important details that we have overlooked on purpose: the existence and the uniqueness of the determinant function. As to the latter, we postpone its analysis to the Appendix. As to the former, the existence of a function that satisfies the axioms (Ax1)–(Ax3) will be dealt with in this section. We give below the definition of the determinant as a 'proper' function through a formula, the Leibniz's formula of the determinant. We will show that this function satisfies the axioms (Ax1)–(Ax3). Once this is done, the existence part is settled and out of the way.

Definition 20 *Let $S = (1, \dots, n)$. A* ***permutation*** *p: $S \to S$ of S is a(ny) bijection from S onto itself. An $n \times n$* ***permutation matrix*** *is a matrix P such that each row and each column of P have exactly one non-zero entry which is equal to* 1.

Notice that the total number of permutations of S is $n!$ Observe also that a permutation matrix is what you obtain by 'scrambling' the rows of the identity matrix.

Example 2.5 *Let $S = (1, 2, 3)$. There are $6 = 3!$ permutations of S:*

$$p_1 = (1,2,3), p_2 = (1,3,2), p_3 = (2,1,3),$$

$$p_4 = (2,3,1), p_5 = (3,1,2), p_6 = (3,2,1).$$

For example,

$$\underbrace{\begin{bmatrix} 1 & 0 & 0 \\ 0 & 0 & 1 \\ 0 & 1 & 0 \end{bmatrix}}_{P_1} \begin{bmatrix} 1 \\ 2 \\ 3 \end{bmatrix} = \begin{bmatrix} 1 \\ 3 \\ 2 \end{bmatrix} = p_2$$

and

$$\underbrace{\begin{bmatrix} 0 & 1 & 0 \\ 0 & 0 & 1 \\ 1 & 0 & 0 \end{bmatrix}}_{P_2} \begin{bmatrix} 1 \\ 2 \\ 3 \end{bmatrix} = \begin{bmatrix} 2 \\ 3 \\ 1 \end{bmatrix} = p_4.$$

The matrices P_1 and P_2 are permutation matrices. To obtain the identity I from matrix P_1, one needs to make a single row exchange between rows 2 and 3. The rows of P_2 must be exchanged twice to obtain I. A possibility is row 2 $\leftrightarrow$ row 3 followed by row 2 $\leftrightarrow$ row 1. One could also opt for row 1 $\leftrightarrow$ row 3 followed by row 2 $\leftrightarrow$ row 3.

As this example suggests, to obtain a permutation of S, we multiply the vector $\begin{bmatrix} 1 & \dots & n \end{bmatrix}^T$, corresponding to S, by an appropriate permutation matrix.

Definition 21 *A permutation of $S = (1, \dots, n)$ is said to be* ***even*** *if the number of row exchanges in the corresponding matrix P required to obtain the identity is even. A permutation is* ***odd*** *if the number of row exchanges in the corresponding matrix P required to obtain the identity is odd.*

The ***sign*** *sign(p) of a permutation p is $+1$, if the permutation is even, and is -1 if the permutation is odd.*

Clearly, the definition of the parity of the permutation could be equivalently expressed in terms of the number of exchanges between the position of the elements in the permutation to obtain S. We saw in Example 2.5 that, in general, the exchanges can be done in more than one way. However, it is possible to show that the parity remains unchanged.

Let $A = [a_{ij}]$ be a $n \times n$ matrix and let P_S be the set of all permutations of S. The Leibniz's formula for the determinant of A is

$$\det A = \sum_{p \in P_S} \operatorname{sign}(p) a_{1p(1)} a_{2p(2)} \dots a_{np(n)}. \tag{2.4}$$

Notice that each summand in (2.4) is a product of n entries such that no column or row is shared by two different entries.

Example 2.6 *We recalculate the determinant of the matrix*

$$A = \begin{bmatrix} 3 & -3 & -3 \\ 0 & 1 & -1 \\ -1 & 0 & 0 \end{bmatrix}.$$

of Example 2.3. We have that the permutations p_1, p_4, p_5 are even and correspond to the products

$$a_{11}a_{22}a_{33} = 0,\ a_{12}a_{23}a_{31} = -3,\ a_{13}a_{21}a_{32} = 0.$$

The rest of the permutations are odd

$$a_{13}a_{22}a_{31} = 3,\ a_{12}a_{21}a_{33} = 0,\ a_{11}a_{23}a_{32} = 0.$$

By Leibniz's formula,

$$\det A = (0 - 3 + 0) - (3 + 0 + 0) = -6.$$

Proposition 2.6 *The function defined on $M_n(\mathbb{K})$ by (2.4) satisfies (Ax1)–(Ax3)*

Proof *The determinant of the identity matrix $I = [u_{ii}]$ is*

$$\det I = \sum_{p \in P_S} sign(p) u_{1p(1)} u_{2p(2)} \dots u_{np(n)} = +u_{11}u_{22} \dots u_{nn} = 1$$

which shows that (Ax1) holds.

As to (Ax2), suppose that two rows of A, say, i and j are exchanged, obtaining a new matrix A'. Hence, for each $p \in P_S$ and the corresponding summand in (2.4), we have a new permutation associated with A', which is

$$a_{1p'(1)} a_{2p'(2)} \dots a_{ip'(i)} \dots a_{jp'(j)} \dots a_{np'(n)},$$

with $p'(i) = p(j)$, $p'(j) = p(i)$ and $p'(k) = p(k)$, for $k \neq i, j$. Moreover, all permutations p' are obtained in this way. Then,

$$\det A' = \sum_{p' \in P_S} sign(p') a_{1p'(1)} a_{2p'(2)} \dots a_{ip'(i)} \dots a_{jp'(j)} \dots a_{np'(n)}.$$

But p coincides with p' except for the extra exchange between $p(i)$ and $p(j)$. Consequently, the parity of the permutations change and we have

$$sign(p') = -sign(p).$$

It follows that

$$
\begin{aligned}
\det A' &= \sum_{p' \in P_S} sign(p') a_{1p'(1)} a_{2p'(2)} \cdots a_{ip'(i)} \cdots a_{jp'(j)} \cdots a_{np'(n)} \\
&= \sum_{p' \in P_S} sign(p') a_{1p(1)} a_{2p(2)} \cdots a_{ip(j)} \cdots a_{jp(i)} \cdots a_{np(n)} \\
&= \sum_{p \in P_S} -sign(p) a_{1p(1)} a_{2p(2)} \cdots a_{ip(i)} \cdots a_{jp(j)} \cdots a_{np(n)} \\
&= -\det A.
\end{aligned}
$$

Finally, we tackle (Ax3). It is clear that the formula (2.4) *satisfies the first equality in (Ax3). Now, let* $A = [a_{ij}], B = [b_{ij}]$*, and* $C = [c_{ij}]$ *be matrices as in* (2.1)*, and* (2.2)*. Then*

$$
\begin{aligned}
\det C &= \sum_{p \in P_S} sign(p) c_{1p(1)} c_{2p(2)} \cdots c_{ip(i)} \cdots c_{np(n)} \\
&= \sum_{p \in P_S} sign(p) c_{1p(1)} c_{2p(2)} \cdots (a_{ip(i)} + b_{ip(i)}) \cdots c_{np(n)} \\
&= \sum_{p \in P_S} sign(p) c_{1p(1)} c_{2p(2)} \cdots a_{ip(i)} \cdots c_{np(n)} \\
&\quad + \sum_{p \in P_S} sign(p) c_{1p(1)} c_{2p(2)} \cdots b_{ip(i)} \cdots c_{np(n)} \\
&= \sum_{p \in P_S} sign(p) a_{1p(1)} a_{2p(2)} \cdots a_{ip(i)} \cdots a_{np(n)} \\
&\quad + \sum_{p \in P_S} sign(p) b_{1p(1)} b_{2p(2)} \cdots b_{ip(i)} \cdots b_{np(n)} \\
&= \det A + \det B,
\end{aligned}
$$

as required.

It is now clear that there exists a determinant function, i.e., a function that satisfies (Ax1)–(Ax3). This was shown by using Leibniz's formula for the determinant. As said before, this way of calculating the determinant is useful, at least to show that there exists in fact a determinant function, but is far from being practical when it comes to actual calculations. It will never be used in the remainder of the book, apart from proving Laplace's formula in the next section.

2.3 Laplace's Formula

In this section, we present yet another formula, the Laplace's formula of the determinant function. This formula is particularly useful when we need to calculate determinants of (small) matrices whose entries have parameters (see Chapter 4). However, due to the quantity of multiplications involved when we deal with large matrices, this is not an effective way of calculating the determinant.

Definition 22 *Let A be a square matrix of order n. Let $[\boldsymbol{A_{ij}}]$ be the submatrix of A obtained by deleting row i and column j. The* ***minor-ij*** $\boldsymbol{M_{ij}}$ *and the* ***cofactor-ij*** $\boldsymbol{C_{ij}}$ *are defined by*

$$M_{ij} = \det[A_{ij}], \qquad C_{ij} = (-1)^{i+j} M_{ij}.$$

Theorem 2.2 (Laplace's formula with expansion along row i) *Let A be a square matrix of order n and let $\mathbf{l}_i$ be a row of A. Then,*

$$|A| = \sum_{j=1}^{n} a_{ij} C_{ij}. \tag{2.5}$$

Proof *This proof follows closely §11.3 of [13]. Given a matrix A, we know that, by* (2.4),

$$\det A = \sum_{p \in P_S} sign(p) a_{1p(1)} a_{2p(2)} \dots a_{ip(i)} \dots a_{np(n)}. \tag{2.6}$$

Notice that, if $\mathbf{l}_i$ is some fixed row of A, then each summand has a single entry from $\mathbf{l}_i$.

Consider firstly the permutations $p \in P_S$ such that $p(1) = 1$ and denote by P'_S this subset of permutations. That is to say, we are speaking of all the summands of the form $a_{11} a_{2p(2)} \dots a_{ip(i)} \dots a_{np(n)}$. In this case, the sum of all this summands gives

$$\begin{aligned}
\sum_{p \in P'_S} sign(p) a_{11} a_{2p(2)} \dots a_{ip(i)} \dots a_{np(n)} &= a_{11} \sum_{q \in S_1} sign(q) a_{2q(2)} \dots a_{nq(n)} \\
&= a_{11} M_{11} \\
&= a_{11} (-1)^{1+1} M_{11} \\
&= a_{11} C_{11}
\end{aligned}$$

where S_1 is the set of permutations of the ordered set $(2, \dots, n)$ and $i = 1 = p(1) = j$.

Consider now the general case where $p(i) = j$ and, therefore, we have all summands in (2.6) containing the entry a_{ij} as a factor. Let B be the matrix

$$B = \begin{bmatrix} a_{i1} & a_{i2} & \cdots & a_{in} \\ a_{11} & a_{12} & \cdots & a_{1n} \\ \vdots & & & \\ a_{i-1,1} & a_{i-1,2} & \cdots & a_{i-1,n} \\ a_{i+1,1} & a_{i+1,2} & \cdots & a_{i+1,n} \\ \vdots & & & \\ a_{n1} & a_{n2} & \cdots & a_{nn} \end{bmatrix}.$$

We have that $\det B = (-1)^{i-1} \det A$*, since we exchanged rows $i-1$ many times. Now exchange columns in B such that we obtain the matrix*

$$B' = \begin{bmatrix} a_{ij} & a_{i1} & a_{i2} & \cdots & a_{in} \\ a_{1j} & a_{11} & a_{12} & \cdots & a_{1n} \\ \vdots & & & & \\ a_{i-1,j} & a_{i-1,1} & a_{i-1,2} & \cdots & a_{i-1,n} \\ a_{i,+1j} & a_{i+1,1} & a_{i+1,2} & \cdots & a_{i+1,n} \\ \vdots & & & & \\ a_{nj} & a_{n1} & a_{n2} & \cdots & a_{nn} \end{bmatrix}.$$

It follows that

$$\det A = (-1)^{i-1} \det B = (-1)^{i-1}(-1)^{j-1} \det B' = (-1)^{i+j} \det B'.$$

Notice that, since transposition does not modify the determinant, when two columns are exchanged, as with rows, the determinant is multiplied by -1.

We have shown above that

$$\det B' = a_{ij} \det[B'_{11}] + \sigma,$$

where σ is the sum of all the summands in the Leibniz's formula for $\det B'$*, which do not have a_{ij} as a factor and $[B'_{11}]$ is the submatrix of B' obtained by deleting row 1 and column 1 of B'. Observing that*

$$\det[B'_{11}] = M_{ij},$$

it follows that

$$\det A = a_{ij}(-1)^{i+j} M_{ij} + \mu = a_{ij} C_{ij} + \mu,$$

where μ is the sum of all summands in (2.6) which do not contain a_{ij} as a factor. This ends the proof.

Example 2.7 *We apply formula (2.5) of Theorem 2.2 to find the determinant of*

$$A = \begin{bmatrix} 1 & 0 & 2 \\ 3 & 0 & -1 \\ 1 & 3 & 1 \end{bmatrix},$$

fixing row 3, i.e., $i = 3$. *It follows*

$$\begin{aligned}
|A| = \begin{vmatrix} 1 & 0 & 2 \\ 3 & 0 & -1 \\ 1 & 3 & 1 \end{vmatrix} &= a_{31}C_{31} + a_{32}C_{32} + a_{33}C_{33} \\
&= 1(-1)^{3+1}\begin{vmatrix} 0 & 2 \\ 0 & -1 \end{vmatrix} + 3(-1)^{3+2}\begin{vmatrix} 1 & 2 \\ 3 & -1 \end{vmatrix} + 1(-1)^{3+3}\begin{vmatrix} 1 & 0 \\ 3 & 0 \end{vmatrix} \\
&= 0 - 3(1 \times (-1) - 2 \times 3) + 0 \\
&= 21
\end{aligned}$$

As is clear in this example, (2.5) is a recursive formula that in each step decreases the order of the minors involved until one has only 2×2 minors to calculate. Moreover, given that transposition does not change the determinant, it is only natural that a Laplace's formula exist with expansion along a column.

Corollary 2.3 (Laplace's formula with expansion along column j**)** *Let* A *be a square matrix of order* n *and let* $\mathbf{c}_j$ *be a column of* A. *Then,*

$$|A| = \sum_{i=1}^{n} a_{ij}C_{ij}. \tag{2.7}$$

Proof *Exercise. (Hint: use the fact that transposition does not change the determinant.)*

Example 2.8 *We re-calculate the determinant of Example 2.7 using an expansion along column 2.*

$$\begin{aligned}
|A| = \begin{vmatrix} 1 & 0 & 2 \\ 3 & 0 & -1 \\ 1 & 3 & 1 \end{vmatrix} &= a_{12}C_{12} + a_{22}C_{22} + a_{32}C_{32} \\
&= 0 \times (-1)^{1+2}\begin{vmatrix} 3 & -1 \\ 1 & 1 \end{vmatrix} + 0 \times (-1)^{2+2}\begin{vmatrix} 1 & 2 \\ 1 & 1 \end{vmatrix} \\
&\quad + 3 \times (-1)^{3+2}\begin{vmatrix} 1 & 2 \\ 3 & -1 \end{vmatrix} \\
&= 0 + 0 - 3 \times (1 \times (-1) - 2 \times 3) \\
&= 21
\end{aligned}$$

It is clear from Examples 2.7 and 2.8 that a right choice of the row or the column may simplify considerably the calculations.

How to calculate the determinant of a $n \times n$ matrix A by recursion

1. Choose a row or a column with as many zeros as possible.
2. Calculate $\det A$ using either (2.5) or (2.7), depending on your choice of a row or a column, respectively.
3. Apply the process described in 1. and 2. to the $n-1 \times n-1$ determinants obtained.
4. Keep on repeating the process until obtaining 2×2 determinants, which you can then calculate directly (see Example 2.4).

Definition 23 *Let A be a $n \times n$ matrix. The* ***matrix of cofactors*** *of A is the $n \times n$ matrix* $\mathbf{cof}\,\boldsymbol{A}$ *defined by*

$$\operatorname{cof} A = [C_{ij}]_{i,j=1,\dots,n}.$$

The ***adjugate matrix*** *of A is the matrix* $\mathbf{adj}\ \boldsymbol{A}$ *defined by*

$$\operatorname{adj} A = (\operatorname{cof} A)^T.$$

Proposition 2.7 *Let A be a square matrix of order n. The following hold.*

(i)

$$A \operatorname{adj} A = (\det A)I = (\operatorname{adj} A)A.$$

(ii) If $\det A \neq 0$, then the inverse matrix of A is

$$A^{-1} = \frac{1}{\det A} \operatorname{adj} A.$$

Proof *(i) We calculate firstly the diagonal entries of $A \operatorname{adj} A$. The entry-ii is*

$$(A \operatorname{adj} A)_{ii} = \sum_{j=1}^{n} a_{ij} C_{ij}.$$

It then follows from Theorem 2.2 that

$$(A \operatorname{adj} A)_{ii} = \det A.$$

On the other hand, an off-diagonal entry $(A \operatorname{adj} A)_{ik}$, with $i \neq k$,

$$(A \operatorname{adj} A)_{ik} = \sum_{j=1}^{n} a_{ij} C_{kj} = 0,$$

since this is the determinant of a matrix whose rows i and k coincide.

The equality (adj $A)A = (\det A)I$ *can be proved similarly.*

(ii) If $\det A \neq 0$ *then, by Theorem 2.1, A is invertible. The fact that in this case*

$$A^{-1} = \frac{1}{\det A} \operatorname{adj} A$$

is a direct consequence of (i).

How to calculate the inverse of a 2×2 matrix

Let A be the matrix $A = \begin{bmatrix} a & b \\ c & d \end{bmatrix}$, where $a, b, c, d \in \mathbb{K}$ and $\det A \neq 0$. By Proposition 2.7, we have

$$A^{-1} = \frac{1}{ad - bc} \begin{bmatrix} d & -b \\ -c & a \end{bmatrix}.$$

Hence, we have now a simple formula to calculate the inverse of a 2×2 square matrix which might come in handy (and actually does). No Gaussian elimination is required!

Let $A\mathbf{x} = \mathbf{b}$ be a system of linear equations with n equations, n unknowns and exactly one solution. As we know, this means that A is an invertible (real or complex) matrix or, in other words, $\det A \neq 0$. Cramer's rule gives a formula which uses the determinant to find the solution of this type of system of linear equations.

Suppose then that we have the following system

$$\begin{bmatrix} a_{11} & a_{12} & \dots & a_{1n} \\ a_{21} & a_{22} & \dots & a_{2n} \\ \vdots & \vdots & & \vdots \\ a_{n1} & a_{n2} & \dots & a_{nn} \end{bmatrix} \begin{bmatrix} x_1 \\ x_2 \\ \vdots \\ x_n \end{bmatrix} = \begin{bmatrix} b_1 \\ b_2 \\ \vdots \\ b_n \end{bmatrix}.$$

That is,

$$x_1 \begin{bmatrix} a_{11} \\ a_{21} \\ \vdots \\ a_{k1} \end{bmatrix} + x_2 \begin{bmatrix} a_{12} \\ a_{22} \\ \vdots \\ a_{k2} \end{bmatrix} + \dots + x_i \begin{bmatrix} a_{1i} \\ a_{2i} \\ \vdots \\ a_{ni} \end{bmatrix} + \dots + x_n \begin{bmatrix} a_{1n} \\ a_{2n} \\ \vdots \\ a_{nn} \end{bmatrix} = \begin{bmatrix} b_1 \\ b_2 \\ \vdots \\ b_n \end{bmatrix}.$$

It follows, for all $i = 1, \dots, n$, that

$$x_1 \begin{bmatrix} a_{11} \\ a_{21} \\ \vdots \\ a_{k1} \end{bmatrix} + x_2 \begin{bmatrix} a_{12} \\ a_{22} \\ \vdots \\ a_{k2} \end{bmatrix} + \dots + \begin{bmatrix} x_i a_{1i} - b_1 \\ x_i a_{2i} - b_2 \\ \vdots \\ x_i a_{ni} - b_n \end{bmatrix} + \dots + x_n \begin{bmatrix} a_{1n} \\ a_{2n} \\ \vdots \\ a_{nn} \end{bmatrix} = \begin{bmatrix} 0 \\ 0 \\ \vdots \\ 0 \end{bmatrix}. \quad (2.8)$$

We begin by supposing that $\mathbf{b} \neq \mathbf{0}$. Notice that, in this case, $\mathbf{x} \neq \mathbf{0}$. Hence, there exists r with $1 \leq r \leq n$ for which $x_r \neq 0$. It follows from equality (2.8) that column r of

$$B = \begin{bmatrix} a_{11} & a_{12} & \dots & x_i a_{1i} - b_1 & \dots & a_{1n} \\ a_{21} & a_{22} & \dots & x_i a_{2i} - b_2 & \dots & a_{2n} \\ \vdots & \vdots & & \vdots & & \vdots \\ a_{n1} & a_{n2} & \dots & x_i a_{ni} - b_n & \dots & a_{nn} \end{bmatrix}$$

is a linear combination of the remaining columns. In other words, using only elementary operations of type (iii) in Definition 2, row r in B^T can be transformed into a zero row, yielding $|B| = |B^T| = 0$. Hence,

$$x_i \det \begin{bmatrix} a_{11} & a_{12} & \dots & a_{1i} & \dots & a_{1n} \\ a_{21} & a_{22} & \dots & a_{2i} & \dots & a_{2n} \\ \vdots & \vdots & & \vdots & & \vdots \\ a_{n1} & a_{n2} & \dots & a_{ni} & \dots & a_{nn} \end{bmatrix} - \det \begin{bmatrix} a_{11} & a_{12} & \dots & b_1 & \dots & a_{1n} \\ a_{21} & a_{22} & \dots & b_2 & \dots & a_{2n} \\ \vdots & \vdots & & \vdots & & \vdots \\ a_{n1} & a_{n2} & \dots & b_n & \dots & a_{nn} \end{bmatrix} = 0.$$

Finally, we have **Cramer's rule**

$$x_i = \frac{\det \begin{bmatrix} a_{11} & a_{12} & \dots & b_1 & \dots & a_{1n} \\ a_{21} & a_{22} & \dots & b_2 & \dots & a_{2n} \\ \vdots & \vdots & & \vdots & & \vdots \\ a_{n1} & a_{n2} & \dots & b_n & \dots & a_{nn} \end{bmatrix}}{\det A}. \tag{2.9}$$

Observe that this formula also holds for $\mathbf{b} = \mathbf{0}$, since, in this case, the determinant in the numerator of (2.9) has a zero column and, therefore, is null.

Example 2.9 *Solve the system of linear equations*

$$\begin{cases} x + y + z = 3 \\ x - y + 2z = 2 \\ 2x + y - z = 2 \end{cases}.$$

Using Cramer's rule,

$$x = \frac{\det \begin{bmatrix} 3 & 1 & 1 \\ 2 & -1 & 2 \\ 2 & 1 & -1 \end{bmatrix}}{\det \begin{bmatrix} 1 & 1 & 1 \\ 1 & -1 & 2 \\ 2 & 1 & -1 \end{bmatrix}} = \frac{7}{7} = 1,$$

$$y = \frac{\det \begin{bmatrix} 1 & 3 & 1 \\ 1 & 2 & 2 \\ 2 & 2 & -1 \end{bmatrix}}{\det \begin{bmatrix} 1 & 1 & 1 \\ 1 & -1 & 2 \\ 2 & 1 & -1 \end{bmatrix}} = 1,$$

$$z = \frac{\det \begin{bmatrix} 1 & 1 & 3 \\ 1 & -1 & 2 \\ 2 & 1 & 2 \end{bmatrix}}{\det \begin{bmatrix} 1 & 1 & 1 \\ 1 & -1 & 2 \\ 2 & 1 & -1 \end{bmatrix}} = 1.$$

2.4 Exercises

EX 2.4.1. Use Gaussian elimination to calculate the determinant of

$$A = \begin{bmatrix} 6 & -12 & 18 \\ -3 & 1 & 4 \\ 6 & 7 & -1 \end{bmatrix}$$

EX 2.4.2. Let A and B be 3×3 matrices such that

$$\det A = 3 \quad \text{and} \quad \det B = -3.$$

Complete:

$$\det(-2A^{-3}) = \_\_\_\_, \quad \det\left(\left(AB^T\right)^2\right) = \_\_\_\_,$$

$$\det(E_{32}(-2)D_3(2)P_{34}A) = \_\_\_\_.$$

EX 2.4.3. Let A be a 4×4 matrix such that $|A| = -2$. Consider the following assertions.

I) The diagonal entries of A might be all equal to zero.
II) $|(2A)^{-1}| = -1/32$.
III) $|(2A)^{-1}| = -1/4$.
IV) $|(-A^T)^2| = 4$.

The complete list of correct assertions is

A) I, III, IV B) I, II C) I, II, IV D) III, IV

EX 2.4.4. Consider the matrices

$$A = \begin{bmatrix} 5 & 3 & 2 & 1 \\ b & 4 & d & 0 \\ a & 6 & c & 0 \\ 0 & 2 & 0 & 0 \end{bmatrix} \qquad B = \begin{bmatrix} a & b \\ -c & -d \end{bmatrix}.$$

where $a, b, c, d \in \mathbb{R}$. Suppose that $\det A = 20$. Find $\det B$.

EX 2.4.5. Use Laplace's rule to calculate the determinant of

$$A = \begin{bmatrix} 1 & 2 & 0 & 1 \\ 5 & 2 & 30 & 1 \\ 0 & 0 & 6 & 7 \\ 30 & 60 & 1 & 29 \end{bmatrix}$$

EX 2.4.6. Consider the matrix

$$A = \begin{bmatrix} -2 & -3 & 9 & 1 \\ -2 & 5 & 13 & 3 \\ 0 & 0 & 1 & 0 \\ -4 & -1 & -4 & 0 \end{bmatrix}$$

and the matrix $B = E_{23}(-1)\,A\,E_{34}(-5)$. Consider also the following assertions.

I) B is invertible.

II) $\det(B) = 52$.

III) $\det(\frac{1}{2}B)^{-1} = -\frac{2}{3}$.

IV) $B^{-1} = E_{23}(1)\,A^{-1}\,E_{34}(5)$.

The complete list of correct assertions is

A) I, II B) I, II, IV C) III, IV D) II, IV

EX 2.4.7. Determine

$$\operatorname{adj} \begin{bmatrix} 1 & -1 & -1 \\ 0 & 1 & 0 \\ 0 & 0 & 1 \end{bmatrix}.$$

EX 2.4.8. Find the cofactor C_{14} and the entry $(A^{-1})_{41}$ of the inverse matrix of

$$A = \begin{bmatrix} 15 & 3 & 6 & -4i \\ -15 & 1 & 1 & 30i \\ 15 & 3 & 7 & 20i \\ 15 & 3 & 8 & 12i \end{bmatrix}.$$

EX 2.4.9. Use Cramer's rule to solve the system

$$\begin{cases} 8x_1 - 6x_3 = -4 \\ -x_1 + 3x_2 - x_3 = -4 \\ 2x_1 - x_2 = -2 \end{cases}$$

EX 2.4.10. Let A_α and $\mathbf{b}_{\alpha,\beta}$ be the matrices

$$A_\alpha = \begin{bmatrix} 1 & \alpha+i & 2 \\ -1 & 1 & -2 \\ 0 & \alpha+i & 1 \end{bmatrix} \qquad \mathbf{b}_{\alpha,\beta} = \begin{bmatrix} \beta \\ 0 \\ \alpha \end{bmatrix},$$

where $\alpha, \beta \in \mathbb{C}$.

(a) Find all $\alpha \in \mathbb{C}$ for which A_α is not invertible.

(b) With $\alpha = -i$ and $\beta = 1$, use Cramer's rule to find the solution of the system $A_\alpha \mathbf{x} = \mathbf{b}_{\alpha,\beta}$.

EX 2.4.11. Let

$$A = \begin{bmatrix} a^2 & 3 & 0 \\ 5 & 0 & a \\ a^2 & a & 0 \end{bmatrix},$$

where a is a real number. Answer the following questions **without** calculating A^{-1}, should it exist.

(a) Find all a for which A is invertible.

(b) Whenever A is invertible, find the entry-23 of A^{-1}.

EX 2.4.12. Show that the determinant of a 3×3 matrix A satisfies **Sarrus' rule**: the determinant of A is a sum of all products of entries having the same shade in (a) minus the sum of all products of entries having the same shade in (b).

(*a*)

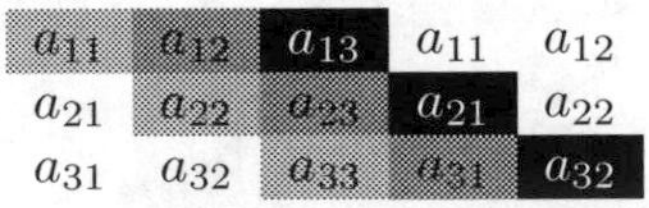

(*b*)

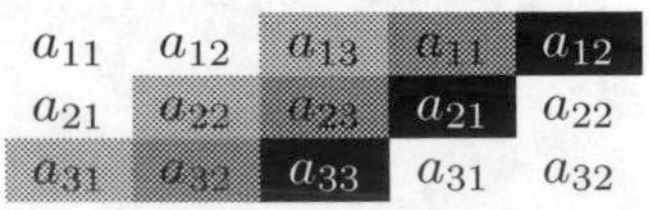

2.5 At a Glance

The determinant is a $\mathbb{K}$-valued function defined on the square matrices in $M_n(\mathbb{K})$. It can be calculated either using Gaussian elimination or a formula, Leibniz's or Laplace's. Easiest to calculate is the determinant of a triangular matrix for it is the product of its diagonal entries.

Matrix multiplication is not commutative but $|AB| = |BA|$. Moreover, the determinant is invariant under transposition.

The determinant can be used as a test for invertibility since invertible matrices are those having a non-zero determinant, and the inverse can be calculated by means of determinants.

Cramer's rule gives an explicit formula to obtain the solution of a system of linear equations as quotients of determinants, under the constraint that the system has a square coefficient matrix and a unique solution.

Chapter 3

Vector Spaces

3.1	Vector Spaces	82
3.2	Linear Independence	87
3.3	Bases and Dimension	93
	3.3.1 Matrix spaces and spaces of polynomials	99
	3.3.2 Existence and construction of bases	101
3.4	Null Space, Row Space, and Column Space	107
	3.4.1 $A\mathbf{x} = \mathbf{b}$	115
3.5	Sum and Intersection of Subspaces	116
3.6	Change of Basis	119
3.7	Exercises	123
3.8	At a Glance	130

If one were to be asked what matrices, polynomials, and vectors have in common, the first answer to spring to mind would be 'nothing'. This chapter is about proving this answer wrong.

What links matrices, polynomials, and vectors is the concept of vector space. We shall see how far reaching it is to look at these apparently far removed entities through the lens of this abstract concept.

Mathematics is the art of giving the same name to different things.
Henri Poincaré

Matrices profit from being regarded as vectors but this is a symbiotic relation: the analysis of vector spaces relies heavily on matrix techniques. Although this chapter might look at first as totally different in nature from the previous ones, we will see that vector spaces and matrices are definitely entangled. This entanglement is so strong that we will introduce in §3.4 four fundamental vector spaces associated with any given matrix . Their relevance is such that these vector spaces will be present in all of the remainder of the book.

To remain true to our purpose, the outstanding number in this chapter is the dimension of a vector space, of which much of the theory revolves around.

DOI: 10.1201/9781351243452-3

3.1 Vector Spaces

A **vector space over** $\mathbb{K}$ or a **linear space over** $\mathbb{K}$ is a non-empty set V endowed with the operations of addition $+$ and scalar multiplication μ

addition
$$+ : V \times V \to V$$
$$(\boldsymbol{u}, \boldsymbol{v}) \mapsto \boldsymbol{u} + \boldsymbol{v}$$

scalar multiplication
$$\mu : \mathbb{K} \times V \to V$$
$$(\alpha, \boldsymbol{u}) \mapsto \alpha \boldsymbol{u}$$

satisfying, for all $\boldsymbol{u}, \boldsymbol{v}, \boldsymbol{w} \in V$ and $\alpha, \beta \in \mathbb{K}$,

(i) $\boldsymbol{u} + \boldsymbol{v} = \boldsymbol{v} + \boldsymbol{u}$

(ii) $\boldsymbol{u} + (\boldsymbol{v} + \boldsymbol{w}) = (\boldsymbol{u} + \boldsymbol{v}) + \boldsymbol{w}$

(iii) There exists an element $\mathbf{0}$ in V, called the **additive identity**, such that
$$\boldsymbol{u} + \mathbf{0} = \boldsymbol{u} = \mathbf{0} + \boldsymbol{u}$$

(iv) Given $\boldsymbol{u} \in V$, there exists an element $-\boldsymbol{u} \in V$, called the **additive inverse** of $\boldsymbol{u}$, such that
$$\boldsymbol{u} + (-\boldsymbol{u}) = \mathbf{0} = (-\boldsymbol{u}) + \boldsymbol{u}$$

(v) $\alpha(\boldsymbol{u} + \boldsymbol{v}) = \alpha\boldsymbol{u} + \alpha\boldsymbol{v}$

(vi) $(\alpha\beta)\boldsymbol{u} = \alpha(\beta\boldsymbol{u})$

(vii) $(\alpha + \beta)\boldsymbol{u} = \alpha\boldsymbol{u} + \beta\boldsymbol{u}$

(viii) $1\boldsymbol{u} = \boldsymbol{u}$

When $\mathbb{K} = \mathbb{R}$ (respectively, $\mathbb{K} = \mathbb{C}$), V is also called a **real vector space** (respectively, a **complex vector space**).

An element of a vector space is said to be a **vector** or **point**. Axioms (i),(ii) say, respectively, that the addition of vectors is commutative and associative. We can also see in (v) and (vii) that the multiplication by a scalar is distributive relative to the addition of vectors and that the multiplication by a vector is distributive relative to the addition of scalars.

The additive identity is unique: if $\mathbf{0}$ and $\tilde{\mathbf{0}}$ were additive identities, then, by (iii) above,
$$\mathbf{0} = \mathbf{0} + \tilde{\mathbf{0}} = \tilde{\mathbf{0}}.$$

Notice also that, given $\boldsymbol{u} \in V$, we have $0\boldsymbol{u} = \mathbf{0}$. Observe that

$$\boldsymbol{u} = 1\boldsymbol{u} = (1+0)\boldsymbol{u} = 1\boldsymbol{u} + 0\boldsymbol{u} = \boldsymbol{u} + 0\boldsymbol{u}.$$

Adding the vector $-\boldsymbol{u}$ to both members of the equality $\boldsymbol{u} = \boldsymbol{u} + 0\boldsymbol{u}$, it follows that $0\boldsymbol{u} = \mathbf{0}$.

Similarly, the additive inverse of a vector $\boldsymbol{u}$ is unique. In fact, should there exist vectors $\boldsymbol{u_1}, \boldsymbol{u_2}$ such that

$$\boldsymbol{u} + \boldsymbol{u}_1 = \mathbf{0} = \boldsymbol{u} + \boldsymbol{u}_2,$$

then

$$\boldsymbol{u}_1 = \boldsymbol{u}_1 + (\boldsymbol{u} + \boldsymbol{u}_2) = (\boldsymbol{u}_1 + \boldsymbol{u}) + \boldsymbol{u}_2 = \boldsymbol{u}_2.$$

Moreover, $-\boldsymbol{u} = (-1)\boldsymbol{u}$, since

$$\boldsymbol{u} + (-1)\boldsymbol{u} = (1 + (-1))\boldsymbol{u} = \mathbf{0}$$

Consider again the set

$$\mathbb{K}^n = \{(a_1, a_2, \dots, a_n) : a_1, a_2, \dots, a_n \in \mathbb{K}\}$$

consisting of the n-tuples of scalars. We shall use throughout the following notation for $\boldsymbol{u} \in \mathbb{K}^n$:

$$\boldsymbol{u} = (a_1, a_2, \dots, a_n) \qquad \text{or} \qquad \mathbf{u} = \begin{bmatrix} a_1 \\ a_2 \\ \vdots \\ a_n \end{bmatrix}.$$

Although less frequently, we shall also use

$$[\boldsymbol{u}] = \begin{bmatrix} a_1 \\ a_2 \\ \vdots \\ a_n \end{bmatrix}.$$

For all $\boldsymbol{u}, \boldsymbol{v} \in \mathbb{K}^n$ and $\alpha \in \mathbb{K}$, define addition and scalar multiplication by

$$\mathbf{u} + \mathbf{v} = \underbrace{\begin{bmatrix} a_1 \\ a_2 \\ \vdots \\ a_n \end{bmatrix}}_{\mathbf{u}} + \underbrace{\begin{bmatrix} b_1 \\ b_2 \\ \vdots \\ b_n \end{bmatrix}}_{\mathbf{v}} = \begin{bmatrix} a_1 + b_1 \\ a_2 + b_2 \\ \vdots \\ a_n + b_n \end{bmatrix}$$

$$\alpha\mathbf{u} = \begin{bmatrix} \alpha a_1 \\ \alpha a_2 \\ \vdots \\ \alpha a_n \end{bmatrix}.$$

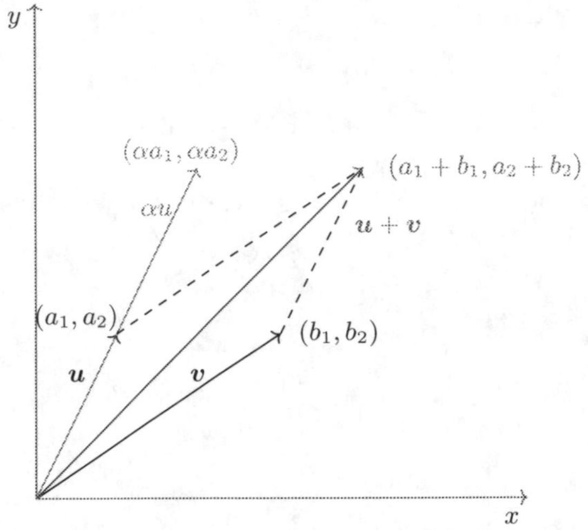

FIGURE 3.1: Addition and multiplication by a scalar for $\mathbb{R}^2$-plane vectors.

It is easily checked that, when endowed with these two operations, $\mathbb{K}^n$ is a vector space over $\mathbb{K}$. That is, these operations satisfy Axioms (i)–(viii). We shall see in §3.3 that $\mathbb{K}^n$, although being a concrete example of a vector space over $\mathbb{K}$, captures the essence of these spaces and will act as a model and a source of insight for the general spaces.

At this point, it is worthwhile noting that the vectors of $\mathbb{R}^n$ are a natural generalisation of the $\mathbb{R}^2$-plane vectors and the $\mathbb{R}^3$-space vectors. The same applying to the operations of addition and scalar multiplication. Figure 3.1 illustrates these operations on the $\mathbb{R}^2$-plane.

Example 3.1 *Let A be a $k \times n$ matrix over $\mathbb{K}$ and let $A\mathbf{x} = \mathbf{0}$ be the associated homogeneous system of linear equations. The solution set $S \neq \emptyset$ of this system is contained in $\mathbb{K}^n$ and, together with the restriction of the addition and scalar multiplication of $\mathbb{K}^n$, is itself a real vector space.*

To see this, we begin by showing that S is closed for the addition of vectors and scalar multiplication, that is, given $\boldsymbol{x}, \boldsymbol{y}, \boldsymbol{z} \in S$ and $\alpha \in \mathbb{R}$, the vectors $\boldsymbol{x} + \boldsymbol{y}$, $\alpha \boldsymbol{z}$ lie in S. In fact,

$$A(\mathbf{x} + \mathbf{y}) = A\mathbf{x} + A\mathbf{y} = \mathbf{0}, \quad A(\alpha\mathbf{z}) = \alpha A\mathbf{z} = \mathbf{0},$$

which shows that $\boldsymbol{x} + \boldsymbol{y}, \alpha\boldsymbol{z} \in S$, i.e., the vectors $\boldsymbol{x} + \boldsymbol{y}, \alpha\boldsymbol{z}$ are solutions of $A\mathbf{x} = \mathbf{0}$.

The above ensures that addition of vectors and scalar multiplication are well-defined in S. Notice also that $\mathbf{0} \in \mathbb{K}^n$ lies in S (Ax. (iii)) and that, for each $\boldsymbol{x} \in S$, its additive inverse $-\boldsymbol{x} = (-1)\boldsymbol{x}$ lies in S (Ax. (iv)).

It would remain to show that Axioms (i),(ii), and (v)–(viii) are satisfied. But that we know for granted, since the operations on S are the restrictions of those in $\mathbb{K}^n$.

Exercise 3.1 *Show that the following are real vector spaces.*

a) The set $M_{k,n}(\mathbb{R})$ *of real* $k \times n$ *matrices with the usual addition of matrices and multiplication of a matrix by a scalar.*

b) The set $\mathbb{P}_n$ *of real polynomials*

$$p(t) = a_0 + a_1 t + \cdots + a_n t^n, \qquad a_0, a_1, \ldots, a_n \in \mathbb{R}$$

of degree less that or equal to n *with the usual addition of real functions and multiplication of a function by a scalar.*

c) The set $\mathbb{P}$ *of real polynomials (of any degree) with the usual addition of real functions and multiplication of a function by a scalar.*

d) The set $C([a,b])$ *of continuous real functions on the real interval* $[a,b]$, *with* $a < b$, *endowed with the usual addition of real functions and multiplication of a function by a scalar.*

Example 3.2 *The solution set of the homogeneous system*

$$\underbrace{\begin{bmatrix} i & -i & 0 \\ -1+i & 1-i & 0 \end{bmatrix}}_{A} \begin{bmatrix} x_1 \\ x_2 \\ x_3 \end{bmatrix} = \begin{bmatrix} 0 \\ 0 \end{bmatrix}$$

is the vector space $S = \{(x_1, x_2, x_3) \in \mathbb{C}^3 : x_1 = x_2\}$. *Notice that* S *is contained in* $\mathbb{C}^3$, *yet another vector space.*

Exercise 3.2 *Show that the following are complex vector spaces.*

a) The set $M_{k,n}(\mathbb{C})$ *of real* $k \times n$ *matrices with the usual addition of matrices and multiplication of a matrix by a scalar.*

b) The set $\mathbb{P}_n$ *of real polynomials*

$$p(t) = a_0 + a_1 z + \cdots + a_n z^n, \qquad a_0, a_1, \ldots, a_n \in \mathbb{C}$$

of degree less that or equal to n *with the usual addition of complex functions and multiplication of a function by a scalar.*

c) The set $\mathbb{P}$ *of complex polynomials (of any degree) with the usual addition of complex functions and multiplication of a function by a scalar.*

Examples 3.1 and 3.2 outline a particularly relevant kind of subset in a vector space V, that that together with the restriction of the operations on V is itself a vector space.

Definition 24 *A non-empty subset* S *of a vector space* V *over* $\mathbb{K}$ *is said to be a* ***vector subspace*** *or a* ***linear subspace*** *of* V *if, together with the restriction of vector addition and scalar multiplication on* V, S *is itself a vector space over* $\mathbb{K}$.

Proposition 3.1 *Let S be a non-empty subset of a vector space V over $\mathbb{K}$ which is* **closed** *for the vector addition and scalar multiplication in V, i.e., for $\boldsymbol{u}, \boldsymbol{v} \in S$, $\alpha \in \mathbb{K}$, the vectors $\boldsymbol{u} + \boldsymbol{v} \in S$ and $\alpha\boldsymbol{u}$ lie in S. Then S is a subspace of V.*

Proof *Exercise (follow Example 3.1).*

Notice that with this proposition we have an equivalent definition of subspace. Indeed, a subspace of V could be equivalently defined as a non-empty substet of V which is closed for vector addition and scalar multiplication.

Example 3.3 *Examples of subspace of $\mathbb{R}^2$ are*

a) $\{(0,0)\}$, *b) a line through* $\{(0,0)\}$, *c)* $\mathbb{R}^2$.

Observe that all of the sets a)–c) contain the point $(0,0)$.

Can you find more subspaces of $\mathbb{R}^2$? Try.

Proposition 3.2 *Let S be a subspace of a vector space V over $\mathbb{K}$. Then the zero vector $\mathbf{0}$ lies in S.*

Proof *Let $\boldsymbol{u}$ be a vector in S (observe that, by definition, $S \neq \emptyset$). Since S is a subspace, then $0\boldsymbol{u} \in S$, i.e., $0\boldsymbol{u} = \mathbf{0} \in S$.*

A consequence of Proposition 3.2 is that any subset of V which does not contain the zero vector cannot be a subspace. For example, it is immediate that the line of equation $y = 2$ is not a subspace of $\mathbb{R}^2$ (see Figure 3.2).

How to see if a subset is a subspace

Let S be a non-empty subset of a vector space V.

1. Check if $\mathbf{0} \in S$. If it does not, then S is not a subspace.
2. If $\mathbf{0} \in S$, then take two arbitrary vectors $\boldsymbol{u}, \boldsymbol{v} \in S$ and an arbitrary scalar α and verify whether $\boldsymbol{u} + \boldsymbol{v} \in S$ and $\alpha\boldsymbol{u} \in S$ (as in Example 3.1). If both conditions are satisfied, then S is a subspace. If at least one of them fails, then S is not a subspace.

We shall see further on that $\mathbb{R}^2$ does not possess any subspaces apart from those in Example 3.3 (cf. Example 3.13). But that requires developing further the theory of vector spaces which, at this point in the book, is not mature enough to answer with confidence the (apparently) simple question

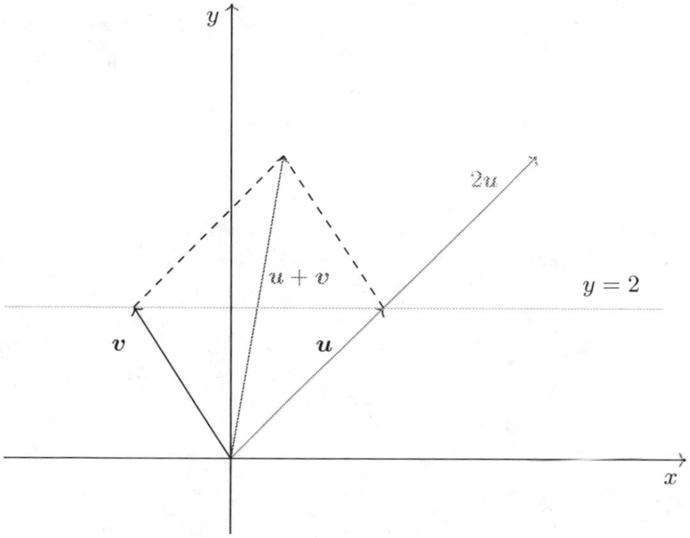

FIGURE 3.2: The straight line $y = 2$ is neither closed for the multiplication by scalars nor for the addition of vectors.

Which subsets of $\mathbb{R}^2$ are subspaces?

3.2 Linear Independence

Definition 25 *Let $\boldsymbol{u}_1, \boldsymbol{u}_2, \ldots, \boldsymbol{u}_k$ be vectors in a vector space V over $\mathbb{K}$. A* ***linear combination*** *of $\boldsymbol{u}_1, \boldsymbol{u}_2, \ldots, \boldsymbol{u}_k$ is any vector which can be presented as*

$$\alpha_1 \boldsymbol{u}_1 + \alpha_2 \boldsymbol{u}_2 + \cdots + \alpha_k \boldsymbol{u}_k,$$

where $\alpha_1, \alpha_2, \ldots, \alpha_k$ are scalars.

The set span$\{\boldsymbol{u}_1, \boldsymbol{u}_2, \ldots, \boldsymbol{u}_k\}$ *consisting of all linear combinations of $\{\boldsymbol{u}_1, \boldsymbol{u}_2, \ldots, \boldsymbol{u}_k\}$ is called the* ***span of the vectors*** *$\boldsymbol{u}_1, \boldsymbol{u}_2, \ldots, \boldsymbol{u}_k$, i.e.,*

$$\text{span}\{\boldsymbol{u}_1, \boldsymbol{u}_2, \ldots, \boldsymbol{u}_k\} = \{\alpha_1 \boldsymbol{u}_1 + \alpha_2 \boldsymbol{u}_2 + \cdots + \alpha_k \boldsymbol{u}_k \colon \alpha_1, \alpha_2, \ldots, \alpha_k \in \mathbb{R}\}.$$

Proposition 3.3 *Let $\boldsymbol{u}_1, \boldsymbol{u}_2, \ldots, \boldsymbol{u}_k$ be vectors in a vector space V over $\mathbb{K}$. Then the set* span$\{\boldsymbol{u}_1, \boldsymbol{u}_2, \ldots, \boldsymbol{u}_k\}$ *is a vector subspace of V.*

Proof *We must show that* span$\{\boldsymbol{u}_1, \boldsymbol{u}_2, \ldots, \boldsymbol{u}_k\}$ *is a (non-empty) set closed under vector addition and scalar multiplication. We start by showing that* span$\{\boldsymbol{u}_1, \boldsymbol{u}_2, \ldots, \boldsymbol{u}_k\}$ *is closed for vector addition.*

Let $\boldsymbol{u}, \boldsymbol{v}$ *be vectors in* $\text{span}\{\boldsymbol{u}_1, \boldsymbol{u}_2, \ldots, \boldsymbol{u}_k\}$, *i.e, there exist scalars* $\alpha_1, \alpha_2, \ldots, \alpha_k$ *and* $\beta_1, \beta_2, \ldots, \beta_k$ *such that*

$$\boldsymbol{u} = \alpha_1 \boldsymbol{u}_1 + \alpha_2 \boldsymbol{u}_2 + \cdots + \alpha_k \boldsymbol{u}_k,$$

and

$$\boldsymbol{v} = \beta_1 \boldsymbol{u}_1 + \beta_2 \boldsymbol{u}_2 + \cdots + \beta_k \boldsymbol{u}_k.$$

Hence, using the properties of the operations $+$ *and* μ,

$$\begin{aligned} \boldsymbol{u} + \boldsymbol{v} &= (\alpha_1 \boldsymbol{u}_1 + \alpha_2 \boldsymbol{u}_2 + \cdots + \alpha_k \boldsymbol{u}_k) + (\beta_1 \boldsymbol{u}_1 + \beta_2 \boldsymbol{u}_2 + \cdots + \beta_k \boldsymbol{u}_k) \\ &= \underbrace{(\alpha_1 + \beta_1)}_{\gamma_1} \boldsymbol{u}_1 + \underbrace{(\alpha_2 + \beta_2)}_{\gamma_2} \boldsymbol{u}_2 + \cdots + \underbrace{(\alpha_k + \beta_k)}_{\gamma_k} \boldsymbol{u}_k. \end{aligned}$$

Consequently, there exist scalrs $\gamma_1, \gamma_2, \ldots, \gamma_k$ *such that*

$$\boldsymbol{u} + \boldsymbol{v} = \gamma_1 \boldsymbol{u}_1 + \gamma_2 \boldsymbol{u}_2 + \cdots + \gamma_k \boldsymbol{u}_k,$$

from which follows that $\boldsymbol{u} + \boldsymbol{v}$ *lies in* $\text{span}\{\boldsymbol{u}_1, \boldsymbol{u}_2, \ldots, \boldsymbol{u}_k\}$. *In other words,* $\text{span}\{\boldsymbol{u}_1, \boldsymbol{u}_2, \ldots, \boldsymbol{u}_k\}$ *is closed under vector addition.*

Now let α *be a scalar and let* $\boldsymbol{u} = \alpha_1 \boldsymbol{u}_1 + \alpha_2 \boldsymbol{u}_2 + \cdots + \alpha_k \boldsymbol{u}_k$ *be a vector in* $\text{span}\{\boldsymbol{u}_1, \boldsymbol{u}_2, \ldots, \boldsymbol{u}_k\}$. *Then, using once again the properties of the operations* $+$ *and* μ,

$$\begin{aligned} \alpha \boldsymbol{u} &= \alpha(\alpha_1 \boldsymbol{u}_1 + \alpha_2 \boldsymbol{u}_2 + \cdots + \alpha_k \boldsymbol{u}_k) \\ &= \alpha(\alpha_1 \boldsymbol{u}_1) + \alpha(\alpha_2 \boldsymbol{u}_2) + \cdots + \alpha(\alpha_k \boldsymbol{u}_k) \\ &= (\alpha\alpha_1)\boldsymbol{u}_1 + (\alpha\alpha_2)\boldsymbol{u}_2 + \cdots + (\alpha\alpha_k)\boldsymbol{u}_k. \end{aligned}$$

Hence, $\alpha \boldsymbol{u}$ *is a linear combination of* $\boldsymbol{u}_1, \boldsymbol{u}_2, \ldots, \boldsymbol{u}_k$. *Hence* $\alpha \boldsymbol{u} \in \text{span}\{\boldsymbol{u}_1, \boldsymbol{u}_2, \ldots, \boldsymbol{u}_k\}$, *as required.*

Example 3.4 *The span of each of the sets below is* $\mathbb{R}^2$:

a) $\{(1,0), (0,1)\}$;

b) $\{(1,1), (-1,0)\}$;

c) $\{(1,1), (-1,0), (2,3)\}$.

This is clear for the set in a). We show now that the span of the set in b) is $\mathbb{R}^2$.

We must prove that any vector $(a_1, a_2) \in \mathbb{R}^2$ *is a linear combination of* $(1,1), (-1,0)$. *That is, there must exist* $\alpha, \beta \in \mathbb{R}$ *such that*

$$\begin{bmatrix} a_1 \\ a_2 \end{bmatrix} = \alpha \begin{bmatrix} 1 \\ 1 \end{bmatrix} + \beta \begin{bmatrix} -1 \\ 0 \end{bmatrix}.$$

This equality corresponds to the system of linear equations whose augmented matrix is

$$[A|\mathbf{b}] = \left[\begin{array}{cc|c} 1 & -1 & a_1 \\ 1 & 0 & a_2 \end{array}\right].$$

Since

$$\operatorname{rank}\left([A|\mathbf{b}]\right) = \operatorname{rank}\left(\left[\begin{array}{cc|c} 1 & -1 & a_1 \\ 1 & 0 & a_2 \end{array}\right]\right) = \operatorname{rank}(A)$$

this system is consistent, that is, (a_1, a_2) lies in $\operatorname{span}\{(1,1),(-1,0)\}$. *Solving the system, we get the unique solution $\alpha = a_2$ and $\beta = a_2 - a_1$.*

In c), the linear combination in question is

$$\begin{bmatrix} a_1 \\ a_2 \end{bmatrix} = \alpha \begin{bmatrix} 1 \\ 1 \end{bmatrix} + \beta \begin{bmatrix} -1 \\ 0 \end{bmatrix} + \gamma \begin{bmatrix} 2 \\ 3 \end{bmatrix},$$

corresponding to the augmented matrix

$$[A|\mathbf{b}] = \left[\begin{array}{ccc|c} 1 & -1 & 2 & a_1 \\ 1 & 0 & 3 & a_2 \end{array}\right].$$

This system is consistent but no longer does it have a unique solution. In fact, all linear combinations of $(1,1),(-1,0),(2,3)$ with coefficients $\alpha = -\gamma + a_1 - a_2, \beta = -\gamma + a_2, \gamma \in \mathbb{R}$ coincide with (a_1, a_2).

Exercise 3.3 *Show that the span of each of the sets in Example 3.4 is $\mathbb{C}^2$. Is this surprising?*

A set $\{\boldsymbol{u}_1, \boldsymbol{u}_2, \dots, \boldsymbol{u}_k\}$ of vectors in a vector space V is said to **span** V or to be a **spanning set** for V if

$$V = \operatorname{span}\{\boldsymbol{u}_1, \boldsymbol{u}_2, \dots, \boldsymbol{u}_k\}.$$

The sets in Example 3.4 are spanning sets for $\mathbb{R}^2$. In a similar vein, it is easily seen that

$$\{(1,0,\dots,0),(0,1,0,\dots,0),(0,0,1,0,\dots,0),\dots,(0,\dots,0,1)\}$$

is a spanning set for $\mathbb{R}^n$ (and also for $\mathbb{C}^n$).

Example 3.5 *The straight line in $\mathbb{R}^3$ defined by the equations $y = -x, z = x$ is spanned by the single vector $\{(1,-1,1)\}$.*

On the other hand, the plane S in $\mathbb{R}^3$ with equation $y = x$ needs two vectors to be spanned. For example, a spanning set for this plane is $\{(1,1,0),(0,0,1)\}$. In fact, any vector $\boldsymbol{u}$ in S is of the form $\boldsymbol{u} = (a,a,b)$, for some $a, b \in \mathbb{R}$. Hence,

$$\boldsymbol{u} = a(1,1,0) + b(0,0,1).$$

Given a vector space V, we have seen that it is possible to have several sets which span V. Moreover, as suggested by Example 3.4, some spanning sets are minimal. The sets in a) and b) span $\mathbb{R}^2$ in a more 'economical' way than the set in c). This set has one vector too many.

At this point, we are left with some questions.

(i) *Can one always find a spanning set for V?*

(ii) *How does one select the minimal spanning sets (avoiding redundancy)?*

(iii) *If, say, two vectors span V, do any two vectors span V?*

Question (iii) has a clear *No* for an answer. Already we can see this in the discussion above. For example, to span a plane in $\mathbb{R}^3$ (containing $(0,0,0)$) we need two vectors but they must not be colinear. To span $\mathbb{R}^3$ we need three vectors but none of them can be in the plane (or the line) spanned by the other two, i.e., the three vectors must not be coplanar.

The point seeming to be that, when selecting a spanning set, we must be sure not to include a vector which is already in the space spanned by the remaining vectors (as this vector does not bring anything new to the set of linear combinations of the other vectors). This still vague idea is conveyed precisely by the notion of linear independence.

Definition 26 *A subset $\{\boldsymbol{u}_1, \boldsymbol{u}_2, \dots, \boldsymbol{u}_k\}$ of a vector space V over $\mathbb{K}$ is said to be a* ***linearly independent set****, or that the vectors $\boldsymbol{u}_1, \boldsymbol{u}_2, \dots, \boldsymbol{u}_k$ are* ***linearly independent****, if*

$$\alpha_1 \boldsymbol{u}_1 + \alpha_2 \boldsymbol{u}_2 + \dots + \alpha_k \boldsymbol{u}_k = \boldsymbol{0} \quad \Rightarrow \quad \alpha_1 = \alpha_2 = \dots = \alpha_k = 0. \tag{3.1}$$

In the definition, we could have used an equivalence instead of the implication. In fact, if we set $\alpha_1 = \alpha_2 = \dots = \alpha_k = 0$, it is clear that

$$\alpha_1 \boldsymbol{u}_1 + \alpha_2 \boldsymbol{u}_2 + \dots + \alpha_k \boldsymbol{u}_k = \boldsymbol{0}.$$

It is evident from the definition that

any subset of V containing the zero vector cannot be linearly independent

because the linear combination whose coefficients are all equal to the scalar 0 except for that of the vector $\boldsymbol{0}$, which can be any given non-zero number, will coincide with $\boldsymbol{0}$.

Why condition (3.1) conveys the idea that $\operatorname{span}\{\boldsymbol{u}_1, \boldsymbol{u}_2, \dots, \boldsymbol{u}_k\}$ loses something if any single vector is removed from the set might not be clear at first. However, before illuminating this part (postponed until Theorem 3.1), we shall explore firstly linear independence *per se*.

Example 3.6 *The vectors $(1,0,1), (0,-1,1), (1,1,1) \in \mathbb{R}^3$ are linearly independent. In other words,*

$$\alpha_1 \begin{bmatrix} 1 \\ 0 \\ 1 \end{bmatrix} + \alpha_2 \begin{bmatrix} 0 \\ -1 \\ 1 \end{bmatrix} + \alpha_3 \begin{bmatrix} 1 \\ 1 \\ 1 \end{bmatrix} = \begin{bmatrix} 0 \\ 0 \\ 0 \end{bmatrix}$$

only when $\alpha_1 = \alpha_2 = \alpha_3 = 0$. *This means that the homogeneous system*

$$\begin{bmatrix} 1 & 0 & 1 \\ 0 & -1 & 1 \\ 1 & 1 & 1 \end{bmatrix} \begin{bmatrix} \alpha_1 \\ \alpha_2 \\ \alpha_3 \end{bmatrix} = \begin{bmatrix} 0 \\ 0 \\ 0 \end{bmatrix}$$

must admit the trivial solution $(\alpha_1, \alpha_2, \alpha_3) = (0, 0, 0)$ *only. For this to hold,*

$$\operatorname{rank}\left(\begin{bmatrix} 1 & 0 & 1 \\ 0 & -1 & 1 \\ 1 & 1 & 1 \end{bmatrix}\right)$$

must be equal to 3 *which is the case. Hence the vectors* $(1, 0, 1), (0, -1, 1)$, $(1, 1, 1)$ *are linearly independent.*

Exercise 3.4 *Show that the sets in Example 3.4 a), b), and Exercise 3.3 a), b) are linearly independent and that the set in c) of the same examples is not linearly independent. Check that removing any single vector from the sets in Example 3.4 c) and Exercise 3.3 c) makes the new set a linearly independent set which still spans* $\mathbb{R}^2$ *or* $\mathbb{C}^2$, *respectively.*

Definition 27 *A subset* $\{\boldsymbol{u}_1, \boldsymbol{u}_2, \ldots, \boldsymbol{u}_k\}$ *of a vector space* V *which is not linearly independent is called* ***linearly dependent*** *or the vectors* $\boldsymbol{u}_1, \boldsymbol{u}_2, \ldots, \boldsymbol{u}_k$ *are called* ***linearly dependent vectors.***

The vectors in Example 3.4 c) are linearly dependent.

The next proposition generalises the 'behaviour' appearing in Example 3.6 and Exercise 3.4.

Proposition 3.4 *Let* $\boldsymbol{u}_1, \boldsymbol{u}_2, \ldots, \boldsymbol{u}_k$ *be vectors in* $\mathbb{K}^n$. *The vectors* $\boldsymbol{u}_1, \boldsymbol{u}_2, \ldots,$ $\boldsymbol{u}_k$ *are linearly independent if and only if*

$$\operatorname{rank}\left(\begin{bmatrix} \mathbf{u}_1 & | & \mathbf{u}_2 & | & \ldots & | & \mathbf{u}_k \end{bmatrix}\right) = k.$$

Proof *Let*

$$A = \begin{bmatrix} \mathbf{u}_1 & | & \mathbf{u}_2 & | & \ldots & | & \mathbf{u}_k \end{bmatrix}, \qquad \mathbf{x} = \begin{bmatrix} \alpha_1 \\ \alpha_2 \\ \vdots \\ \alpha_k \end{bmatrix}.$$

The vectors $\boldsymbol{u}_1, \boldsymbol{u}_2, \ldots, \boldsymbol{u}_k$ *are linearly independent if, and only, if the equation* $A\mathbf{x} = \mathbf{0}$ *admits the trivial solution only. That is, if and only if the corresponding homogeneous system has a unique (trivial) solution. Consequently,* $\boldsymbol{u}_1, \boldsymbol{u}_2, \ldots, \boldsymbol{u}_k$ *are linearly independent if and only if*

$$\operatorname{rank}\left(\begin{bmatrix} \mathbf{u}_1 & | & \mathbf{u}_2 & | & \ldots & | & \mathbf{u}_k \end{bmatrix}\right) = k.$$

An important consequence of this proposition is the next corollary.

Corollary 3.1 *Let $\boldsymbol{u}_1, \boldsymbol{u}_2, \ldots, \boldsymbol{u}_k$ be vectors in $\mathbb{K}^n$. If $k > n$, then the vectors $\boldsymbol{u}_1, \boldsymbol{u}_2, \ldots, \boldsymbol{u}_k$ are linearly dependent.*

Proof *The rank of the matrix $A = \begin{bmatrix} \mathbf{u}_1 & | & \mathbf{u}_2 & | & \ldots & | & \mathbf{u}_k \end{bmatrix}$ can never be k, since $\operatorname{rank} A \leq n < k$. It follows from Proposition 3.4 that $\boldsymbol{u}_1, \boldsymbol{u}_2, \ldots, \boldsymbol{u}_k$ are linearly dependent.*

How to find if a set of vectors in $\mathbb{K}^n$ is linearly independent or linearly dependent

Let $\boldsymbol{u}_1, \boldsymbol{u}_2, \ldots, \boldsymbol{u}_k$ be vectors in $\mathbb{K}^n$.

1. Build the matrix whose columns consist of these vectors, i.e.,

$$A = \begin{bmatrix} \mathbf{u}_1 & | & \mathbf{u}_2 & | & \ldots & | & \mathbf{u}_k \end{bmatrix},$$

and use Gaussian elimination to find its rank;

2. If $\operatorname{rank} A = k$, then the vectors are linearly independent, otherwise they are linearly dependent.

Theorem 3.1 *Let V be a vector space and let $\boldsymbol{u}_1, \boldsymbol{u}_2, \ldots, \boldsymbol{u}_k$ be vectors in V. The vectors $\boldsymbol{u}_1, \boldsymbol{u}_2, \ldots, \boldsymbol{u}_k$ are linearly dependent if and only if one of the vectors is a linear combination of the others.*

Proof *We show firstly that, if some vector in $\{\boldsymbol{u}_1, \boldsymbol{u}_2, \ldots, \boldsymbol{u}_k\}$ is a linear combination of the others, then $\boldsymbol{u}_1, \boldsymbol{u}_2, \ldots, \boldsymbol{u}_k$ are linearly dependent.*

Suppose, without loss of generality, that there exist scalars $\alpha_2, \ldots, \alpha_k$ such that $\boldsymbol{u}_1$ is a linear combination of the other vectors, i.e.,

$$\boldsymbol{u}_1 = \alpha_2 \boldsymbol{u}_2 + \cdots + \alpha_k \boldsymbol{u}_k.$$

Then

$$\underbrace{(-1)}_{\neq 0} \boldsymbol{u}_1 + \alpha_2 \boldsymbol{u}_2 + \cdots + \alpha_k \boldsymbol{u}_k = \mathbf{0},$$

which shows that $\boldsymbol{u}_1, \boldsymbol{u}_2, \ldots, \boldsymbol{u}_k$ are linearly dependent.

Conversely, suppose now that $\boldsymbol{u}_1, \boldsymbol{u}_2, \ldots, \boldsymbol{u}_k$ are linearly dependent, i.e., for some scalars $\alpha_1, \ldots, \alpha_k$, not all equal to zero,

$$\alpha_1 \boldsymbol{u}_1 + \alpha_2 \boldsymbol{u}_2 + \cdots + \alpha_k \boldsymbol{u}_k = \mathbf{0}. \tag{3.2}$$

Suppose, without loss of generality, that $\alpha_1 \neq 0$. It follows from (3.2) that

$$\boldsymbol{u}_1 = -\frac{\alpha_2}{\alpha_1} \boldsymbol{u}_2 - \cdots - \frac{\alpha_k}{\alpha_1} \boldsymbol{u}_k. \tag{3.3}$$

Hence, $\boldsymbol{u}_1$ is a linear combination of the other vectors, which ends the proof.

Suppose that $\boldsymbol{u}_1$ is a linear combination of $\boldsymbol{u}_2, \ldots, \boldsymbol{u}_k$ and let $\boldsymbol{v}$ be a vector in the space $\operatorname{span}\{\boldsymbol{u}_1, \boldsymbol{u}_2, \ldots, \boldsymbol{u}_k\}$, i.e., for some scalars $\beta_1, \beta_2, \ldots, \beta_k$,

$$\boldsymbol{v} = \beta_1 \boldsymbol{u}_1 + \beta_2 \boldsymbol{u}_2 + \cdots + \beta_k \boldsymbol{u}_k.$$

Using (3.3), we have

$$\begin{aligned} \beta_1 \boldsymbol{u}_1 + \beta_2 \boldsymbol{u}_2 + \cdots + \beta_k \boldsymbol{u}_k &= \beta_1 \left(-\frac{\alpha_2}{\alpha_1} \boldsymbol{u}_2 - \cdots - \frac{\alpha_k}{\alpha_1} \boldsymbol{u}_k \right) + \beta_2 \boldsymbol{u}_2 + \cdots + \beta_k \boldsymbol{u}_k \\ &= \left(-\frac{\alpha_2}{\alpha_1} \beta_1 + \beta_2 \right) \boldsymbol{u}_2 + \cdots + \left(-\frac{\alpha_k}{\alpha_1} \beta_1 + \beta_k \right) \boldsymbol{u}_k, \end{aligned}$$

from which follows that

$$\operatorname{span}\{\boldsymbol{u}_1, \boldsymbol{u}_2, \ldots, \boldsymbol{u}_k\} = \operatorname{span}\{\boldsymbol{u}_2, \ldots, \boldsymbol{u}_k\}. \tag{3.4}$$

Hence the span remains unchanged if one eliminates from the spanning set a vector which is already a linear combination of the others. In other words, if one is to choose a minimal set to span a vector space, one needs to focus on linearly independent sets. The next section is devoted to this kind of sets, those linearly independent sets that span a vector space V.

3.3 Bases and Dimension

A crucial notion is that of a basis of a vector space. Intuitively, a basis can be thought of as a system of coordinate axes with respect to which the vectors are described, much like what happens in $\mathbb{R}^2$ or $\mathbb{R}^3$, for example. Roughly speaking, the number of axes in the system is the dimension of the space.

Definition 28 *A **basis** $\mathcal{B}$ of a vector space V is a (any) linearly independent set that spans V.*

Example 3.7 *Here are some examples of basis of $\mathbb{R}^2$.*

*a) The set $\mathcal{E}_2 = \{(1,0),(0,1)\}$ is a basis of $\mathbb{R}^2$, said the **standard basis** of $\mathbb{R}^2$.*

b) The set $\mathcal{B} = \{(1,1),(-1,0)\}$ is another basis of $\mathbb{R}^2$.

c) Another basis of $\mathbb{R}^2$ is $\mathcal{B}_1 = \ldots$ (find one).

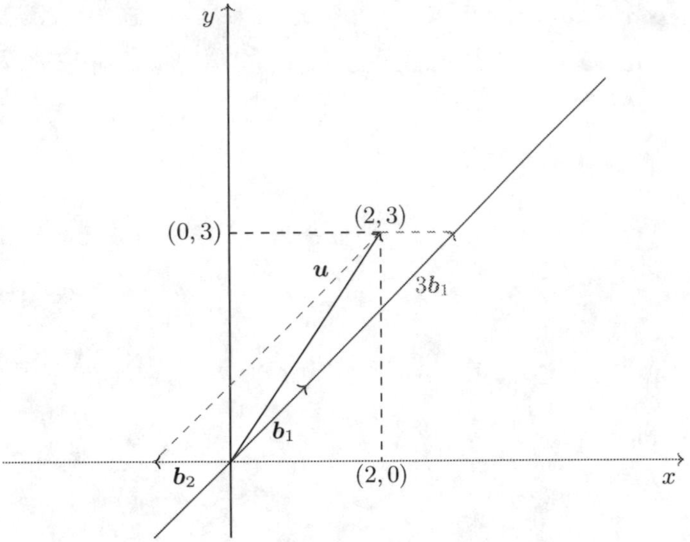

FIGURE 3.3: The coordinate vectors of $\boldsymbol{u} = (2, 3)$ relative to the standard basis and relative to $\mathcal{B} = (\boldsymbol{b}_1, \boldsymbol{b}_2)$ with $\boldsymbol{b}_1 = (1, 1), \boldsymbol{b}_2 = (-1, 0)$.

Exercise 3.5 *Express the vector* $(2, 3)$ *as a linear combination of the vector of the basis* $\mathcal{B}$ *in b) above.*

Solution. One wants to find $\alpha_1, \alpha_2 \in \mathbb{R}$ *such that*

$$\alpha_1 \begin{bmatrix} 1 \\ 1 \end{bmatrix} + \alpha_2 \begin{bmatrix} -1 \\ 0 \end{bmatrix} = \begin{bmatrix} 2 \\ 3 \end{bmatrix}.$$

This leads to solving the system whose augmented matrix is

$$\left[\begin{array}{cc|c} 1 & -1 & 2 \\ 1 & 0 & 3 \end{array}\right],$$

yielding $\alpha_1 = 3$ *e* $\alpha_2 = 1$ *(see Figure 3.3).*

The subset $\mathcal{E}_n = \{(1, 0, \dots, 0), (0, 1, 0, \dots, 0), \dots, (0, \dots, 0, 1)\}$ of $\mathbb{R}^n$ is the **standard basis** of $\mathbb{R}^n$.

Example 3.8

a) The set $\mathcal{B} = \{(1, -i), (1, 2i)\}$ *is a basis of* $\mathbb{C}^2$*. Since*

$$\operatorname{rank} \underbrace{\begin{bmatrix} 1 & 1 \\ -i & 2i \end{bmatrix}}_{A} = 2,$$

by Proposition 3.4, the vectors $(1, -i), (1, 2i)$ *are linearly independent. Moreover, given* $(a, b) \in \mathbb{C}^2$, *the system*

$$A \begin{bmatrix} \alpha \\ \beta \end{bmatrix} = \begin{bmatrix} a \\ b \end{bmatrix}$$

is clearly consistent, showing that $\{(1, -i), (1, 2i)\}$ *spans* $\mathbb{C}^2$.

b) The set $\mathcal{E}_3 = \{(1, 0, 0), (0, 1, 0), (0, 0, 1)\}$ *is a basis of* $\mathbb{C}^3$.

More generally, the set $\mathcal{E}_n = \{(1, 0, \dots, 0), (0, 1, 0, \dots, 0), \dots, (0, \dots, 0, 1)\}$ is the **standard basis** of $\mathbb{C}^n$.

Theorem 3.2 *Let* $\mathcal{B} = \{\boldsymbol{b_1}, \boldsymbol{b_2}, \cdots, \boldsymbol{b_k}\}$ *be a basis of a vector space* V *over* $\mathbb{K}$. *Then every vector in* V *is* <u>*uniquely*</u> *expressed as a linear combination of the vectors of the basis* $\mathcal{B}$.

Proof *Let* $\boldsymbol{u} \in V$ *and let* $\alpha_1, \alpha_2, \dots, \alpha_k, \beta_1, \beta_2, \dots, \beta_k$ *scalars such that*

$$\boldsymbol{u} = \alpha_1 \boldsymbol{b}_1 + \alpha_2 \boldsymbol{b}_2 + \cdots + \alpha_k \boldsymbol{b}_k,$$

and

$$\boldsymbol{u} = \beta_1 \boldsymbol{b}_1 + \beta_2 \boldsymbol{b}_2 + \cdots + \beta_k \boldsymbol{b}_k.$$

Subtracting the corresponding members of the equalities above, we have

$$\boldsymbol{0} = \boldsymbol{u} - \boldsymbol{u} = \alpha_1 \boldsymbol{b}_1 + \alpha_2 \boldsymbol{b}_2 + \cdots + \alpha_k \boldsymbol{b}_k - (\beta_1 \boldsymbol{b}_1 + \beta_2 \boldsymbol{b}_2 + \cdots + \beta_k \boldsymbol{b}_k).$$

Hence, using the properties of the operations $+$ *and* μ,

$$(\alpha_1 - \beta_1)\boldsymbol{b}_1 + (\alpha_2 - \beta_2)\boldsymbol{b}_2 + \cdots + (\alpha_k - \beta_k)\boldsymbol{b}_k = \boldsymbol{0}.$$

Since the set $\{\boldsymbol{b_1}, \boldsymbol{b_2}, \cdots, \boldsymbol{b_k}\}$ *is a basis of* V, *the vectors* $\boldsymbol{b_1}, \boldsymbol{b_2}, \cdots, \boldsymbol{b_k}$ *are linearly independent and, therefore,*

$$\alpha_1 - \beta_1 = \alpha_2 - \beta_2 = \cdots = \alpha_k - \beta_k = 0,$$

i.e.,

$$\alpha_1 = \beta_1, \alpha_2 = \beta_2, \quad \dots, \quad \alpha_k = \beta_k,$$

as required.

Let $\mathcal{B} = (\boldsymbol{b_1}, \boldsymbol{b_2}, \cdots, \boldsymbol{b_k})$ be an **ordered basis** of a vector space V and let

$$\boldsymbol{u} = \alpha_1 \boldsymbol{b_1} + \alpha_2 \boldsymbol{b_2} + \cdots + \alpha_k \boldsymbol{b_k}$$

be a vector in V. Then the **coordinate vector** of u relative to the basis $\mathcal{B}$ is the vector $\boldsymbol{u}_{\mathcal{B}}$ (ou $(\boldsymbol{u})_{\mathcal{B}}$) in $\mathbb{K}^k$ defined by

$$\boldsymbol{u}_{\mathcal{B}} = (\alpha_1, \alpha_2, \cdots, \alpha_k).$$

Whenever writing the coordinate vector of $\boldsymbol{u}$ as a column, we shall adopt the notation

$$\mathbf{u}_{\mathcal{B}} = \begin{bmatrix} \alpha_1 \\ \alpha_2 \\ \vdots \\ \alpha_k \end{bmatrix} \qquad \text{or} \qquad [\boldsymbol{u}]_{\mathcal{B}} = \begin{bmatrix} \alpha_1 \\ \alpha_2 \\ \vdots \\ \alpha_k \end{bmatrix}.$$

Example 3.9 *The coordinate vectors of* $(2,3) \in \mathbb{R}^2$ *relative to the basis* $\mathcal{B} = ((1,1),(-1,0))$ *and to the standard basis* $\mathcal{E}_2 = ((1,0),(0,1))$ *are, respectively,* $(2,3)_{\mathcal{B}} = (3,1)$ *e* $(2,3)_{\mathcal{E}_2} = (2,3)$*. What about* $(2,3)_{\mathcal{B}_1}$ *for* $\mathcal{B}_1$ *in Example 3.7 c)?*

How to calculate the coordinates of a vector in $\mathbb{K}^n$ relative to a basis

Let $(\boldsymbol{b}_1, \boldsymbol{b}_2, \ldots, \boldsymbol{b}_n)$ be a basis of $\mathbb{K}^n$, and let $\boldsymbol{x}$ be a vector in $\mathbb{K}^n$. To calculate the coordinate vector $\boldsymbol{x}_{\mathcal{B}} = (\alpha_1, \alpha_2, \ldots, \alpha_n)$ of $\boldsymbol{x}$ relative to the basis $\mathcal{B}$, i.e., to calculate the coefficients of the linear combination

$$\boldsymbol{x} = \alpha_1 \boldsymbol{b}_1 + \alpha_2 \boldsymbol{b}_2 + \cdots + \alpha_n \boldsymbol{b}_n$$

1) Solve the system whose augmented matrix is $\left[\mathbf{b}_1 \quad \mathbf{b}_2 \quad \ldots \quad \mathbf{b}_k \;\middle|\; \mathbf{x}\right]$;

2) The coordinate vector $\boldsymbol{x}_{\mathcal{B}} = (\alpha_1, \alpha_2, \ldots, \alpha_n) \in \mathbb{K}^n$ is the (unique) solution of the system in 1) above.

Example 3.10 *Find a basis for the subspace of* $\mathbb{R}^3$

$$S = \operatorname{span}\{(1,1,0),(0,0,1)\}.$$

Find also a vetor equation, parametric equations and cartesian equations for S*.*

By Proposition 3.4, we know that the vectors in S *are linearly independent if and only if*

$$\operatorname{rank}\left(\begin{bmatrix} 1 & 0 \\ 1 & 0 \\ 0 & 1 \end{bmatrix}\right) = 2.$$

Since the rank of the matrix is indeed 2, we can conclude that the vectors are linearly independent. That is, the set $S = \{(1,1,0),(0,0,1)\}$ *is a basis for* S*.*

Hence, the vectors (x, y, z) in S are linear combinations of the vectors in the basis of S. In other words, the vectors (x, y, z) of S satisfy the ***vector equation*** *of S*

$$(x, y, z) = t(1, 1, 0) + s(0, 0, 1), \tag{3.5}$$

where $t, s \in \mathbb{R}$.

The ***parametric equations*** *of S can be immediately obtained from (3.5):*

$$\begin{cases} x = t \\ y = t \\ z = s \end{cases}$$

where t, s are real parameters.

Cartesian equations *can be obtained in the following way. Let (x, y, z) be an arbitrary vector of S. This vector is a linear combination of the basis vectors of S and, therefore, the set*

$$\{(1, 1, 0), (0, 0, 1), (x, y, z)\}$$

is linearly dependent (cf. Theorem 3.1). That is, $(x, y, z) \in S$ if and only if the matrix

$$A = \begin{bmatrix} 1 & 0 & x \\ 1 & 0 & y \\ 0 & 1 & z \end{bmatrix}$$

has rank less than 3. (Hence rank A *must be equal to 2.) Keeping this in mind, we reduce A to a row echelon form using Gaussian elimination:*

$$\begin{bmatrix} 1 & 0 & x \\ 1 & 0 & y \\ 0 & 1 & z \end{bmatrix} \xrightarrow[L_2 - L_1]{} \begin{bmatrix} 1 & 0 & x \\ 0 & 0 & -x+y \\ 0 & 1 & z \end{bmatrix} \xrightarrow[L_2 \leftrightarrow L_3]{} \begin{bmatrix} 1 & 0 & x \\ 0 & 1 & z \\ 0 & 0 & -x+y \end{bmatrix}$$

Observe that A must be the augmented matrix of a consistent system. In fact, when consistent, it has a unique solution, given the linear independence of the two first columns of A. It follows that $-x + y$ must coincide with 0. Hence we obtain the cartesian equation $x = y$ for the plane S. In this very simple example, this equation could be obtained just by 'looking at' the set $\{(1, 1, 0), (0, 0, 1)\}$. However, it served to illustrate a general method to solve this kind of problem.

Proposition 3.5 *Let $\mathcal{B} = (\boldsymbol{b_1}, \boldsymbol{b_2}, \cdots, \boldsymbol{b_k})$ be a basis of a vector space V over $\mathbb{K}$. Given any vectors $\boldsymbol{u}, \boldsymbol{v} \in V$ and a scalar α, then*

$$(\boldsymbol{u} + \boldsymbol{v})_{\mathcal{B}} = \boldsymbol{u}_{\mathcal{B}} + \boldsymbol{v}_{\mathcal{B}} \qquad (\alpha\boldsymbol{u})_{\mathcal{B}} = \alpha\boldsymbol{u}_{\mathcal{B}}. \tag{3.6}$$

Moreover, the function $T\colon V \to \mathbb{K}^k$, defined by $T(\boldsymbol{u}) = \boldsymbol{u}_{\mathcal{B}}$, is bijective.

Proof *The proof of (3.6) is left as an exercise. The function T is surjective because, given a vector $(\alpha_1, \alpha_2, \dots, \alpha_k) \in \mathbb{K}^k$, the vector $\boldsymbol{u} \in V$ defined by $\boldsymbol{u} = \alpha_1 \boldsymbol{b}_1 + \alpha_2 \boldsymbol{b}_2 + \cdots + \alpha_k \boldsymbol{b}_k$ is such that*

$$T(\boldsymbol{u}) = \boldsymbol{u}_{\mathcal{B}} = (\alpha_1, \alpha_2, \dots, \alpha_k).$$

Suppose now that $\boldsymbol{u}, \boldsymbol{v} \in V$ are such that $T(\boldsymbol{u}) = T(\boldsymbol{v})$. Then, by (3.6), we have

$$T(\boldsymbol{u}) = T(\boldsymbol{v}) \Leftrightarrow T(\boldsymbol{u} - \boldsymbol{v}) = \boldsymbol{0}_{\mathbb{K}^k} \Leftrightarrow T(\boldsymbol{u} - \boldsymbol{v}) = (\underbrace{0, 0, \dots, 0}_{k}).$$

Hence

$$T(\boldsymbol{u} - \boldsymbol{v}) = (\boldsymbol{u} - \boldsymbol{v})_{\mathcal{B}} = (\underbrace{0, 0, \dots, 0}_{k}),$$

from which follows that

$$\boldsymbol{u} - \boldsymbol{v} = 0\boldsymbol{b}_1 + 0\boldsymbol{b}_2 + \cdots + 0\boldsymbol{b}_k = \boldsymbol{0}_V.$$

Consequently, $\boldsymbol{u} = \boldsymbol{v}$. Thus it has been shown that

$$T(\boldsymbol{u}) = T(\boldsymbol{v}) \Rightarrow \boldsymbol{u} = \boldsymbol{v},$$

that is, T is injective.

Given a basis $\mathcal{B} = (\boldsymbol{b}_1, \boldsymbol{b}_2, \cdots, \boldsymbol{b}_k)$ of a vector space V over $\mathbb{K}$, let T be the function defined by

$$T\colon V \to \mathbb{K}^k$$
$$\boldsymbol{u} \mapsto \boldsymbol{u}_{\mathcal{B}}$$

which sends each vector $\boldsymbol{u} \in V$ to its coordinate vector $\boldsymbol{u}_{\mathcal{B}}$ relative to the basis $\mathcal{B}$. By Proposition 3.5, T transforms vector sums in vector sums and transforms scalar multiplication in scalar multiplication.
Functions between vector spaces having this two properties are called linear transformations and will be the subject of Chapter 5.

Proposition 3.6 *Let $\mathcal{B} = (\boldsymbol{b}_1, \boldsymbol{b}_2, \cdots, \boldsymbol{b}_k)$ be a basis of a vector space V over $\mathbb{K}$ and let $\boldsymbol{u}_1, \dots, \boldsymbol{u}_p$ be vectors in V. The vectors $\boldsymbol{u}_1, \dots, \boldsymbol{u}_p$ are linearly independent if and only if the vectors $(\boldsymbol{u}_1)_{\mathcal{B}}, \dots, (\boldsymbol{u}_p)_{\mathcal{B}}$ in $\mathbb{K}^k$ are linearly independent.*

Proof *Since the function $\boldsymbol{u} \mapsto \boldsymbol{u}_{\mathcal{B}}$ is bijective (cf. Proposition 3.5),*

$$\alpha_1 \boldsymbol{u}_1 + \cdots + \alpha_p \boldsymbol{u}_p = \boldsymbol{0}_V \tag{3.7}$$

if and only if

$$(\alpha_1 \boldsymbol{u}_1 + \cdots + \alpha_p \boldsymbol{u}_p)_{\mathcal{B}} = \mathbf{0}_{\mathbb{K}^k} \,.$$

By Proposition 3.5, we have

$$\alpha_1 (\boldsymbol{u}_1)_{\mathcal{B}} + \cdots + \alpha_p (\boldsymbol{u}_p)_{\mathcal{B}} = \mathbf{0}_{\mathbb{K}^k} \,. \tag{3.8}$$

It follows that (3.7) *has a unique solution (i.e.,* $\alpha_1 = \cdots = \alpha_p = 0$*) if and only if the same applies to* (3.8)*. Hence,* $\boldsymbol{u}_1, \ldots, \boldsymbol{u}_p \in V$ *are linearly independent if and only if* $(\boldsymbol{u}_1)_{\mathcal{B}}, \ldots, (\boldsymbol{u}_p)_{\mathcal{B}} \in \mathbb{R}^k$ *are linearly independent.*

3.3.1 Matrix spaces and spaces of polynomials

Up to this point and although our definitons are general, in practical terms we have been focusing on $\mathbb{K}^n$. It is now the time to see what we get when addressing these concepts in more general vector spaces.

The ordered **standard basis** of the space $\mathrm{M}_{k,n}(\mathbb{K})$ of the $k \times n$ real matrices is the ordered set consisting of the real $k \times n$ matrices having all entries but one equal to zero which takes value 1; the ordering is such that the non-zero entry in the first matrix is entry-11 and it 'circulates' along the lines from left to right.

For example, in the case of $\mathrm{M}_2(\mathbb{K})$, the standard basis is

$$\mathcal{B}_c = \left(\begin{bmatrix} 1 & 0 \\ 0 & 0 \end{bmatrix}, \begin{bmatrix} 0 & 1 \\ 0 & 0 \end{bmatrix}, \begin{bmatrix} 0 & 0 \\ 1 & 0 \end{bmatrix}, \begin{bmatrix} 0 & 0 \\ 0 & 1 \end{bmatrix} \right).$$

Given a matrix $A = [a_{ij}]$, we have

$$A = \begin{bmatrix} a_{11} & a_{12} \\ a_{21} & a_{22} \end{bmatrix} = a_{11} \begin{bmatrix} 1 & 0 \\ 0 & 0 \end{bmatrix} + a_{12} \begin{bmatrix} 0 & 1 \\ 0 & 0 \end{bmatrix} + a_{21} \begin{bmatrix} 0 & 0 \\ 1 & 0 \end{bmatrix} + a_{22} \begin{bmatrix} 0 & 0 \\ 0 & 1 \end{bmatrix}.$$

Hence it is clear that A is a linear combination of the vectors, i.e., of the matrices, in $\mathcal{B}_c$. It is also easy to see that $\mathcal{B}_c$ is a linearly independent set. In fact,

$$\alpha_1 \begin{bmatrix} 1 & 0 \\ 0 & 0 \end{bmatrix} + \alpha_2 \begin{bmatrix} 0 & 1 \\ 0 & 0 \end{bmatrix} + \alpha_3 \begin{bmatrix} 0 & 0 \\ 1 & 0 \end{bmatrix} + \alpha_4 \begin{bmatrix} 0 & 0 \\ 0 & 1 \end{bmatrix} = \begin{bmatrix} 0 & 0 \\ 0 & 0 \end{bmatrix},$$

yields $\alpha_1 = \alpha_2 = \alpha_3 = \alpha_4 = 0$, which shows that $\mathcal{B}_c$ is linearly independent.

We see that $\mathcal{B}_c$ is a basis of $\mathrm{M}_2(\mathbb{K})$ and that, for a matrix A as above, the coordinate vector $A_{\mathcal{B}_c}$ of A relative to the basis $\mathcal{B}_c$ is $(a_{11}, a_{12}, a_{21}, a_{22})$ which lies in $\mathbb{K}^4$. We have then that any matrix

$$A = \begin{bmatrix} a_{11} & a_{12} \\ a_{21} & a_{22} \end{bmatrix}$$

has an image in $\mathbb{K}^4$ according to

$$\begin{aligned} T \colon \mathrm{M}_2(\mathbb{K}) &\to \mathbb{K}^4 \\ A &\mapsto A_{\mathcal{B}} = (a_{11}, a_{12}, a_{21}, a_{22}). \end{aligned}$$

Observe that Proposition 3.5 guarantees that T is bijective.

Example 3.11 *Find a basis $\mathcal{B}$ for the subspace S of $\mathrm{M}_2(\mathbb{R})$ spanned by the matrices*

$$A = \begin{bmatrix} 1 & 1 \\ 1 & 1 \end{bmatrix} \qquad B = \begin{bmatrix} 1 & 1 \\ 1 & 0 \end{bmatrix} \qquad C = \begin{bmatrix} 0 & 0 \\ 0 & -5 \end{bmatrix}.$$

Find also the coordinate vector $A_{\mathcal{B}}$ of A relative to the basis $\mathcal{B}$.

We need to decide whether the set $\{A, B, C\}$ is linearly independent because, if this were the case, $\{A, B, C\}$ would be a basis, i.e., a linearly independent set which spans the space. On the other hand, we know that a set of matrices in $\mathrm{M}_2(\mathbb{R})$ is linearly independent if and only if their coordinate vectors relative to the standard basis is a linearly independent set in $\mathbb{R}^4$ (cf. Proposition 3.6). Hence we shall see if the vectors

$$A_{\mathcal{B}_c} = (1,1,1,1), \qquad B_{\mathcal{B}_c} = (1,1,1,0), \qquad C_{\mathcal{B}_c} = (0,0,0,-5)$$

are linearly independent. We use Gaussian elimination to reduce $[[A_{\mathcal{B}_c}]\,[B_{\mathcal{B}_c}]\,[C_{\mathcal{B}_c}]]$ to a row echelon matrix:

$$\begin{bmatrix} 1 & 1 & 0 \\ 1 & 1 & 0 \\ 1 & 1 & 0 \\ 1 & 0 & -5 \end{bmatrix} \to \begin{bmatrix} 1 & 1 & 0 \\ 0 & 0 & 0 \\ 0 & 0 & 0 \\ 0 & -1 & -5 \end{bmatrix} \to \begin{bmatrix} 1 & 1 & 0 \\ 0 & -1 & -5 \\ 0 & 0 & 0 \\ 0 & 0 & 0 \end{bmatrix}.$$

We see that $A_{\mathcal{B}} = (1,1,1,1), B_{\mathcal{B}} = (1,1,1,0)$ are linearly independent, whilst $C_{\mathcal{B}} = (0,0,0,-5)$ lies in the $\mathrm{span}\{A_{\mathcal{B}}, B_{\mathcal{B}}\}$. Consequently, a basis of S is $\mathcal{B} = (A, B)$.

The coordinate vector $A_{\mathcal{B}}$ is

$$A_{\mathcal{B}} = \left(\begin{bmatrix} 1 & 1 \\ 1 & 1 \end{bmatrix}\right)_{\mathcal{B}} = (1,0).$$

Notice that no calculations are required to obtain this coordinate vector since $A = 1A + 0B$.

Observe that to solve this problem, we transferred it to a problem in $\mathbb{R}^4$, a 'copy/ mirror image' of $\mathrm{M}_2(\mathbb{R})$ obtained using the coordinate vector of each matrix relative to the standard basis. A crucial fact is that this copy is 1-to-1 and onto.
This is a trait of the book: whenever possible, we shall convert problems in some general real vector space (respectively, complex vector space) having a basis with n vectors in problems in $\mathbb{R}^n$ (respectively, $\mathbb{C}^n$) using Proposition 3.6. That is to say, transferring the problem from the space to its space of coordinate vectors relative to the given basis.

Let $\mathbb{P}_n$, and let

$$p(t) = a_o + a_1 t + a_2 t^2 + \cdots + a_{n-1} t^{n-1} + a_n t^n, \qquad a_o, a_1, a_2, \ldots, a_{n-1}, a_n \in \mathbb{R}.$$

denote the generic polynomial.

The **standard basis** of $\mathbb{P}_n$ is the ordered set $\mathcal{P}_n = (1, t, \ldots, t^n)$. As we have seen for matrix spaces, each polynomial $p(t) = a_o + a_1 t + a_2 t^2 + \cdots + a_{n-1}t^{n-1} + a_n t^n$ has an image in $\mathbb{R}^n$ consisting of the coordinate vector $(a_o, a_1, a_2, \ldots, a_{n-1}, a_n)$ of $p(t)$ relative to the basis $\mathcal{P}_n$.

Example 3.12 *Find a basis for the subspace S of $\mathbb{P}_3$ spanned by $X = \{1,\ 1+t, 1+t+t^2, 1+t+t^2+t^3\}$.*

Similarly to Example 3.11, we shall resort to the coordinate vectors of the polynomials relative to the basis $\mathcal{P}_3$. We obtain the matrix

$$\begin{bmatrix} 1 & 1 & 1 & 1 \\ 0 & 1 & 1 & 1 \\ 0 & 0 & 1 & 1 \\ 0 & 0 & 0 & 1 \end{bmatrix}$$

which is already a row echelon matrix having rank 4.

Hence the set X is linearly independent and, therefore, a basis of S. We shall see later that this forces S to coincide with $\mathbb{P}_3$ (see Theorem 3.5 (i)).

3.3.2 Existence and construction of bases

Lately we have seen how important bases are: they act as a 'system of coordinates' with respect to which the space is described. This even allows for treating any space having a basis with n vectors like $\mathbb{K}^n$ (cf. §3.3.1).

One might ask however whether this is always possible. Given a space, does it always have a basis? And if it has two bases, say, is there a relation between their cardinality?

The next two theorems answer these questions for spaces having a spanning set. But before going into that, it should be pointed out that not all spaces have a spanning set, that is, a finite set whose span coincides with the space. For example, if one considers the set $\mathbb{P}$ of real polynomials, it is impossible to find such a set for $\mathbb{P}$. (Why?)

Theorem 3.3 *Every vector space over $\mathbb{K}$ with a spanning set has a basis.*

Here we adopt the convention that the empty set $\emptyset$ is a basis of $V = \{\mathbf{0}\}$.

Proof *The case $V = \{\mathbf{0}\}$ holds trivially. Let $V \neq \{\mathbf{0}\}$ and let X be a spanning set of V. We show next that X contains a maximal linearly independent set Y, that is, any other subset of X which contains Y properly is linearly dependent.*

Let $\boldsymbol{y}_1$ be a non-zero vector in X, and observe that $\{\boldsymbol{y}_1\}$ is linearly independent. Now two situations can occur: either (a) every other vector of X lies in the subspace spanned by $\boldsymbol{y}_1$, or (b) we can find $\boldsymbol{y}_2 \in X$ such that $\{\boldsymbol{y}_1, \boldsymbol{y}_2\}$ is linearly independent.

In the case (a), $Y = \{\boldsymbol{y}_1\}$. *In case (b), one keeps adjoining vectors of* X *to* $\{\boldsymbol{y}_1, \boldsymbol{y}_2\}$, *one at a time, until obtaining a linearly independent set* Y *which cannot be enlarged, either because there are no more vectors in* X, *or because all the remaining vectors in* $X \backslash Y$ *lie in* $\operatorname{span} Y$ *(see Theorem 3.1).*

Let now $Y \subseteq X$ *be a maximal linearly independent set. It follows by the above reasoning that every* $\boldsymbol{x} \in X$ *is a linear combination of the elements of* Y. *Hence, we have that each element of* V *is a linear combination of the elements of* X, *which are in turn all linear combinations of the vectors in* Y. *It follows that* Y *spans* V *(see* (3.4)*) and, being linearly independent, is a basis of* V.

Theorem 3.4 *Let* $\mathcal{B}_1$ *e* $\mathcal{B}_2$ *be bases of a vector space* V *over* $\mathbb{K}$. *Then* $\mathcal{B}_1$ *and* $\mathcal{B}_2$ *have the same cardinality.*

Proof *Suppose, without loss of generality, that*

$$k = \#\mathcal{B}_1 < \#\mathcal{B}_2 = n.$$

Since every vector in $\mathcal{B}_2$ *is in the span of* $\mathcal{B}_1$, *one can construct a* $k \times n$ *matrix* A *whose columns are the coordinate vectors of the vectors in* $\mathcal{B}_2$ *relative to basis* $\mathcal{B}_1$. *It follows that* $\operatorname{rank}(A) \leq k < n$. *Hence, by Propositions 3.4 and 3.6,* $\mathcal{B}_2$ *is a linearly dependent set, yielding a contradiction.*

The **dimension** of a vector space V, denoted by $\dim V$, is the cardinality of a (any) basis of V.

For example, the standard basis $\mathcal{E}_n$ of $\mathbb{R}^n$ has n vectors, from which we see that $\dim \mathbb{R}^n = n$. Obviously, we have also $\dim \mathbb{C}^n = n$.

Going back to the examples in §3.3.1, we have that $\dim M_2(\mathbb{R}) = 4$, $\dim \mathbb{P}_n = n + 1$, and $\dim S = 2$, where S is the subspace in Example 3.11.

Now we know that any vector space with a spanning set has always a basis but also that not every vector space has a spanning set. We make here a distinction between these spaces: a vector space with a spanning set is called a **finite dimensional** vector space, whilst those vector spaces without such a set are called **infinite dimensional**. In the sequel, all vector spaces are supposed to be finite dimensional unless stated otherwise.

Recapping: now that we know that a finite dimensional vector space always possesses a basis (and, therefore, infinitely many basis, if $V \neq \{\mathbf{0}\}$), it would be desirable to have a way to find it (them). This will be accomplished in Theorem 3.5 below which gives a way of obtaining bases from subsets of vectors in V.

We begin with the following lemma.

Lemma 3.1 *Let* $X = \{\boldsymbol{u_1}, \boldsymbol{u_2}, \cdots, \boldsymbol{u_k}\}$ *be a linearly independent set of* $\mathbb{K}^k$. *Then* X *is a basis of* $\mathbb{K}^k$.

Proof *It is enough to show that X spans $\mathbb{K}^k$, i.e, to show that, for all $\boldsymbol{v} = (c_1, c_2, \ldots, c_k) \in \mathbb{K}^k$, there exist $\alpha_1, \alpha_2, \ldots, \alpha_k \in \mathbb{K}$ such that*

$$\boldsymbol{v} = \alpha_1 \boldsymbol{u_1} + \alpha_2 \boldsymbol{u_2} + \cdots + \alpha_k \boldsymbol{u_k}.$$

But the system of linear equations

$$\begin{bmatrix} \mathbf{u_1} & | & \mathbf{u_2} & | & \cdots & | & \mathbf{u_k} \end{bmatrix} \begin{bmatrix} \alpha_1 \\ \alpha_2 \\ \vdots \\ \alpha_k \end{bmatrix} = \begin{bmatrix} c_1 \\ c_2 \\ \vdots \\ c_k \end{bmatrix},$$

is consistent (and has a unique solution), since $\operatorname{rank}\left(\begin{bmatrix} \mathbf{u_1} & | & \mathbf{u_2} & | & \cdots & | & \mathbf{u_k} \end{bmatrix}\right) = k$ *(compare with Proposition 3.4).*

It has thus been shown that each vector $\boldsymbol{v} \in \mathbb{R}^k$ is a linear combination of $\boldsymbol{u_1}, \boldsymbol{u_2}, \cdots, \boldsymbol{u_k}$, i.e., X spans $\mathbb{K}^k$.

Theorem 3.5 *Let V be a vector space over $\mathbb{K}$ of positive dimension k.*

(i) Any k linearly independent vectors span V (i.e., form a basis of V).

(ii) Any subset of V containing m vectors, with $m > k$, is linearly dependent.

(iii) Any linearly independent subset of V consisting of p vectors, where $p < k$, is contained in a basis of V.

(iv) Any subset spanning V contains a basis of V.

Informally, we can say that (iii) asserts that any linearly independent subset of V can be 'augmented' to yield a basis of V. Similarly, it is stated in (iv) that any subset that spans V can be cut down in order to obtain a basis of V.

Proof *Let $\mathcal{B} = (\boldsymbol{b_1}, \boldsymbol{b_2}, \cdots, \boldsymbol{b_k})$ be a basis of V.*

(i) Let $X = \{\boldsymbol{u_1}, \boldsymbol{u_2}, \cdots, \boldsymbol{u_k}\}$ be a linearly independent set of vectors in V. By Proposition 3.6, the set $\{(\boldsymbol{u_1})_\mathcal{B}, (\boldsymbol{u_2})_\mathcal{B}, \cdots, (\boldsymbol{u_k})_\mathcal{B}\}$ is a linearly independent subset of $\mathbb{K}^k$. Hence, by Lemma 3.1, the set

$$\{(\boldsymbol{u_1})_\mathcal{B}, (\boldsymbol{u_2})_\mathcal{B}, \cdots, (\boldsymbol{u_k})_\mathcal{B}\}$$

is a basis of $\mathbb{K}^k$.

Let $\boldsymbol{v} = \alpha_1 \boldsymbol{b}_1 + \alpha_2 \boldsymbol{b}_2 + \cdots + \alpha_k \boldsymbol{b}_k$ be a vector of V. We want to show that $\boldsymbol{v}$ lies in $\operatorname{span} X$. *Then, since $((\boldsymbol{u_1})_\mathcal{B}, (\boldsymbol{u_2})_\mathcal{B}, \cdots, (\boldsymbol{u_k})_\mathcal{B})$ is an ordered basis of $\mathbb{K}^k$, the vector $(\alpha_1, \alpha_2, \ldots, \alpha_k) \in \mathbb{K}^k$ is a linear combination of the vectors $(\boldsymbol{u_1})_\mathcal{B}, (\boldsymbol{u_2})_\mathcal{B}, \cdots, (\boldsymbol{u_k})_\mathcal{B}$. That is,*

$$\boldsymbol{v}_\mathcal{B} = (\alpha_1, \alpha_2, \ldots, \alpha_k) = \beta_1 (\boldsymbol{u_1})_\mathcal{B} + \beta_2 (\boldsymbol{u_2})_\mathcal{B} + \cdots + \beta_k (\boldsymbol{u_k})_\mathcal{B}.$$

Hence, by Proposition 3.5,

$$\begin{aligned} v_{\mathcal{B}} &= \beta_1(\boldsymbol{u_1})_{\mathcal{B}} + \beta_2(\boldsymbol{u_2})_{\mathcal{B}} + \cdots + \beta_k(\boldsymbol{u_k})_{\mathcal{B}} \\ &= (\beta_1\boldsymbol{u_1} + \beta_2\boldsymbol{u_2} + \cdots + \beta_k\boldsymbol{u_k})_{\mathcal{B}} \end{aligned}$$

Using Proposition 3.5 again, we have

$$\boldsymbol{v} = \beta_1\boldsymbol{u_1} + \beta_2\boldsymbol{u_2} + \cdots + \beta_k\boldsymbol{u_k},$$

as required.

(ii) Let $X = \{\boldsymbol{u_1}, \boldsymbol{u_2}, \cdots, \boldsymbol{u_m}\}$ *be a subset of* V *and suppose that* X *is linearly independent. It follows that* $X_k = \{\boldsymbol{u_1}, \boldsymbol{u_2}, \cdots, \boldsymbol{u_k}\}$ *is also linearly independent.*[1]

Assertion (i) of this theorem guarantees that X_k *is a basis of* V*, from which follows that all vectors* $\boldsymbol{u}_{k+1}, \ldots, \boldsymbol{u_m}$ *lie in the span of the basis* X_k*. Consequently, by Theorem 3.1, the set* X *is linearly dependent, contradicting the initial hypothesis of* X *being linearly independent. We have thus shown that* X *is a linearly dependent set.*

(iii) Let X *consist of a linearly independent subset of* V *having* p *vectors and let* $\mathcal{B}$ *be a basis of* V*. Let* $\{\boldsymbol{u}_1, \ldots, \boldsymbol{u}_r\}$ *be a maximal subset of vectors in* $\mathcal{B}$ *such that* $\{\boldsymbol{u}_1, \ldots, \boldsymbol{u}_r\} \cup X$ *is linearly independent.*

We claim that $r = k - p$*. Observe that, if this is the case, then* $\{\boldsymbol{u}_1, \ldots, \boldsymbol{u}_r\} \cup X$ *is a linearly independent set of* k *vectors which, by (i) of this theorem, spans* V*. In other words,* $\{\boldsymbol{u}_1, \ldots, \boldsymbol{u}_r\} \cup X$ *is a basis of* V*.*

We show now that $r = k - p$*. Clearly* $r \leq k - p$*, since otherwise, by (ii) above,* $\{\boldsymbol{u}_1, \ldots, \boldsymbol{u}_r\} \cup X$ *would be linearly dependent.*

Suppose that $r < k - p$*. Then the cardinality of* $\{\boldsymbol{u}_1, \ldots, \boldsymbol{u}_r\} \cup X$ *is less than* k*. It is also the case that, by Theorem 3.1,* $\mathcal{B} \backslash \{\boldsymbol{u}_1, \ldots, \boldsymbol{u}_r\} \subseteq \operatorname{span}(\{\boldsymbol{u}_1, \ldots, \boldsymbol{u}_r\} \cup X)$*. Hence, as in* (3.4),

$$V = \operatorname{span}(\{\boldsymbol{u}_1, \ldots, \boldsymbol{u}_r\} \cup X),$$

from which follows that $\dim V = r < k$*, yielding a contradiction.*

(iv) Let X *be a subset of* V *which spans* V*. We claim that there exists a linearly independent subset* Y *of* X *with* k *elements. Observe that, since* Y *is linearly independent then, by (ii) of this theorem, it cannot have more than* k *elements.*

Suppose that this was not the case, i.e, any linearly independent set $Y \subseteq X$ *has less than* k *elements. Let* $r = \max\{\#Y \colon Y \subseteq X, Y \textit{is linearly independent}\}$ *and let* Y_r *be a linearly independent set contained in* X *consisting of* r *elements. It follows that any other element of* X *lies in* $\operatorname{span} Y_r$ *and, therefore,* $\operatorname{span} Y_r = \operatorname{span} X = V$*. Hence,* Y_r *is a basis of* V*, which contradicts the hypothesis of* $\dim V = k > r$*.*

Proposition 3.7 *Every subspace of a vector space* V *with* $\dim V = k$ *has dimension less than or equal to* k*.*

---

[1] Observe that any subset of a linearly independent set is necessarily also lineraly independent because, if it were not then... (Exercise).

Proof *Let S be a subspace of V. If $S = \{\mathbf{0}\}$, then the assertion holds trivially.*

Suppose now that $S \neq \{\mathbf{0}\}$, and let $\boldsymbol{y}_1 \in S$ be a non-zero vector. Then the set $\boldsymbol{y}_1$ is linearly independent. Two situations can occur: either (a) every other vector of S lies in the subspace spanned by $\boldsymbol{y}_1$ or (b) we can find $\boldsymbol{y}_2 \in S$ such that $\{\boldsymbol{y}_1, \boldsymbol{y}_2\}$ is linearly independent. Now we only have to continue mimicking the proof of Theorem 3.3 to get the desired conclusion. Notice that, by Theorem 3.5 (ii), any linearly independent subset of S contains k vectors, at most.

Example 3.13 *We are now in a position to classify the subspaces of $\mathbb{R}^2$. Since $\dim \mathbb{R}^2 = 2$, we know, by Theorem 3.5 (ii), that any linearly independent subset of $\mathbb{R}^2$ has, at most, two vectors. Hence, given a subspace S of $\mathbb{R}^2$, its basis can have zero vectors, one vector or two vectors, being S, respectively, $\{(0,0)\}$, a straight line through $(0,0)$, or $\mathbb{R}^2$. Notice that, by Theorem 3.5 (i), any two linearly independent vectors span $\mathbb{R}^2$.*

Example 3.14 *This example is an application of Theorem 3.5 (iv).*

Find a basis of the subspace S of $\mathbb{R}^3$ spanned by

$$X = \{(1,2,6), (1,1,1), (2,3,7), (0,1,5)\}.$$

The vectors $(1,2,6), (1,1,1), (2,3,7), (0,1,5)$ are linearly dependent because, since the dimension of $\mathbb{R}^3$ is equal to 3, any set with four vectors cannot be linearly independent (cf. Theorem 3.5 (ii)). Observe that, if these vectors were linearly independent, the dimension of $\mathbb{R}^3$ would have to be greater than or equal to 4, which is impossible.

Since X spans S, Theorem 3.5 (iv) guarantees that the set

$$\{(1,2,6), (1,1,1), (2,3,7), (0,1,5)\}$$

contains a basis of S. We must then find a maximal linearly independent subset of X, in the sense that it is not strictly contained in another linearly independent set contained in X.

Having in mind Proposition 3.4, we reduce the matrix

$$\begin{bmatrix} 1 & 1 & 2 & 0 \\ 2 & 1 & 3 & 1 \\ 6 & 1 & 7 & 5 \end{bmatrix} \tag{3.9}$$

to a row echelon matrix. We have then

$$\begin{bmatrix} 1 & 1 & 2 & 0 \\ 2 & 1 & 3 & 1 \\ 6 & 1 & 7 & 5 \end{bmatrix} \to \begin{bmatrix} 1 & 1 & 2 & 0 \\ 0 & -1 & -1 & 1 \\ 0 & -5 & -5 & 5 \end{bmatrix} \to \begin{bmatrix} \boxed{1} & 1 & 2 & 0 \\ 0 & \boxed{-1} & -1 & 1 \\ 0 & 0 & 0 & 0 \end{bmatrix}$$

Observing that the pivots (in grey) are located in the first and second columns, Proposition 3.4 yields that $(1,2,6), (1,1,1)$ are linearly independent.

On the other hand, using the row echelon matrix above, we know that the homogeneous system associated with the matrix (3.9) has two free variables. Denoting by $(\alpha, \beta, \gamma, \delta)$ *the elements of the solution set of this system, we have that the free variables are* γ *and* δ.

If, for example, we let $\gamma = 1$ *and* $\delta = 0$, *then there exist* $\alpha_1, \beta_1 \in \mathbb{R}$ *such that*

$$\alpha_1(1,2,6) + \beta_1(1,1,1) + (2,3,7) + 0(0,1,5) = 0.$$

Hence

$$\alpha_1(1,2,6) + \beta_1(1,1,1) + (2,3,7) = 0,$$

which shows that $(2,3,7)$ *is a linear combination of* $(1,2,6), (1,1,1)$.

Analogously it could be shown that $(0,1,5)$ *is a linear combination of* $(1,2,6), (1,1,1)$, *being enough to set* $\gamma = 0$ *and* $\delta = 1$.

We conclude thus that the set $\{(1,2,6),(1,1,1)\}$ *is a basis of* S.

It follows from the solution of this problem that a vector equation for S *is*

$$(x,y,z) = t(1,2,6) + s(1,1,1),$$

with $t, s \in \mathbb{R}$, *and that parametric equations of* S *are*

$$\begin{cases} x = t + s \\ y = 2t + s \\ z = 6t + s, \end{cases}$$

with $t, s \in \mathbb{R}$. *Find a cartesian equation of* S *(see Example 3.10).*

Example 3.15 *This is an example of application of Theorem 3.5 (iii).*

Show that it is possible to obtain a basis of $\mathbb{R}^4$ *containing* $\{\boldsymbol{v}_1, \boldsymbol{v}_2, \boldsymbol{v}_3\}$, *where*

$$\boldsymbol{v}_1 = (1,-1,-2,2) \qquad \boldsymbol{v}_2 = (-3,5,5,-6) \qquad \boldsymbol{v}_3 = (1,-1,0,2).$$

We begin by reducing the matrix $[\mathbf{v}_1 \mid \mathbf{v}_2 \mid \mathbf{v}_3]$ *to row echelon form:*

$$\begin{bmatrix} 1 & -3 & 1 \\ -1 & 5 & -1 \\ -2 & 5 & 0 \\ 2 & -6 & 2 \end{bmatrix} \rightarrow \begin{bmatrix} 1 & -3 & 1 \\ 0 & 2 & 0 \\ 0 & -1 & 2 \\ 0 & 0 & 0 \end{bmatrix} \rightarrow \begin{bmatrix} 1 & -3 & 1 \\ 0 & 2 & 0 \\ 0 & 0 & 2 \\ 0 & 0 & 0 \end{bmatrix}$$

By Proposition 3.4, we see that the vectors $\{\mathbf{v}_1, \mathbf{v}_2, \mathbf{v}_3\}$ *are linearly independent. Theorem 3.5 (iii) guarantees it is possible to add vectors to this set in order to construct a basis for* $\mathbb{R}^4$.

Having in mind the location of the grey pivots, we see that, for example, adding $\mathbf{e}_4$ *of the standard basis of* $\mathbb{R}^4$ *to the set* $\{\boldsymbol{v}_1, \boldsymbol{v}_2, \boldsymbol{v}_3\}$, *we shall have a matrix*

$$\begin{bmatrix} 1 & -3 & 1 & 0 \\ 2 & -6 & 2 & 0 \\ -1 & 5 & -1 & 0 \\ -2 & 5 & 0 & 1 \end{bmatrix}$$

whose rank is necessarily 4. Notice that, in the above Gaussian elimination, all elementary operations used leave the fourth column unchanged.

Now by Proposition 3.4 and Theorem 3.5 (i), we have immediately that $\{\boldsymbol{v}_1, \boldsymbol{v}_2, \boldsymbol{v}_3, \boldsymbol{e}_4\}$ *is a basis of* $\mathbb{R}^4$.

3.4 Null Space, Row Space, and Column Space

Although this book has been written in the spirit of binding together matrices and any other concept developed, it might be the case that this section shows it like no other. Here we shall define four fundamental subspaces associated with each matrix, definitely 'entangling' matrices and vector spaces. These are the null space, the row space, and the column space of a matrix and the null space of its transpose.

Definition 29 *The* ***null space*** $N(A)$ *of a* $k \times n$ *matrix* A *over* $\mathbb{K}$ *is the solution set of the homogeneous system of linear equations* $A\mathbf{x} = \mathbf{0}$.

We proved in Example 3.1 the following proposition.

Proposition 3.8 *Let* A *be a* $k \times n$ *matrix over* $\mathbb{K}$. *The null space*

$$N(A) = \{\mathbf{x} \in \mathbb{K}^n : A\mathbf{x} = \mathbf{0}\}$$

is a vector subspace of $\mathbb{K}^n$.

Example 3.16 *Part I: The null space of* A.
Consider the matrix

$$A = \begin{bmatrix} 1 & 1 & 2 & 2 \\ 1 & 2 & 1 & 1 \\ 2 & 3 & 3 & 3 \end{bmatrix}.$$

To find the null space $N(A)$, *we must solve the homogeneous system* $A\mathbf{x} = \mathbf{0}$. *Reducing* A *to a row echelon matrix through Gaussian elimination, one has*

$$\begin{bmatrix} 1 & 1 & 2 & 2 \\ 1 & 2 & 1 & 1 \\ 2 & 3 & 3 & 3 \end{bmatrix} \to \begin{bmatrix} 1 & 1 & 2 & 2 \\ 0 & 1 & -1 & -1 \\ 0 & 1 & -1 & -1 \end{bmatrix} \to \begin{bmatrix} 1 & 1 & 2 & 2 \\ 0 & 1 & -1 & -1 \\ 0 & 0 & 0 & 0 \end{bmatrix}.$$

The solution set of $A\mathbf{x} = \mathbf{0}$ *is then*

$$N(A) = \{(x, y, z, w) \in \mathbb{R}^4 : x = -3z - 3w, y = z + w\}.$$

Since, for all $(x, y, z, w) \in N(A)$, *we have*

$$\begin{bmatrix} x \\ y \\ z \\ w \end{bmatrix} = \begin{bmatrix} 3z - 3w \\ z + w \\ z \\ w \end{bmatrix} = \begin{bmatrix} 3z \\ z \\ z \\ 0 \end{bmatrix} + \begin{bmatrix} -3w \\ w \\ 0 \\ w \end{bmatrix} = z \begin{bmatrix} 3 \\ 1 \\ 1 \\ 0 \end{bmatrix} + w \begin{bmatrix} -3 \\ 1 \\ 0 \\ 1 \end{bmatrix},$$

it follows that $N(A) = \text{span}\{(3,1,1,0),(-3,1,0,1)\}$. *Hence, a basis of* $N(A)$ *is* $\{(3,1,1,0),(-3,1,0,1))\}$, *since this set is linearly independent. We have now that* $\dim N(A) = 2$.

Notice that the way we constructed the spanning vectors of $N(A)$ makes them automatically linearly independent due to the 'strategic' placement of zero in each vector. In fact, the only way to span the zero vector is by making $z = 0 = w$. Hence, when finding a basis for $N(A)$, if the method above is used, then one does not have to verify whether the spanning vectors are linearly independent: they always are.

We can extrapolate from this example that, given a matrix A, the dimension of $N(A)$ coincides with the number of independent variables of the system $A\mathbf{x} = \mathbf{0}$.

The **nullity** $\text{nul}(A)$ of a matrix A is the dimension of its null space.

Definition 30 *Let* A *be a* $k \times n$ *matrix over* $\mathbb{K}$. *The* ***row space*** $L(A)$ *of* A *is the subspace of* $\mathbb{K}^n$ *spanned by the rows of* A. *Supposing that* A *is presented in terms of its rows, i.e.,*

$$A = \begin{bmatrix} \mathbf{l}_1 \\ \mathbf{l}_2 \\ \vdots \\ \mathbf{l}_k \end{bmatrix}, \qquad \mathbf{l}_1^T, \mathbf{l}_2^T, \dots, \mathbf{l}_k^T \in \mathbb{K}^n\ ,$$

we have

$$L(A) = \{\alpha_1 \mathbf{l}_1^T + \alpha_2 \mathbf{l}_2^T + \cdots + \alpha_k \mathbf{l}_k^T : \alpha_1, \alpha_2, \dots, \alpha_k \in \mathbb{K}\}$$

The ***column space*** $C(A)$ *of* A *the subspace of* $\mathbb{K}^k$ *by the columns of* A. *Supposing* A *presented in terms of its columns, i.e.,*

$$A = \begin{bmatrix} \mathbf{c}_1 & | & \mathbf{c}_2 & | & \cdots & | & \mathbf{c}_n \end{bmatrix} \qquad (\mathbf{c}_1, \mathbf{c}_2, \dots, \mathbf{c}_k \in \mathbb{K}^k)\ ,$$

then

$$C(A) = \{\beta_1 \mathbf{c}_1 + \beta_2 \mathbf{c}_2 + \cdots + \beta_n \mathbf{c}_n : \beta_1, \beta_2, \dots, \beta_n \in \mathbb{K}\}$$

Proposition 3.9 *Let* A *be a* $k \times n$ *matrix over* $\mathbb{K}$ *and let* B *be a matrix which is obtained from* A *through a single elementary operation. Then the row space of* B *coincides with the row space of* A.

Proof *Consider the matrix*

$$A = \begin{bmatrix} \mathbf{l}_1 \\ \mathbf{l}_2 \\ \vdots \\ \mathbf{l}_k \end{bmatrix} \qquad (\mathbf{l}_1^T, \mathbf{l}_2^T, \dots, \mathbf{l}_k^T \in \mathbb{K}^n)\ ,$$

The row space $L(A)$ of A is the subspace of $\mathbb{K}^n$ spanned by the vectors $\mathbf{l}_1^T, \mathbf{l}_2^T, \dots, \mathbf{l}_k^T \in \mathbb{K}^n$, i.e.,

$$L(A) = \{\alpha_1 \mathbf{l}_1^T + \alpha_2 \mathbf{l}_2^T + \cdots + \alpha_k \mathbf{l}_k^T : \alpha_1, \alpha_2, \dots, \alpha_k \in \mathbb{K}\}.$$

If B is obtained by exchanging two rows of A, e.g., $\mathbf{l}_i \leftrightarrow \mathbf{l}_j$, it is obvious that $L(A) = L(B)$.

Suppose now that B is obtained by multiplying row $\mathbf{l}_i$ of A by the scalar $\alpha \neq 0$, i.e.,

$$A \xrightarrow[\alpha \mathbf{l}_i]{} B\,.$$

A linear combination of the rows of B is a vector of the form

$$\alpha_1 \mathbf{l}_1^T + \alpha_2 \mathbf{l}_2^T + \cdots + \alpha_i(\alpha \mathbf{l}_i^T) + \cdots + \alpha_k \mathbf{l}_k^T,$$

from which follows that

$$\begin{aligned}\alpha_1 \mathbf{l}_1^T + \alpha_2 \mathbf{l}_2^T + \cdots + \alpha_i(\alpha \mathbf{l}_i^T) + \cdots + \alpha_k \mathbf{l}_k^T \\ = \alpha_1 \mathbf{l}_1^T + \alpha_2 \mathbf{l}_2^T + \cdots + (\alpha_i \alpha)\mathbf{l}_i^T + \cdots + \alpha_k \mathbf{l}_k^T.\end{aligned} \tag{3.10}$$

It is now clear that (3.10) is also a linear combination of the rows of A. Hence we showed that $L(B) \subseteq L(A)$.

Conversely, let $\mathbf{x}$ be a vector in $L(A)$. That is,

$$\mathbf{x} = \alpha_1 \mathbf{l}_1^T + \alpha_2 \mathbf{l}_2^T + \cdots + \alpha_i \mathbf{l}_i^T + \cdots + \alpha_k \mathbf{l}_k^T.$$

However,

$$\begin{aligned}\alpha_1 \mathbf{l}_1^T + \alpha_2 \mathbf{l}_2^T + \cdots + \alpha_i \mathbf{l}_i^T + \cdots + \alpha_k \mathbf{l}_k^T \\ = \alpha_1 \mathbf{l}_1^T + \alpha_2 \mathbf{l}_2^T + \cdots + \frac{\alpha_i}{\alpha}(\alpha \mathbf{l}_i^T) + \cdots + \alpha_k \mathbf{l}_k^T,\end{aligned} \tag{3.11}$$

which shows that $L(A) \subseteq L(B)$, since the vector (3.11) lies in $L(B)$.

Finally, suppose that B is obtained from A by replacing row $\mathbf{l}_i$ by $\mathbf{l}_i + \alpha \mathbf{l}_j$, where $i \neq j$ and α is a scalar. Observing that

$$\begin{aligned}\alpha_1 \mathbf{l}_1^T + \alpha_2 \mathbf{l}_2^T + \cdots + \alpha_i(\mathbf{l}_i^T + \alpha \mathbf{l}_j^T) + \cdots + \alpha_j \mathbf{l}_j^T + \cdots + \alpha_k \mathbf{l}_k^T \\ = \alpha_1 \mathbf{l}_1^T + \alpha_2 \mathbf{l}_2^T + \cdots + \alpha \mathbf{l}_i^T + \cdots + (\alpha + \alpha_j)\mathbf{l}_j^T + \cdots + \alpha_k \mathbf{l}_k^T\end{aligned}$$

and using a reasoning similar to that above, it is easy to see that $L(A) = L(B)$.

Proposition 3.10 *The non-zero rows of a row echelon matrix are linearly independent.*

Proof *Let R be $m \times n$ row echelon matrix which we supposed presented in terms of its rows, i.e.,*

$$R = \begin{bmatrix} \mathbf{l}_1 \\ \mathbf{l}_2 \\ \vdots \\ l_m \end{bmatrix}, \qquad \mathbf{l}_1^T, \mathbf{l}_2^T, \dots, \mathbf{l}_m^T \in \mathbb{K}^n ,$$

and consider the equality

$$\alpha_1 \mathbf{l}_1^T + \alpha_2 \mathbf{l}_2^T + \cdots + \alpha_k \mathbf{l}_k^T = \mathbf{0}, \qquad \alpha_1, \alpha_2, \dots, \alpha_k \in \mathbb{K}, \tag{3.12}$$

where $k \leq m$ is the index of the bottom most non-zero row of R. If a_{1j} is the first non-zero entry of $\mathbf{l}_1$, since R is a row echelon matrix, we have that $a_{2j} = a_{3j} = \cdots = a_{kj} = 0$ (i.e, all entries of R in the j column are zero). Hence, for (3.12) to hold it is necessary that $\alpha_1 = 0$. Thus we obtain

$$\alpha_2 \mathbf{l}_2^T + \cdots + \alpha_k \mathbf{l}_k^T = \mathbf{0}, \qquad \alpha_2, \dots, \alpha_k \in \mathbb{K}. \tag{3.13}$$

The submatrix consisting only of the rows $\mathbf{l}_2, \dots, \mathbf{l}_k$ is also a row echelon matrix to which we can apply again the preceding reasoning, concluding that $\alpha_2 = 0$. Repeating this procedure sufficiently many times, we have

$$\alpha_1 = \alpha_2 = \cdots = \alpha_k = 0$$

and, therefore, the rows of R are linearly independent.

Example 3.17 *Part II: The row and column spaces of A.*
Consider again the matrix

$$A = \begin{bmatrix} 1 & 1 & 2 & 2 \\ 1 & 2 & 1 & 1 \\ 2 & 3 & 3 & 3 \end{bmatrix}$$

and its reduction to a row echelon matrix done in Part I (see Example 3.16).

By Proposition 3.9, the row space of A and the row space of any matrix obtained from A using elementary operations coincide. Hence the row space of

$$\begin{bmatrix} 1 & 1 & 2 & 2 \\ 0 & 1 & -1 & -1 \\ 0 & 0 & 0 & 0 \end{bmatrix}$$

is the row space $L(A)$ of A. Observing that, on the other hand, the non-zero rows of a row echelon matrix are linearly independent (cf. Proposition 3.10), we have that the set $\{(1,1,2,2), (0,1,-1,-1)\}$ is a basis of $L(A)$.

The set consisting of the columns of A corresponding to the dependent variables (i.e., the columns corresponding to pivots) is linearly independent. Notice *that, if one considers the matrix A' obtained by removing the grey columns of A, this matrix is a row echelon matrix. Hence, since the number of columns is equal to the number of pivots, the homogeneous system $A'\mathbf{x} = \mathbf{0}$ admits only the trivial solution. Since $A'\mathbf{x}$ is a linear combination of the columns of A', it is now clear que that the columns of A' are linearly independent.*

On the other hand, if one adds any of the grey columns to the columns of A', i.e., any columns corresponding to an independent variable, this new set is linearly dependent. In fact, these columns form an augmented matrix of a system which is consistent and has a unique solution, showing that this grey column is a linear combination of the columns of A'. We can now conclude that $\{(1,1,2),(1,2,3)\}$ is a basis of $C(A)$.

We have finally that

$$\dim L(A) = \textit{number of pivots} = \operatorname{rank}(A) = \dim C(A)$$

$$\begin{aligned} \dim N(A) &= \textit{number of independent variables} \\ &= \textit{number of columns} - \textit{number of pivots} \\ &= \textit{number of columns} - \operatorname{rank}(A), \end{aligned}$$

from which follows that

$$\dim N(A) + \dim L(A) = \textit{number of columns.}$$

A common mistake. When choosing the columns in the basis of $C(A)$, one must go back to the original matrix. One has to choose the columns in matrix A corresponding to those having pivots in the row echelon matrix. It is a common mistake to select those of the row echelon matrix. **This is wrong**.

How to find bases for the spaces associated with a matrix

Let A be a $k \times n$ matrix over $\mathbb{K}$.

1. Use Gaussian elimination to reduce A to a row echelon matrix R.
2. The set of the rows of R having pivots is a basis of the row space $L(A)$.
3. The set of the columns of A corresponding to those in R having pivots is a basis of $C(A)$.
4. Solve the system $R\mathbf{x} = \mathbf{0}$. Find a basis for this solution set (cf. Example 3.16). This is also a basis of $N(A)$.

One has only to reduce A to a row echelon matrix!

In 2., above, alternatively, one could go back to A and choose the corresponding rows. However, this might be tricky if row exchange was involved in the Gaussian elimination. Moreover, the rows of A are more 'complicated' than those of R, since the latter have more zero entries, in general. Hence, there is nothing to be gained from going back to A. Why do it then?

Proposition 3.11 *Let A be a $k \times n$ matrix over $\mathbb{K}$. Then,*

(i) $\dim N(A) = n - \operatorname{rank}(A)$;

(ii) $\dim L(A) = \operatorname{rank}(A)$;

(iii) $\dim C(A) = \operatorname{rank}(A)$.

Proof *(i) The dimension of the null space is the number of independent variables in $A\mathbf{x} = \mathbf{0}$ and, consequently, coincides with $n - \operatorname{rank}(A)$.*
(ii) This is a consequence of Propositions 3.9 and 3.10.
(iii) Removing from A the columns corresponding to those without pivots in the row echelon matrix, we obtain a matrix A' whose columns are linearly independent, since he system $A'\mathbf{x} = \mathbf{0}$ has only the trivial solution. It is also the case that these columns correspond exactly to the maximum number of linearly independent columns in A, yielding $\dim C(A) = \operatorname{rank}(A)$.

The next theorem is an immediate consequence of this proposition.

Theorem 3.6 (Rank-nullity theorem) *Let A be a $k \times n$ matrix over $\mathbb{K}$. Then,*

$$n = \dim N(A) + \dim L(A) = \operatorname{nul}(A) + \operatorname{rank}(A) \tag{3.14}$$

Example 3.18 *Find the spaces $N(A), L(A), C(A)$, and check that Theorem 3.6 holds for*

$$A = \begin{bmatrix} 1 & i & 0 \\ -i & 1 & 2i \end{bmatrix}.$$

We begin by finding $N(A)$. Solving the homogeneous system $A\mathbf{x} = \mathbf{0}$, we have

$$A = \begin{bmatrix} 1 & i & 0 \\ -i & 1 & 2i \end{bmatrix} \xrightarrow[l_2 + il_1]{} \begin{bmatrix} 1 & i & 0 \\ 0 & 0 & 2i \end{bmatrix}. \tag{3.15}$$

The null space is, therefore,

$$N(A) = \{(x_1, x_2, x_3) \in \mathbb{C}^3 \colon x_1 = -ix_2, x_3 = 0\}.$$

Hence, every vector $(x_1, x_2, x_3) \in N(A)$ can be written as

$$(x_1, x_2, x_3) = (-ix_2, x_2, 0) = x_2(-i, 1, 0)$$

from which follows that

$$\mathcal{B}_{N(A)} = \{(-i, 1, 0)\}$$

is a basis of $N(A)$. *Consequently,* $\dim N(A) = 1$.

Using (3.15), *we see that*

$$L(A) = \mathcal{L}(\{(1, i, 0), (0, 0, 2i)\}), \qquad C(A) = \mathcal{L}(\{(1, -i), (0, 2i)\},$$

and that bases for these spaces are

$$\mathcal{B}_{L(A)} = \{(1, i, 0), (0, 0, 2i)\}, \qquad \mathcal{B}_{C(A)} = \{(1, -i), (0, 2i)\}.$$

Hence, $\dim L(A) = 2 = \dim C(A)$.

We see now that Theorem 3.6 holds, since

$$\underbrace{\textit{number of columns of} A}_{3} = \underbrace{\dim N(A)}_{1} + \underbrace{\dim L(A)}_{2}.$$

Corollary 3.2 *Let* A *be a* $k \times n$ *matrix over* $\mathbb{K}$. *Then,*

$$k = \dim N(A^T) + \dim C(A),$$

$\operatorname{rank}(A) = \operatorname{rank}(A^T)$, *and*

$$n - k = \dim N(A) - \dim N(A^T).$$

Proof *Applying Theorem 3.6 to* A^T, *we have*

$$k = \dim N(A^T) + \dim L(A^T) = \dim N(A^T) + \dim C(A),$$

since $L(A^T) = C(A)$. *The remaining assertions are immediate, since, by Proposition 3.11,* $\dim L(A) = \operatorname{rank}(A) = \dim C(A)$.

The next example shows how the row space of a matrix can be used to obtain a basis of a subspace of $\mathbb{K}^n$.

Example 3.19 *We want to show that the vectors*

$$\boldsymbol{v}_1 = (1, 2, -1, -2, 1), \quad \boldsymbol{v}_2 = (-3, -6, 5, 5, -8), \quad \boldsymbol{v}_3 = (1, 2, -1, 0, 4)$$

are linearly independent and to find a basis of $\mathbb{R}^5$ *containing these vectors.*

The vectors $\boldsymbol{v}_1, \boldsymbol{v}_2, \boldsymbol{v}_3$ *are linearly independent if and only if the rows of the matrix*

$$\begin{bmatrix} 1 & 2 & -1 & -2 & 1 \\ -3 & -6 & 5 & 5 & -8 \\ 1 & 2 & -1 & 0 & 4 \end{bmatrix}$$

are also linearly independent. Using Gaussian elimination,

$$\begin{bmatrix} 1 & 2 & -1 & -2 & 1 \\ -3 & -6 & 5 & 5 & -8 \\ 1 & 2 & -1 & 0 & 4 \end{bmatrix} \to \underbrace{\begin{bmatrix} 1 & 2 & -1 & -2 & 1 \\ 0 & 0 & 2 & -1 & -5 \\ 0 & 0 & 0 & 2 & 3 \end{bmatrix}}_{R},$$

from which follows that $\boldsymbol{v}_1, \boldsymbol{v}_2, \boldsymbol{v}_3$ are linearly independent and, therefore, the set

$$\{\underbrace{(1,2,-1,-2,1)}_{\boldsymbol{u}_1}, \underbrace{(0,0,2,-1,-5)}_{\boldsymbol{u}_2}, \underbrace{(0,0,0,2,3)}_{\boldsymbol{u}_3}\}$$

is a basis of $\operatorname{span}\{\boldsymbol{v}_1, \boldsymbol{v}_2, \boldsymbol{v}_3\}$. *Notice that here we applied Proposition 3.9.*

Since the pivots of R are situated in the columns $1, 3$, and 4, it is clear that $\{\boldsymbol{u}_1, \boldsymbol{u}_2, \boldsymbol{u}_3, (0,1,0,0,0), (0,0,0,0,1)\}$ is a basis of $\mathbb{R}^5$. Hence, $\{\boldsymbol{v}_1, \boldsymbol{v}_2, \boldsymbol{v}_3, (0,1,0,0,0), (0,0,0,0,1)\}$ is also a basis of $\mathbb{R}^5$.

How to find a basis for the space spanned by a set of vectors using the row space of a matrix

Let $\boldsymbol{u}_1, \boldsymbol{u}_2, \ldots, \boldsymbol{u}_k$ be vectors in $\mathbb{K}^n$.

1. Construct the matrix whose rows consist of these vectors, i.e.,

$$A = \begin{bmatrix} \mathbf{u}_1^T \\ \mathbf{u}_2^T \\ \vdots \\ \mathbf{u}_k^T \end{bmatrix}.$$

   Observe that the vectors are column vectors and that to obtain row vectors one has, therefore, to use transposition.

2. Use Gaussian elimination to reduce A to a row echelon matrix R.

3. A basis of $\operatorname{span}\{\boldsymbol{u}_1, \boldsymbol{u}_2, \ldots, \boldsymbol{u}_k\} = L(A) = L(R)$ is formed by the rows of R with pivots.

Alternatively, we could use Proposition 3.4 to find a maximal subset of linearly independent vectors of $\{\boldsymbol{u}_1, \boldsymbol{u}_2, \ldots, \boldsymbol{u}_k\}$, that is, beginning with a matrix B whose columns are $\boldsymbol{u}_1, \boldsymbol{u}_2, \ldots, \boldsymbol{u}_k$ (see also Example 3.10). But this is more complicated inasmuch as we have always to go back to the initial matrix B to get our answers. Again, why complicate things?

The next theorem characterises matrix invertibility in terms of the its null space, row space, and column space.

Theorem 3.7 (Necessary and sufficient conditions of invertibility (III)) *Let A square matrix of order n with entries in $\mathbb{K}$. The following assertions are equivalent.*

(i) A *is invertible.*

(ii) $N(A) = \{\mathbf{0}\}$.

(iii) The rows of A are linearly independent.

(iv) The rows of A are a basis of $\mathbb{K}^n$.

(v) $\dim L(A) = n$.

(vi) The columns of A are linearly independent.

(vii) The columns of A are a basis of $\mathbb{K}^n$.

(viii) $\dim C(A) = n$.

Proof *The equivalence (i) $\Leftrightarrow$ (ii) has already been proved (cf. Theorem 1.1).*

(ii) $\Leftrightarrow$ (v) The null space $N(A)$ coincides with $\{\mathbf{0}\}$ if and only if the homogeneous system $A\mathbf{x} = \mathbf{0}$ is consistent and has a unique solution. Hence $N(A) = \{\mathbf{0}\}$ if and only if $\operatorname{rank} A = n$ *and, therefore, if and only if* $\dim L(A) = n$.

The equivalences (iii) $\Leftrightarrow$ (iv) $\Leftrightarrow$ (v) are immediate. Observing that $L(A^T) = C(A)$ and that A is invertible if and only if A^T is invertible (cf. Proposition 1.18 (v)), it is obvious that(i) $\Leftrightarrow$ (vi) $\Leftrightarrow$ (vii) $\Leftrightarrow$ (viii).

3.4.1 $A\mathbf{x} = \mathbf{b}$

Let A be a matrix over $\mathbb{K}$ of size $k \times n$. Let $\mathbf{x}_0$ be a solution of the homogeneous system $A\mathbf{x} = \mathbf{0}$ and let $\mathbf{x}_p$ be a particular solution of the system $A\mathbf{x} = \mathbf{b}$. In other words, we are supposing that $\mathbf{x}_0, \mathbf{x}_p$ are vectors in $\mathbb{K}^n$ such that

$$A\mathbf{x}_0 = \mathbf{0} \qquad\qquad A\mathbf{x}_p = \mathbf{b}.$$

Example 3.20 *Consider the system of linear equations*

$$\underbrace{\begin{bmatrix} 1 & -1 & 0 \\ -2 & 2 & 0 \end{bmatrix}}_{A} \begin{bmatrix} x_1 \\ x_2 \\ x_3 \end{bmatrix} = \begin{bmatrix} 1 \\ -2 \end{bmatrix}.$$

The solution set of $A\mathbf{x} = \mathbf{0}$ is $\{(x_1, x_2, x_3) \in \mathbb{R}^3 : x_1 = x_2\}$. Hence, for example, $\mathbf{x}_0 = (3, 3, -17)$ is a solution of $A\mathbf{x} = \mathbf{0}$.

Check that $\mathbf{x}_p = (1, 0, 3)$ *is a solution of the non-homogeneous system.*

The vector $\mathbf{x} = \mathbf{x}_0 + \mathbf{x}_p$ is a solution of $A\mathbf{x} = \mathbf{b}$ because

$$\begin{aligned} A\mathbf{x} &= A(\mathbf{x}_0 + \mathbf{x}_p) \\ &= A\mathbf{x}_0 + A\mathbf{x}_p \\ &= \mathbf{0} + \mathbf{b} \\ &= \mathbf{b}. \end{aligned}$$

Conversely, it is easy to see that any solution of $A\mathbf{x} = \mathbf{b}$ has this form. In fact, if $\mathbf{x}_1$ is a solution of $A\mathbf{x} = \mathbf{b}$, we have

$$\begin{aligned} A(\mathbf{x}_1 - \mathbf{x}_p) &= A\mathbf{x}_1 - A\mathbf{x}_p \\ &= \mathbf{b} - \mathbf{b} \\ &= \mathbf{0}, \end{aligned}$$

from which follows that $\mathbf{x}_0' = \mathbf{x}_1 - \mathbf{x}_p$ is a solution of the homogeneous system $A\mathbf{x} = \mathbf{0}$. Hence, we have once again that $\mathbf{x}_1 = \mathbf{x}_0' + \mathbf{x}_p$, that is, $\mathbf{x}_1$ is the sum of a solution of the homogeneous system with the particular solution $\mathbf{x}_p$ of $A\mathbf{x} = \mathbf{b}$.

We can thus conclude that the solution set $\mathcal{S}$ of $A\mathbf{x} = \mathbf{b}$ can be presented as

$$\mathcal{S} = \mathbf{x_p} + N(A),$$

where

$$\mathbf{x_p} + N(A) := \{\mathbf{x_p} + \mathbf{x} \colon \mathbf{x} \in N(A)\}$$

(see Figure 3.4).

Proposition 3.12 *Let* A *be a* $k \times n$ *matrix over* $\mathbb{K}$ *and let* $\mathbf{b}$ *be a* $k \times 1$ *vector. The system* $A\mathbf{x} = \mathbf{b}$ *is consistent if and only if* $\mathbf{b} \in C(A)$. *Moreover, if* $A\mathbf{x} = \mathbf{b}$ *is consistent, then its solution set* $\mathcal{S}$ *is*

$$\mathcal{S} = \mathbf{x_p} + N(A),$$

where $\mathbf{x_p}$ *a particular solution of* $A\mathbf{x} = \mathbf{b}$.

Proof *Since* $A\mathbf{x}$ *is a linear combination of the columns of* A, *that is,* $A\mathbf{x}$ *lies in* $C(A)$, *it is clear that the system is consistent if and only if* $\mathbf{b}$ *lies in* $C(A)$.

The remaining assertion has just been proved above.

3.5 Sum and Intersection of Subspaces

In this section we construct new spaces out of old ones.

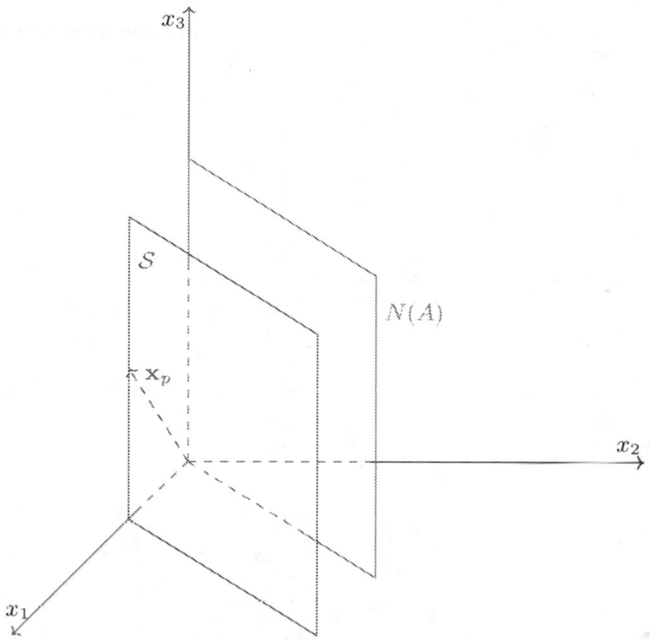

FIGURE 3.4: The solution set $\mathcal{S}$ is obtained adding the particular solution $\mathbf{x}_p = (1, 0, 3)$ to the solution set of the homogeneous system.

Definition 31 *Let U and W be subspaces of a vector space V over $\mathbb{K}$. The* ***sum of the subspaces** U **and** W*, *denoted $U + W$, is defined by*

$$U + W = \{\boldsymbol{x} + \boldsymbol{y} \colon \boldsymbol{x} \in U \wedge \boldsymbol{y} \in W\}.$$

Proposition 3.13 *Let U and W be subspaces of a vector space V. Then $U+W$ and $U \cap W$ are vector subspaces of V.*

Proof *Notice that $U+W$ and $U\cap W$ are non-empty sets. Let α be a scalar and let $\boldsymbol{z} \in U + W$. Then there exist $\boldsymbol{x} \in U$ and $\boldsymbol{y} \in W$ such that $\boldsymbol{z} = \boldsymbol{x} + \boldsymbol{y}$ and*

$$\alpha\boldsymbol{z} = \alpha(\boldsymbol{x} + \boldsymbol{y}) = \alpha\boldsymbol{x} + \alpha\boldsymbol{y}.$$

Since $\alpha\boldsymbol{x} \in U$ and $\alpha\boldsymbol{y} \in W$, it follows that $\alpha\boldsymbol{z} \in U + W$. Hence, $U + W$ is closed under scalar multiplication.
To see that $U + W$ is closed for vector addition, let $\boldsymbol{z}_1$ and $\boldsymbol{z}_2$ be vectors in $U + W$. Then there exist $\boldsymbol{x}_1, \boldsymbol{x}_2 \in U$ and $\boldsymbol{y}_1, \boldsymbol{y}_2 \in W$ such that $\boldsymbol{z}_1 = \boldsymbol{x}_1 + \boldsymbol{y}_1$ and $\boldsymbol{z}_2 = \boldsymbol{x}_2 + \boldsymbol{y}_2$. Hence

$$\begin{aligned}\boldsymbol{z}_1 + \boldsymbol{z}_2 &= \boldsymbol{x}_1 + \boldsymbol{y}_1 + \boldsymbol{x}_2 + \boldsymbol{y}_2\\ &= (\boldsymbol{x}_1 + \boldsymbol{x}_2) + (\boldsymbol{y}_1 + \boldsymbol{y}_2),\end{aligned}$$

which shows that $U + W$ is closed for the operation $+$.

As to showing that $U \cap W$ is a subspace, we begin with the scalar multiplication. If $\boldsymbol{x} \in U \cap W$ then $\alpha\boldsymbol{x} \in U$, $\alpha\boldsymbol{x} \in W$ and, therefore, $U \cap W$ is closed under scalar multiplication.

Let $\boldsymbol{x}, \boldsymbol{y} \in U \cap W$, then $\boldsymbol{x}+\boldsymbol{y} \in U$ and $\boldsymbol{x}+\boldsymbol{y} \in W$. Hence, $\boldsymbol{x}+\boldsymbol{y} \in U \cap W$, from which follows that $U \cap W$ is closed for vector addition.

Example 3.21 *Find bases for $U + W$ and $U \cap W$, where U is the plane $x + 2y + z = 0$ and $W = \operatorname{span}\{(1, -1, 1), (0, 0, 1)\}$.*

Solution. The cartesian equation of U leads to

$$\begin{bmatrix} x \\ y \\ z \end{bmatrix} = y \begin{bmatrix} -2 \\ 1 \\ 0 \end{bmatrix} + z \begin{bmatrix} -1 \\ 0 \\ 1 \end{bmatrix},$$

from which follows that a basis for U is

$$B_U = \left\{ \begin{bmatrix} -2 \\ 1 \\ 0 \end{bmatrix}, \begin{bmatrix} -1 \\ 0 \\ 1 \end{bmatrix} \right\}.$$

Hence $U+W = \operatorname{span}\{(1, -1, 1), (0, 0, 1), (-2, 1, 0), (-1, 0, 1)\}$. To obtain a basis of $U+W$, we only have to find a maximal set of linearly independent vectors contained in $\operatorname{span}\{(1, -1, 1), (0, 0, 1), (-2, 1, 0), (-1, 0, 1)\}$. It is easy to see that we can find three linearly independent vectors and, therefore, $U+W = \mathbb{R}^3$.

A cartesian equation for W is $x+y = 0$ and, consequently, $U \cap W$ consists of the vectors in $\mathbb{R}^3$ such that

$$\begin{cases} x + 2y + z = 0 \\ x + y = 0 \end{cases}.$$

Hence, a basis for $U \cap W$ is $B_{U \cap W} = \{(-1, 1, -1)\}$.

Theorem 3.8 *Let U and W be subspaces of a finite dimensional vector space V. Then*

$$\dim U + \dim W = \dim(U + W) + \dim(U \cap W).$$

Proof *Let $B_{U \cap W}$ be a basis of the intersection. Then, by Theorem 3.5 (iii), we can find bases B_U and B_W such that $B_{U \cap W} \subseteq B_U, B_W$. Hence, $B_U \cup B_W$ is a basis for B_{U+W}. It is clear that $B_U \cup B_W$ is a spanning set for $U + W$. We show now that $B_U \cup B_W$ is linearly independent. Let*

$$B_U \cup B_W = \{\boldsymbol{u}_1, \dots, \boldsymbol{u}_r, \boldsymbol{v}_1, \dots, \boldsymbol{v}_m, \boldsymbol{w}_1, \dots \boldsymbol{w}_k\},$$

where $\{\boldsymbol{u}_1, \dots, \boldsymbol{u}_r\} \subseteq U \backslash (U \cap W)$, $\{\boldsymbol{v}_1, \dots, \boldsymbol{v}_m\} \subseteq U \cap W$, and $\{\boldsymbol{w}_1, \dots \boldsymbol{w}_k\} \subseteq W \backslash (U \cap W)$.

If

$$\sum_{i=1}^{r} \alpha_i \boldsymbol{u}_i + \sum_{j=1}^{m} \beta_j \boldsymbol{v}_j + \sum_{l=1}^{k} \gamma_l \boldsymbol{w}_l = \boldsymbol{0}, \tag{3.16}$$

then

$$\sum_{i=1}^{r} \alpha_i \boldsymbol{u}_i + \sum_{j=1}^{m} \beta_j \boldsymbol{v}_j = -\sum_{l=1}^{k} \gamma_l \boldsymbol{w}_l,$$

from which follows that $-\sum_{l=1}^{k} \gamma_l \boldsymbol{w}_l \in U$. *Consequently,* $\sum_{l=1}^{k} \gamma_l \boldsymbol{w}_l \in U \cap W$ *is a linear combination of* $\{\boldsymbol{v}_1, \dots, \boldsymbol{v}_m\}$, *yielding that* $\sum_{l=1}^{k} \gamma_l \boldsymbol{w}_l = \boldsymbol{0}$, *since the set* $\{\boldsymbol{v}_1, \dots, \boldsymbol{v}_m, \boldsymbol{w}_1, \dots \boldsymbol{w}_k\}$ *is the linearly independent set* B_W. *It now follows that all scalars in* (3.16) *are* 0 *and, therefore,* $B_U \cup B_W$ *is linearly independent.*

Notice that

$$\#B_{U+W} = \#B_U + \#B_W - \#B_{U\cap W},$$

since the vectors in $B_{U\cap W}$ *appear twice in* $\#B_U + \#B_W$. *Now, it is immediate that*

$$\dim U + \dim W = \dim(U + W) + \dim(U \cap W).$$

Definition 32 *Let* V *be a vector space and let* U, W *be subspaces of* V. *The space* V *is said to be a* **direct sum** *of* U *and* W *if* $V = U + W$ *and* $U \cap W = \{\boldsymbol{0}\}$. *We denote by*

$$V = U \oplus W$$

the direct sum of U *and* W.

The sum $U + W = \mathbb{R}^3$ in Example 3.21 is not a direct sum. However, if U is the plane $x = 0$ and W is the straight line $x = y = z$, then $\mathbb{R}^3 = U \oplus W$.

Example 3.22 *Prove that, in a direct sum* $V = U \oplus W$, *each* $\boldsymbol{x} \in V$ *decomposes uniquely as* $\boldsymbol{x} = \boldsymbol{u} + \boldsymbol{w}$, *with* $\boldsymbol{u} \in U, \boldsymbol{w} \in W$.

Solution. Suppose that $\boldsymbol{x} = \boldsymbol{u} + \boldsymbol{w}$ *and* $\boldsymbol{x} = \boldsymbol{u}' + \boldsymbol{w}'$. *Then,* $\boldsymbol{u} + \boldsymbol{w} = \boldsymbol{x} = \boldsymbol{u}' + \boldsymbol{w}'$. *It follows that*

$$\boldsymbol{u} - \boldsymbol{u}' = \boldsymbol{w} - \boldsymbol{w}'$$

and, consequently, $\boldsymbol{u} - \boldsymbol{u}' = \boldsymbol{0} = \boldsymbol{w} - \boldsymbol{w}'$, *since* $U \cap W = \{\boldsymbol{0}\}$. *That is,* $\boldsymbol{u} = \boldsymbol{u}'$ *and* $\boldsymbol{w} = \boldsymbol{w}'$.

3.6 Change of Basis

Sometimes a problem becomes more tractable if one chooses an appropriate basis for the space in question. In the problem tackled in Example 5.12 of Chapter 5, we will see that it is more convenient to use a particular basis of $\mathbb{R}^2$ and then go back to the standard basis. Is there, however, an easy way to relate the coordinate vectors in both bases? The answer is *yes*. We will justify our answer in this section.

Let V be a vector space over $\mathbb{K}$ of dimension k, let $\mathcal{B}_1$ and $\mathcal{B}_2$ be bases of V and let $\boldsymbol{u} \in V$ be a vector. One might ponder whether there exists a relation between the coordinate vectors $\boldsymbol{u}_{\mathcal{B}_1}, \boldsymbol{u}_{\mathcal{B}_2} \in \mathbb{K}^k$ of $\boldsymbol{u}$ relative to $\mathcal{B}_1$ and $\mathcal{B}_2$, respectively. We shall see that there exists a matrix linking these two coordinate vectors making it easy to 'pass' from one basis to the other.

Consider the bases $\mathcal{B}_1 = \big(\boldsymbol{b_1}, \boldsymbol{b_2}, \cdots, \boldsymbol{b_k}\big)$ and $\mathcal{B}_2$ of V. We have that

$$\boldsymbol{u} = \alpha_1 \boldsymbol{b_1} + \alpha_2 \boldsymbol{b_2} + \cdots + \alpha_k \boldsymbol{b_k}.$$

Hence, by Proposition 3.5,

$$\begin{aligned}(\boldsymbol{u})_{\mathcal{B}_2} &= (\alpha_1 \boldsymbol{b_1} + \alpha_2 \boldsymbol{b_2} + \cdots + \alpha_k \boldsymbol{b_k})_{\mathcal{B}_2} \\ &= (\alpha_1 \boldsymbol{b_1})_{\mathcal{B}_2} + (\alpha_2 \boldsymbol{b_2})_{\mathcal{B}_2} + \cdots + (\alpha_k \boldsymbol{b_k})_{\mathcal{B}_2} \\ &= \alpha_1 (\boldsymbol{b_1})_{\mathcal{B}_2} + \alpha_2 (\boldsymbol{b_2})_{\mathcal{B}_2} + \cdots + \alpha_k (\boldsymbol{b_k})_{\mathcal{B}_2}.\end{aligned}$$

This equality can also be written in a matrix form as

$$\mathbf{u}_{\mathcal{B}_2} = \big[(\mathbf{b_1})_{\mathcal{B}_2} \quad | \quad (\mathbf{b_2})_{\mathcal{B}_2} \quad | \quad \cdots \quad | \quad (\mathbf{b_k})_{\mathcal{B}_2}\big] \begin{bmatrix}\alpha_1\\ \alpha_2\\ \vdots \\ \alpha_k\end{bmatrix}.$$

That is,

$$\mathbf{u}_{\mathcal{B}_2} = \underbrace{\big[(\mathbf{b_1})_{\mathcal{B}_2} \quad | \quad (\mathbf{b_2})_{\mathcal{B}_2} \quad | \quad \cdots \quad | \quad (\mathbf{b_k})_{\mathcal{B}_2}\big]}_{M_{\mathcal{B}_2 \leftarrow \mathcal{B}_1}} \mathbf{u}_{\mathcal{B}_1}.$$

The matrix whose columns are the coordinate vectors of the vectors of basis $\mathcal{B}_1$ relative to the basis $\mathcal{B}_2$ is called the **change of basis matrix** from basis $\mathcal{B}_1$ to basis $\mathcal{B}_2$, and is denoted by $M_{\mathcal{B}_2 \leftarrow \mathcal{B}_1}$. Hence,

$$\mathbf{u}_{\mathcal{B}_2} = M_{\mathcal{B}_2 \leftarrow \mathcal{B}_1} \mathbf{u}_{\mathcal{B}_1}.$$

Example 3.23 *Consider the basis $\mathcal{B} = ((1,1),(-1,0))$ and the standard basis $\mathcal{E}_2 = (\mathbf{e}_1, \mathbf{e}_2)$ of $\mathbb{R}^2$. Use the change of basis matrix $M_{\mathcal{B} \leftarrow \mathcal{E}_2}$ to find the coordinate vector $(2,3)_{\mathcal{B}}$ (cf. Figure 3.3).*

We have just seen that

$$\begin{bmatrix}2\\3\end{bmatrix}_{\mathcal{B}} = \underbrace{\big[(\mathbf{e}_1)_{\mathcal{B}} \quad (\mathbf{e}_2)_{\mathcal{B}}\big]}_{M_{\mathcal{B} \leftarrow \mathcal{E}_2}} \begin{bmatrix}2\\3\end{bmatrix}.$$

Hence we have to find the coordinate vectors $(\mathbf{e}_1)_{\mathcal{B}}, (\mathbf{e}_2)_{\mathcal{B}}$, that is, solve the systems whose augmented matrices are

$$\left[\begin{array}{cc|c}1 & -1 & 1\\ 1 & 0 & 0\end{array}\right], \quad \left[\begin{array}{cc|c}1 & -1 & 0\\ 1 & 0 & 1\end{array}\right].$$

Since the coefficient matrices are the same, to save time we shall solve them simultaneously. Then, using Gauss–Jordan elimination, we have

$$\left[\begin{array}{cc|cc} 1 & -1 & 1 & 0 \\ 1 & 0 & 0 & 1 \end{array}\right] \xrightarrow[l_2 - l_1]{} \left[\begin{array}{cc|cc} 1 & -1 & 1 & 0 \\ 0 & 1 & -1 & 1 \end{array}\right] \xrightarrow[l_1 + l_2]{} \left[\begin{array}{cc|cc} 1 & 0 & 0 & 1 \\ 0 & 1 & -1 & 1 \end{array}\right].$$

We have now that

$$M_{\mathcal{B}\leftarrow\mathcal{E}_2} = \begin{bmatrix} 0 & 1 \\ -1 & 1 \end{bmatrix},$$

from which follows that

$$\begin{bmatrix} 2 \\ 3 \end{bmatrix}_{\mathcal{B}} = \begin{bmatrix} 0 & 1 \\ -1 & 1 \end{bmatrix} \begin{bmatrix} 2 \\ 3 \end{bmatrix} = \begin{bmatrix} 3 \\ 1 \end{bmatrix}.$$

The next proposition asserts that the change of basis matrix is unique. More precisely,

Proposition 3.14 *Let V be a vector space over $\mathbb{K}$ of dimension n and let $\mathcal{B}_1, \mathcal{B}_2$ be bases of V. Then, there exists uniquely a matrix M such that, for all $\boldsymbol{x} \in V$, $\mathbf{x}_{\mathcal{B}_2} = M\mathbf{x}_{\mathcal{B}_1}$.*

Proof *We prove the uniqueness part, since the existence has been taken care of in the discussion above.*

Suppose that, for all $\boldsymbol{x} \in V$,

$$\mathbf{x}_{\mathcal{B}_2} = M\mathbf{x}_{\mathcal{B}_1} = M'\mathbf{x}_{\mathcal{B}_1}.$$

It follows that $(M - M')\mathbf{x}_{\mathcal{B}_1} = \mathbf{0}$.

Choosing $\boldsymbol{x}_{\mathcal{B}_1} = (1, 0, \dots, 0)$, we see that the first column of M and M' coincide. Letting $\boldsymbol{x}_{\mathcal{B}_1}$ be any vector in $\mathcal{E}_n$, we will obtain the uniqueness of the change of basis matrix.

Example 3.24 *Consider the ordered basis $\mathcal{B} = \big((0,1,0), (1,0,1), (2,1,0)\big)$ of $\mathbb{R}^3$. Find the change of basis matrix from the standard basis of $\mathbb{R}^3$ to the basis $\mathcal{B}$.*

Setting $\mathcal{E}_3 = (\mathbf{e}_1, \mathbf{e}_2, \mathbf{e}_3)$, the change of basis matrix $M_{\mathcal{B}\leftarrow\mathcal{E}_3}$ is

$$M_{\mathcal{B}\leftarrow\mathcal{E}_3} = \left[(\mathbf{e}_1)_{\mathcal{B}} \mid (\mathbf{e}_2)_{\mathcal{B}} \mid (\mathbf{e}_3)_{\mathcal{B}}\right].$$

We must find the coordinate vectors of $\mathbf{e}_1, \mathbf{e}_2$, and $\mathbf{e}_3$ relative to $\mathcal{B}$. That is, we will solve three systems of linear which we will do simultaneously, as above. Hence,

$$\left[\begin{array}{ccc|ccc} 0 & 1 & 2 & 1 & 0 & 0 \\ 1 & 0 & 1 & 0 & 1 & 0 \\ 0 & 1 & 0 & 0 & 0 & 1 \end{array}\right] \rightarrow \left[\begin{array}{ccc|ccc} 1 & 0 & 1 & 0 & 1 & 0 \\ 0 & 1 & 2 & 1 & 0 & 0 \\ 0 & 1 & 0 & 0 & 0 & 1 \end{array}\right] \rightarrow \dots$$

$$\cdots \to \left[\begin{array}{ccc|ccc} 1 & 0 & 1 & 0 & 1 & 0 \\ 0 & 1 & 2 & 1 & 0 & 0 \\ 0 & 0 & -2 & -1 & 0 & 1 \end{array}\right] \to \left[\begin{array}{ccc|ccc} 1 & 0 & 0 & -1/2 & 1 & 1/2 \\ 0 & 1 & 0 & 0 & 0 & 1 \\ 0 & 0 & -2 & -1 & 0 & 1 \end{array}\right] \to \cdots$$

$$\cdots \to \left[\begin{array}{ccc|ccc} 1 & 0 & 0 & -1/2 & 1 & 1/2 \\ 0 & 1 & 0 & 0 & 0 & 1 \\ 0 & 0 & 1 & 1/2 & 0 & -1/2 \end{array}\right].$$

It follows that

$$M_{\mathcal{B}\leftarrow\mathcal{E}_3} = \begin{bmatrix} -1/2 & 1 & 1/2 \\ 0 & 0 & 1 \\ 1/2 & 0 & -1/2 \end{bmatrix}.$$

How to find the coordinate vectors after a change of basis

Let V be a vector of dimension k over $\mathbb{K}$ and let $\mathcal{B}_1 = \big(\boldsymbol{b_1}, \boldsymbol{b_2}, \cdots, \boldsymbol{b_k}\big)$ and $\mathcal{B}_2 = \big(\boldsymbol{u_1}, \boldsymbol{u_2}, \cdots, \boldsymbol{u_k}\big)$ be a basis of V.

1. Find the change of basis matrix $M_{\mathcal{B}_2\leftarrow\mathcal{B}_1}$ as follows:

   (a) for each vector $\mathbf{b_i}$, with $i = 1, \ldots, k$, determine the scalars $\alpha_1^{(i)}, \ldots, \alpha_k^{(i)} \in \mathbb{K}$ such that

   $$\mathbf{b_i} = \alpha_1^{(i)}\mathbf{u_1} + \cdots + \alpha_k^{(i)}\mathbf{u_k};$$

   (b) the coordinate vector $(\mathbf{b_i})_{\mathcal{B}_2}$ of each $\mathbf{b_i}$ relative to basis $\mathcal{B}_2$ is

   $$(\mathbf{b_i})_{\mathcal{B}_2} = \begin{bmatrix} \alpha_1^{(i)} \\ \vdots \\ \alpha_k^{(i)} \end{bmatrix};$$

   (c) build the change of basis matrix

   $$M_{\mathcal{B}_2\leftarrow\mathcal{B}_1} = \big[(\mathbf{b_1})_{\mathcal{B}_2} \;\big|\; \cdots \;\big|\; (\mathbf{b_k})_{\mathcal{B}_2}\big].$$

2. If $\boldsymbol{x}$ is a vector in V, then

   $$\mathbf{x}_{\mathcal{B}_2} = \underbrace{\big[(\mathbf{b_1})_{\mathcal{B}_2} \;\big|\; \cdots \;\big|\; (\mathbf{b_k})_{\mathcal{B}_2}\big]}_{M_{\mathcal{B}_2\leftarrow\mathcal{B}_1}} \mathbf{x}_{\mathcal{B}_1}.$$

Given a vector space V with $\dim V = k$ and two bases $\mathcal{B}_1, \mathcal{B}_2$ of V, the change of basis matrix $M_{\mathcal{B}_2\leftarrow\mathcal{B}_1}$ is always non-singular. In fact, since

$\text{rank}\,(M_{\mathcal{B}_2 \leftarrow \mathcal{B}_1}) = k$ (cf. Proposition 3.4 and Proposition 3.6), this matrix is invertible. Hence, using the equality

$$\mathbf{u}_{\mathcal{B}_2} = M_{\mathcal{B}_2 \leftarrow \mathcal{B}_1} \mathbf{u}_{\mathcal{B}_1},$$

we have

$$\mathbf{u}_{\mathcal{B}_1} = \left(M_{\mathcal{B}_2 \leftarrow \mathcal{B}_1}\right)^{-1} \mathbf{u}_{\mathcal{B}_2}.$$

By Proposition 3.14, we obtain

$$M_{\mathcal{B}_1 \leftarrow \mathcal{B}_2} = (M_{\mathcal{B}_2 \leftarrow \mathcal{B}_1})^{-1}.$$

Example 3.25 *Consider the ordered basis* $\mathcal{B} = \big(\underbrace{(0,1,0)}_{\boldsymbol{b}_1}, \underbrace{(1,0,1)}_{\boldsymbol{b}_2}, \underbrace{(2,1,0)}_{\boldsymbol{b}_3}\big)$ *of* $\mathbb{R}^3$*. Find the change of basis matrix from* $\mathcal{B}$ *to the standard basis of* $\mathbb{R}^3$*.*

The matrix $M_{\mathcal{E}_3 \leftarrow \mathcal{B}}$ *is*

$$\begin{aligned} M_{\mathcal{E}_3 \leftarrow \mathcal{B}} &= \left[(\mathbf{b}_1)_{\mathcal{E}_3} \;\middle|\; (\mathbf{b}_2)_{\mathcal{E}_3} \;\middle|\; (\mathbf{b}_3)_{\mathcal{E}_3}\right] \\ &= \begin{bmatrix} 0 & 1 & 2 \\ 1 & 0 & 1 \\ 0 & 1 & 0 \end{bmatrix}. \end{aligned}$$

Observe that we can now easily obtain $M_{\mathcal{B} \leftarrow \mathcal{E}_3}$*. Indeed, we need only to calculate the inverse of* $M_{\mathcal{B} \leftarrow \mathcal{E}_3}$*, that is,* $M_{\mathcal{B} \leftarrow \mathcal{E}_3} = (M_{\mathcal{E}_3 \leftarrow \mathcal{B}})^{-1}$ *(compare with Example 3.24).*

When calculating a change of basis matrix between two bases, one should ponder which change of basis matrix is the easiest to obtain. If there is one, then find that change of basis matrix first and calculate its inverse, if necessary.

3.7 Exercises

EX 3.7.1. Which vectors are a linear combination of $\boldsymbol{u} = (0,-1,1)$ e $\boldsymbol{v} = (-1,-3,1)$?

(a) $(1,1,1)$

(b) $(-6,-2,-10)$

(c) $(0, 2, \frac{5}{2})$

(d) $(0, 0, 0)$

EX 3.7.2. Consider the vectors $\boldsymbol{u} = (\frac{2}{3}, \frac{1}{3}, \frac{4}{3})$, $\boldsymbol{v} = (\frac{1}{3}, \frac{-1}{3}, 1)$ and $\boldsymbol{w} = (1, \frac{2}{3}, \frac{5}{3})$. Write each of the following vectors as a linear combination of $\boldsymbol{u}, \boldsymbol{v}$, and $\boldsymbol{w}$.

(a) $(-3, -\frac{7}{3}, -5)$

(b) $(2, \frac{11}{3}, 2)$

(c) $(0, 0, 0)$

(d) $(\frac{7}{3}, \frac{8}{3}, 3)$

EX 3.7.3. Let

$$\boldsymbol{v_1} = (-3, 1, -5, -2) \qquad \boldsymbol{v_2} = (-2, 0, 4, 2) \qquad \boldsymbol{v_3} = (6, 3, 0, 9)$$

be vectors in $\mathbb{R}^4$. Find which of the vectors below lie in $\operatorname{span}\{\boldsymbol{v_1}, \boldsymbol{v_2}, \boldsymbol{v_3}\}$.

(a) $(-2, -3, 7, -3)$

(b) $(0, 0, 0, 0)$

(c) $(2, 2, 2, 2)$

(d) $(-6, 3, -6, 1)$

EX 3.7.4. Find which of the subsets of $\mathbb{R}^2, \mathbb{R}^3$, and $\mathbb{R}^4$ below are vector subspaces when endowed with the vector addition and multiplication of vectors by scalars induced by those of the relevant vector space.

(a) $\{(0, a) \colon a \in \mathbb{R}\}$

(b) $\{(a, 1) \colon a \in \mathbb{R}\}$

(c) $\{(a, b, c) \in \mathbb{R}^3 \colon b = a - c\}$

(d) $\{(a, b, c) \in \mathbb{R}^3 \colon a, b, c \in \mathbb{N}\}$

(e) $\{(a, b, c, d) \in \mathbb{R}^4 \colon b = a - c + d\}$

EX 3.7.5. Find which of the subsets below are vector subspaces when endowed with the vector addition and multiplication of vectors by scalars induced by those of the relevant $\mathbb{R}^n$.

(a) $\operatorname{span}\{(-1, 0, 1)\} \cup \{(x, y, z) \in \mathbb{R}^3 : x - y = z\}$

(b) $\operatorname{span}\{(-1, 0, 1)\} \cup \{(x, y, z) \in \mathbb{R}^3 : x - y = -z\}$

(c) $\{(x, y, z) \in \mathbb{R}^3 : x + z = 3 + y\}$

(d) $\{(x, y, z) \in \mathbb{R}^2 : xz = 0\}$

(e) $\{(x, y, z) \in \mathbb{R}^3 : x = y \wedge z + x = 2\}$

(f) $\{(x, y, z, w) \in \mathbb{R}^4 : x + 2z - w = 0 \wedge x - 2y - w = 0\}$

EX 3.7.6. Show that $M_{k,n}(\mathbb{R}), \mathbb{P}_n, \mathbb{P}$, and $C[a, b]$ (with $a < b$) are real vector spaces and $M_{k,n}(\mathbb{C})$ is a complex vector space.

EX 3.7.7. Consider the following subsets of the space $\mathbb{P}_2$ of the real polynomials of degree less than or equal to 2:

(a) $\{3at^2 - at + 3a \colon a \in \mathbb{R}\}$

(b) $\{-5at^2 - 3t^2 + 3a - 4 \colon a \in \mathbb{R}\}$

(c) $\{at^2 - 2at^2 + 3a \colon a \in \mathbb{R}\}$

(d) $\{-5at + 3a - 1 \colon a \in \mathbb{R}\}$

Which of the subsets above are vector subspaces of $\mathbb{P}_2$?

EX 3.7.8. Suppose that $\mathbb{C}^2$ is endowed with the operations

- $(x, y) + (x', y') = (x + x', y + y')$ for $(x, y), (x', y') \in \mathbb{C}^2$
- $\alpha(x, y) = (\alpha y, -\alpha x)$ for $(x, y) \in \mathbb{C}^2$ and $\alpha \in \mathbb{C}$

Is $\mathbb{C}^2$ a vector space together with this operations?

EX 3.7.9. Which of the sets below are linearly independent? Find a basis for the subspace spanned by each of the sets.

(a) $\{(1, -1, 0), (0, 0, 2)\}$

(b) $\{(2, 4, 12), (-1, -1, -1), (2, 4, 12), (0, 1, 5)\}$

(c) $\{(1, 2, 3, 4), (0, 0, 0, 0), (0, 1, 1, 0)\}$

(d) $\{(1 + i, 2i, 4 - i), (2 + 2i, 4i, 8 - 2i)\}$

(e) $\{(1, 2, 6, 0), (3, 4, 1, 0), (4, 3, 1, 0), (3, 3, 1, 0)\}$

EX 3.7.10. Consider the subspace

$$W = \{(x, y, z) \in \mathbb{R}^3 : x - 2y + 3z = 0\}.$$

(a) Find an ordered basis $\mathcal{B}$ for W and $\dim W$.

(b) Show that the vector $\boldsymbol{v} = (-4, -1, 2)$ lies in W, and determine the coordinate vector $(\boldsymbol{v})_{\mathcal{B}}$ of $\boldsymbol{v}$ relative to $\mathcal{B}$.

EX 3.7.11. Consider the subspace of $\mathbb{C}^4$

$$W = \{(x, y, z, w) \in \mathbb{C}^4 : -4y + z = 0 \wedge x - y + z = w \wedge x = w\}$$

(a) Find an ordered basis $\mathcal{B}$ of W. What is $\dim W$?

(b) Does the vector $\boldsymbol{v} = (1, 0, 0, 1)$ lie in W? If it does, find the coordinate vector $\mathbf{v}_{\mathcal{B}}$ of $\boldsymbol{v}$ relative to $\mathcal{B}$.

EX 3.7.12. Let W be the subspace of $\mathbb{R}^4$ spanned by the vectors $\boldsymbol{u} = (1,0,0,1)$, $\boldsymbol{v} = (2,2,0,1)$, and $\boldsymbol{w} = (4,2,0,3)$.

(a) Show that $S = \{\boldsymbol{u}, \boldsymbol{v}, \boldsymbol{w}\}$ is not a basis for W.

(b) Determine a basis $\mathcal{B}_W$ for W and its dimension.

(c) Find vector, parametric and cartesian equations for W.

EX 3.7.13. Add a vector of the standard basis of $\mathbb{C}^3$ to the set $S = \{(-i, 2i, 3i), (1, -2, -2)\}$ to form a basis of $\mathbb{C}^3$.

EX 3.7.14. Consider the vectors $\boldsymbol{v}_1 = (1, 0, -2, 1), \boldsymbol{v}_2 = (-3, 0, 2, -1)$ in $\mathbb{R}^4$. Add vectors of the standard basis of $\mathbb{R}^4$ to $S = \{\boldsymbol{v}_1, \boldsymbol{v}_2\}$ in order to form a basis of $\mathbb{R}^4$.

EX 3.7.15. Let A be the matrix

$$A = \begin{bmatrix} 1 & 1 & -1 & 1 \\ 0 & -1 & 3 & 0 \\ -1 & 0 & -2 & 0 \\ 0 & -1 & 3 & 1 \end{bmatrix}.$$

(a) Determine bases for the null space, the row space, and column space of A.

(b) Verify the solution (a) using the Rank-nullity Theorem.

(c) If C is an invertible 4×4 matrix, what is the dimension of the null space of CA^T? What is the dimension of the column space of CA^T?

EX 3.7.16. Determine a basis and the dimension of each of the spaces $N(A), L(A)$, and $C(A)$ where

$$A = \begin{bmatrix} 1 & 1 & 0 \\ -3i & i & 0 \\ 0 & 0 & 0 \\ 1 & -1 & 0 \end{bmatrix}.$$

EX 3.7.17. Let A be a real square matrix such that its column space is

$$C(A) = \{(x, y, z) \in \mathbb{R}^3 \colon x + y + z = 0 \wedge x - y = z\}.$$

(a) Is the system

$$A \begin{bmatrix} x \\ y \\ z \end{bmatrix} = \begin{bmatrix} 2 \\ 1 \\ -1 \end{bmatrix}$$

consistent?

(b) Suppose that B is a square matrix whose null space is

$$N(B) = \text{span}\{(1,1,1),(1,-1-1)\}.$$

If $(1,2,3)$ is a particular solution of

$$B\begin{bmatrix} x \\ y \\ z \end{bmatrix} = \begin{bmatrix} 0 \\ -1 \\ 1 \end{bmatrix},$$

what is the solution set of this system?

(c) Consider the matrix B above. What is the minimum value of $\dim N(B^T B)$?

EX 3.7.18. Consider the subspaces of $\mathbb{R}^4$

$$U = \text{span}\{(6,6,2,-4),(1,1,1,0)\},$$

$$V = \{(x,y,z,w) \in \mathbb{R}^4 :\ -x = 2y\}.$$

(a) Find a basis $\mathcal{B}_{U\cap V}$ for the subspace $U \cap V$.

(b) What is the dimension of $\mathbb{R}^4 + (U \cap V)$?

EX 3.7.19. Let U and W be the subspaces of $\mathbb{C}^4$

$$U = \text{span}\{(0,0,3-i,0),(1-2i,0,0,1-2i)\},$$

$$W = \{(x,y,z,w) \in \mathbb{R}^4\ :\ ix + 2y - z - iw = w - x \wedge x - w = 0\}.$$

Find a basis and the dimension of each of the subspaces $U \cap W$ and $U + W$. Verify that the formula

$$\dim U + \dim W = \dim(U \cap W) + \dim(U + W)$$

holds.

EX 3.7.20. Write the polynomial $5 + 9t + 3t^2 + 5t^3$ as a linear combination of $p_1 = 2 + t + t^2 + 4, p_2 = 1 - t + 3t^3$, and $p_3 = 3 + 2t + 5t^3$.

EX 3.7.21. Find if the subsets of polynomials are linearly independent or linearly dependent.

(a) $\{1 + 2t, t^2 - 1 + t, t\}$.

(b) $\{1 + t - t^2 + t^3, 2t + 2t^3, 1 + 3t - t^2 + 3t^3\}$.

(c) $\{t^5 - t^4, t^2, t^3 - 2\}$.

EX 3.7.22. Let $\mathcal{B} = (\frac{1}{3} - \frac{1}{3}t, \frac{1}{3} + \frac{2}{3}t)$ be an ordered basis of the vector space $\mathbb{P}_1$ of the real polynomials of degree less than or equal to 1. Find the coordinate vectors $(3-2t)_{\mathcal{B}}$ and $(3-2t)_{\mathcal{P}_1}$.

EX 3.7.23. Let $\mathcal{B} = (t^3, 1+t, t+t^2, 2-2t)$ be a basis of $\mathbb{P}_3$ and let S be the subspace of $\mathbb{P}_3$ tal que

$$\{(p(t))_{\mathcal{B}} \colon p(t) \in S\} = \{(x,y,z,w) \in \mathbb{R}^4 \colon x+y+z=0\},$$

where $(p(t))_{\mathcal{B}}$ is the coordinate vector of the polynomial $p(t)$ relative to $\mathcal{B}$.

Find a basis $\mathcal{B}_S$ for S and the coordinate vector $(3+t^2-2t^3)_{\mathcal{B}_S}$.

EX 3.7.24. Write

$$M = \begin{bmatrix} -9 & -7 \\ 4 & 0 \end{bmatrix}$$

as a linear combination of

$$A = \begin{bmatrix} 2 & 1 \\ 4 & 1 \end{bmatrix}, \quad B = \begin{bmatrix} 1 & -1 \\ 3 & 2 \end{bmatrix}, \quad C = \begin{bmatrix} 3 & 2 \\ 5 & 0 \end{bmatrix}.$$

EX 3.7.25. Consider the matrices

$$\begin{bmatrix} 0 & 4 \\ -2 & -2 \end{bmatrix} \quad \begin{bmatrix} -1 & 1 \\ 3 & 2 \end{bmatrix} \quad \begin{bmatrix} 2 & 0 \\ 4 & 1 \end{bmatrix}.$$

Find which matrices are a linear combination of the matrices above.

(a) $\begin{bmatrix} -8 & 6 \\ -8 & -1 \end{bmatrix}$

(b) $\begin{bmatrix} 0 & 0 \\ 0 & 0 \end{bmatrix}$

(c) $\begin{bmatrix} 0 & 6 \\ 8 & 3 \end{bmatrix}$

(d) $\begin{bmatrix} 5 & -1 \\ 1 & 7 \end{bmatrix}$

EX 3.7.26. Find the coordinate vectors of

$$A = \begin{bmatrix} 1-2i & 1+2i \\ -1-i & 2 \end{bmatrix}$$

relative to the basis $\mathcal{B} = (A_1, A_2, A_3, A_4)$ of $\mathrm{M}_2(\mathbb{C})$ where

$$A_1 = \begin{bmatrix} -i & i \\ 0 & 0 \end{bmatrix}, \quad A_2 = \begin{bmatrix} 1 & 1 \\ 0 & 0 \end{bmatrix}, \quad A_3 = \begin{bmatrix} 0 & 0 \\ 1+i & 0 \end{bmatrix}, \quad A_4 = \begin{bmatrix} 0 & 0 \\ 0 & 1-i \end{bmatrix}.$$

EX 3.7.27. Let

$$\mathcal{B} = \left(\begin{bmatrix} 2 & -2 \\ 0 & 0 \end{bmatrix}, \begin{bmatrix} 0 & 1 \\ 1 & 0 \end{bmatrix}, \begin{bmatrix} 1 & 1 \\ 0 & 0 \end{bmatrix}, \begin{bmatrix} 0 & 0 \\ 0 & 1 \end{bmatrix} \right)$$

be a basis of $M_2(\mathbb{R})$. Let S be the subspace of $M_2(\mathbb{R})$ such that

$$\{A_{\mathcal{B}} \colon A \in S\} = \{(x, y, z, w) \in \mathbb{R}^4 \colon x + z - w = 0\},$$

where $A_{\mathcal{B}}$ is the coordinate vector of matrix A relative to $\mathcal{B}$. Find a basis B_S for S.

EX 3.7.28.

(a) Find the change of basis matrix $M_{\mathcal{B} \leftarrow \mathcal{E}_2}$ from the standard basis of $\mathbb{R}^2$ to the basis $\mathcal{B} = \left((-\frac{1}{2}, 0), (-\frac{1}{2}, \frac{1}{2})\right)$. Find the coordinate vector $(1,1)_{\mathcal{B}}$.

(b) Find the change of basis matrix $M_{\mathcal{E}_2 \leftarrow \mathcal{B}'}$ from the basis $\mathcal{B}' = \left((\frac{1}{2}, 1), (-1, \frac{1}{2})\right)$ to the standard basis.

(c) Use the matrices above to obtain $M_{\mathcal{B} \leftarrow \mathcal{B}'}$.

EX 3.7.29. Let $\mathcal{B} = \left((0, \frac{1}{2}, 0), (\frac{1}{2}, 0, \frac{1}{2}), (1, \frac{1}{2}, 0)\right)$ be an ordered basis of $\mathbb{R}^3$.

(a) Find the change of basis matrix $M_{\mathcal{B} \leftarrow \mathcal{E}_3}$.

(b) Use the matrix above to find $\boldsymbol{v}$ such that $\mathbf{v}_{\mathcal{B}} = \begin{bmatrix} 2 \\ -2 \\ 4 \end{bmatrix}$.

(c) Find $M_{\mathcal{B}' \leftarrow \mathcal{B}}$, where $\mathcal{B}' = \left((-\frac{1}{2}, 0, -\frac{1}{2}), (1, \frac{1}{2}, 0), (0, \frac{1}{2}, 0)\right)$.

(d) Use the two change of basis matrices to determine $M_{\mathcal{B}' \leftarrow \mathcal{E}_3}$.

EX 3.7.30. Let $\mathcal{B}$ be an ordered basis of the space $\mathbb{P}_1$ consisting of the real polynomials of degree less than or equal to 1. Let

$$M_{\mathcal{B} \leftarrow \mathcal{P}_1} = \begin{bmatrix} 3 & 1 \\ -1 & 5 \end{bmatrix}$$

be the change of basis matrix from the standard basis $\mathcal{P}_1$ to basis the $\mathcal{B}$.

(a) Find $(1 - t)_{\mathcal{B}}$.

(b) Determine the basis $\mathcal{B}$.

EX 3.7.31. Let S be the subspace of the space V consisting of 2×2 upper triangular real matrices having zero trace. Find a basis $\mathcal{B}_S$ of S and a basis $\mathcal{B}$ of V containing $\mathcal{B}_S$. Determine the change of basis matrix $M_{\mathcal{B}_1 \leftarrow \mathcal{B}}$, where $\mathcal{B}_1$ is the basis of V

$$\mathcal{B} = \left(\begin{bmatrix} -2 & 0 \\ 0 & 2 \end{bmatrix}, \begin{bmatrix} 0 & 1 \\ 0 & 0 \end{bmatrix}, \begin{bmatrix} 1 & 0 \\ 0 & 0 \end{bmatrix} \right).$$

3.8 At a Glance

A vector space over the field of scalars $\mathbb{K}$ is a non-empty set endowed with two operations, addition and scalar multiplication, satisfying some fixed axioms. Examples of vector spaces are $\mathbb{R}^n, \mathbb{C}^n, M_{k,n}(\mathbb{K})$, and $\mathbb{P}_n$. Elements of a vector space are called vectors.

In EX 3.7.8 we give an example of a set which satisfies all but one of these axioms imposed on the operations. Although the property that is not verified seems innocent enough, the ending result is that we do not have a vector space in this case. In other words, all axioms matter.

Important subsets of a vector space are its subspaces, that is, non-empty subsets that are closed under addition and scalar multiplication.

Two crucial concepts pervade the theory of vector spaces: linear combination and linear independence. Some vector spaces have a spanning set, that is, a finite subset of vectors such that every vector in the space is a linear combination of those vectors. Some vector spaces do not have a spanning set, e.g., the space $\mathbb{P}$ of real polynomials. These are called infinite dimensional vector spaces, as opposed to the former which are called finite dimensional vector spaces. In the book, we deal almost exclusively with finite dimensional vector spaces.

Minimal spanning sets must be linearly independent and they are called bases of the vector space. All bases of a vector space have the same number of vectors called the dimension of the space.

The dimension classifies finite dimensional vector spaces in the sense that an n dimensional vector space over $\mathbb{K}$ can be essentially identified with $\mathbb{K}^n$, from a purely algebraic point of view. More precisely, vectors can be given by their coordinates relative to a basis, thereby allowing for an 'identification' of an n dimensional vector space over $\mathbb{K}$ with $\mathbb{K}^n$.

Each matrix has four vector subspaces associated: its null space, row space, and column space, and the null space of its transpose. These spaces are fundamental in the analysis of the properties of the matrix. The Rank-nullity Theorem gives a formula relating the dimensions of these spaces. The general solution of a system of linear equations is obtained in terms of a particular solution and the null space of the coefficient matrix of the system.

In a vector space, coordinates relative to two different bases can be related through the so-called change of basis matrix.

Chapter 4

Eigenvalues and Eigenvectors

4.1	Spectrum of a Matrix	131
4.2	Spectral Properties	134
4.3	Similarity and Diagonalisation	139
4.4	Jordan Canonical Form	149
4.4.1	Nilpotent matrices	149
4.4.2	Generalised eigenvectors	156
4.4.3	Jordan canonical form	160
4.5	Exercises	171
4.6	At a Glance	173

This chapter is about the spectrum of a matrix. The spectrum is a set of numbers, called the eigenvalues, each of which has an associated eigenspace containing the eigenvectors. The so-called generalised eigenvectors allow for establishing a similarity relation between the matrix in hand and a particular upper triangular matrix which, in some cases, is in fact diagonal. By means of this similarity relation, both matrices share many properties which might be easier to come by analysing the triangular matrix instead of the general matrix we started with.

The spectrum of a matrix is extremely important in applications such as Quantum Mechanics, Biology, or Atomic Physics, to name a few. The applications in Chapter 7 will be mostly related with the spectra of matrices.

4.1 Spectrum of a Matrix

Definition 33 *Let A be a square matrix in $\mathrm{M}_n(\mathbb{K})$. A non-zero vector $\boldsymbol{x} \in \mathbb{K}^n$ is said to be an* ***eigenvector*** *of A if there exists $\lambda \in \mathbb{K}$ such that*

$$A\mathbf{x} = \lambda\mathbf{x}. \tag{4.1}$$

Under these conditions, λ is called an ***eigenvalue*** *of A associated with $\boldsymbol{x}$. The* ***spectrum of*** *A, denoted by $\sigma(A)$, is the set of eigenvalues of matrix A.*

DOI: 10.1201/9781351243452-4

To find the spectrum of A, it is necessary to solve the equation (4.1) or, equivalently, to solve

$$(A - \lambda I)\mathbf{x} = \mathbf{0}. \qquad (4.2)$$

Since we want to find $\lambda \in \mathbb{K}$ for which there exist non-zero vectors $\mathbf{x}$ satisfying (4.1), the homogeneous system (4.2) must have non-zero solutions. Hence, the coefficient matrix $A - \lambda I$ must be singular. By Theorem 2.1, this holds if and only if the determinant $\det(A - \lambda I)$ is equal to zero.

The polynomial $p(\lambda) = \det(A - \lambda I)$ has degree n and is called the **characteristic polinomyal** of A. The equation

$$\det(A - \lambda I) = 0 \qquad (4.3)$$

is the **characteristic equation**. The eigenvalues of matrix A are, therefore, the roots of the characteristic polynomial of A.

Given an eigenvalue λ, the **eigenspace** $E(\lambda)$ associated with the eigenvalue λ is the solution set of (4.2). In other words, $E(\lambda)$ is the null space of $A - \lambda I$, i.e.,

$$E(\lambda) = N(A - \lambda I).$$

Example 4.1 *Find the spectrum and bases for the eigenspaces of*

$$A = \begin{bmatrix} 0 & -1 & 0 \\ -1 & 0 & 0 \\ 0 & 0 & -1 \end{bmatrix}.$$

We begin by finding the eigenvalues of A. The characteristic polynomial of A is

$$p(\lambda) = \det(A - \lambda I) = \begin{vmatrix} -\lambda & -1 & 0 \\ -1 & -\lambda & 0 \\ 0 & 0 & -1-\lambda \end{vmatrix} = (-1-\lambda)\begin{vmatrix} -\lambda & -1 \\ -1 & -\lambda \end{vmatrix}$$

$$= (-1-\lambda)(\lambda^2 - 1) = (-1-\lambda)^2(1-\lambda).$$

The roots of $p(\lambda) = (-1-\lambda)^2(1-\lambda)$, that is, the solutions of the characteristic equation (4.3) are $\lambda_1 = 1$ and $\lambda_2 = -1$, being the latter a double root. Hence, the eigenvalues of A are $\lambda_1 = 1$ and $\lambda_2 = -1$, and the spectrum of A is $\sigma(A) = \{-1, 1\}$.

The eigenspace $E(1)$ consists of $\mathbf{0}$ and the eigenvectors corresponding the eigenvalue $\lambda_1 = 1$ and, therefore, $E(1)$ is the null space $N(A - I)$ of $A - I$. Using Gaussian elimination to solve the corresponding homogeneous system, we have

$$A - I = \begin{bmatrix} -1 & -1 & 0 \\ -1 & -1 & 0 \\ 0 & 0 & -2 \end{bmatrix} \to \begin{bmatrix} -1 & -1 & 0 \\ 0 & 0 & 0 \\ 0 & 0 & -2 \end{bmatrix} \to \begin{bmatrix} -1 & -1 & 0 \\ 0 & 0 & -2 \\ 0 & 0 & 0 \end{bmatrix}.$$

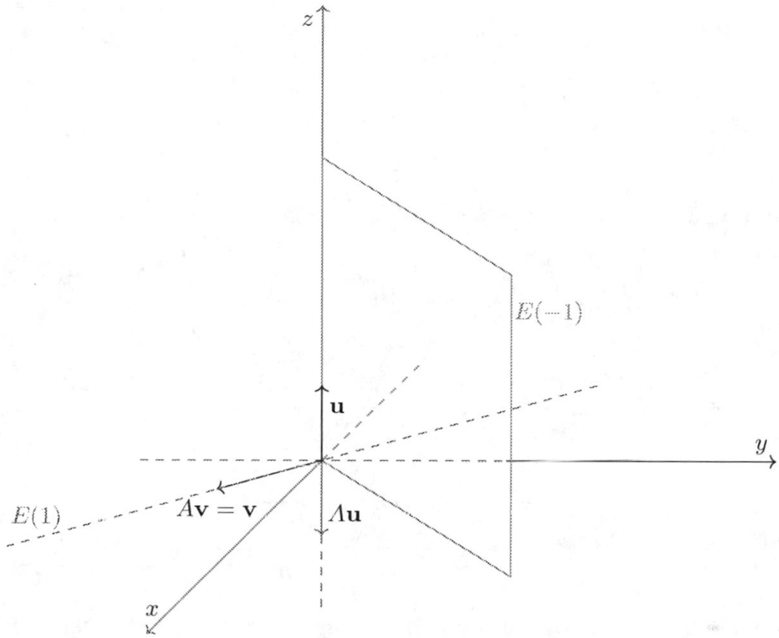

FIGURE 4.1: Vectors **u** and **v** in the eigenspaces $E(-1)$ and $E(1)$, respectively, and the vectors $A\mathbf{u}$ and $A\mathbf{v}$.

The eigenspace $E(1)$ *is the solution set associated with the matrix* $A - I$ *and, therefore,*

$$E(1) = \{(x, y, 0) \in \mathbb{R}^3 \colon x = -y\}.$$

The eigenspace $E(1)$ *is, consequently, the straight line satisfying the equations* $z = 0$, $x = -y$. *Observe that* $E(1)$ *is the solution set of a consistent system having a single independent variable.*

The eigenspace $E(-1)$ *consists of (*$\mathbf{0}$ *and) the eigenvectors associated with the eigenvalue* $\lambda_2 = -1$, *i.e.,* $E(-1) = N(A + I)$. *Similarly to what has been done above, we have*

$$A + I = \begin{bmatrix} 1 & -1 & 0 \\ -1 & 1 & 0 \\ 0 & 0 & 0 \end{bmatrix} \rightarrow \begin{bmatrix} 1 & -1 & 0 \\ 0 & 0 & 0 \\ 0 & 0 & 0 \end{bmatrix},$$

yielding

$$E(-1) = \{(x, y, z) \in \mathbb{R}^3 \colon x = y\}.$$

The eigenspace $E(-1)$ *is the plane* $x = y$ *corresponding to the solution of a consistent system having two independent variables (see Figure 4.1).*

If λ is an eigenvalue, then the eigenspace $E(\lambda)$ is always the solution set of a consistent system with, at least, one independent variable. Indeed, $E(\lambda)$

is the solution set of a homogeneous, hence consistent, system. On the other hand, by the very definition of eigenvalue, the system (4.2) must have non-zero solutions, that is, (4.2) has infinitely many solutions (cf. Proposition 1.3). It follows, that there exists, at least, one independent variable.

Definition 34 *Let λ be an eigenvalue of a square matrix A of order n over $\mathbb{K}$. The **algebraic multiplicity** $m_a(\lambda)$ of λ is the multiplicity of λ as a root of the characteristic polynomial $p(\lambda)$. The **geometric multiplicity** $m_g(\lambda)$ of λ is the dimension of the eigenspace $E(\lambda)$.*

In Example 4.1, we have

Eigenvalue	$m_a(\lambda)$	$m_g(\lambda)$
$\lambda_1 = 1$	1	1
$\lambda_2 = -1$	2	2

Although in this example the algebraic and geometric multiplicities of each eigenvalue do coincide, this is not the case in general (see Example 4.4).

Proposition 4.1 *Let λ be an eigenvalue of a matrix. Then, the geometric multiplicity of λ is less than or equal to its algebraic multiplicity.*

We will discuss the proof of this in Corollary 4.6 (see §4.4).

4.2 Spectral Properties

Proposition 4.2 *Let A be a square matrix of order n over $\mathbb{K}$, let*

$$p(\lambda) = a_0 + a_1\lambda + \cdots + a_{n-1}\lambda^{n-1} + (-1)^n\lambda^n$$

be the characteristic polynomial of A, and let $\lambda_1, \ldots, \lambda_n$ be the eigenvalues of A (possibly repeated). The following hold.

(i) $|A| = \lambda_1\lambda_2\cdots\lambda_n$.

(ii) $a_{n-1} = (-1)^{n-1}\operatorname{tr} A$.

(iii) $\operatorname{tr} A = \sum_{i=1}^{n} \lambda_i$.

Proof *(i) Let*

$$p(\lambda) = (\lambda_1 - \lambda)(\lambda_2 - \lambda)\cdots(\lambda_n - \lambda) \tag{4.4}$$

be the characteristic polynomial of A (where the roots may not be all distinct).

Observing that

$$p(0) = |A - 0I| = |A|$$

and letting $\lambda = 0$ *in* (4.4), *we have*

$$|A| = \lambda_1 \lambda_2 \cdots \lambda_n.$$

(ii) This will be shown by induction. For $n = 1$, *the statement is clear. Suppose now that* $n \geq 2$ *and that the assertion holds for* $n - 1$. *Let* $A = [a_{ij}]$ *be an* $n \times n$ *matrix and let* $p(\lambda) = |A - \lambda I|$ *be its characteristic polynomial. Hence, using the Laplace's expansion along the first row, we have*

$$p(\lambda) = (a_{11} - \lambda I) \det[(A - \lambda I)_{11}] + \sum_{j=2}^{n} a_{1j} C_{1j}.$$

Recall that $[(A - \lambda I)_{11}]$ *is the* $n - 1 \times n - 1$ *matrix obtained by deleting the first row and the first column of* $A - \lambda I$. *It then follows from the induction hypothesis that the coefficient corresponding to* λ^{n-1} *is*

$$a_{11}(-1)^{n-1} - (-1)^{n-2} \operatorname{tr}[A_{11}] = (-1)^{n-1} \operatorname{tr} A.$$

Assertion (iii) follows immediately from (4.4) and assertion (ii).

A useful fact to keep in mind is that, for an $n \times n$ matrix A over $\mathbb{K}$ with characteristic polynomial

$$p(\lambda) = (\lambda_1 - \lambda)^{r_1} (\lambda_2 - \lambda)^{r_2} \cdots (\lambda_k - \lambda)^{r_k}, \tag{4.5}$$

where $\lambda_1, \lambda_2, \ldots, \lambda_k$ are all distinct, we have

$$r_1 + r_2 + \cdots + r_k = n.$$

Before the next proposition, we need to make a definition.

Definition 35 *Given a* $k \times n$ *complex matrix* A, *define the* $k \times n$ *matrix* $\bar{\mathbf{A}}$ *by* $\bar{A}_{ij} = \overline{A_{ij}}$. *In other words, the entries of* $\bar{A}$ *are the complex conjugates of the entries of* A.

Example 4.2 *Let* A *be the matrix*

$$A = \begin{bmatrix} 1+i & -2 & -5+3i \\ 7 & 6 & 1-2i \end{bmatrix}.$$

The matrix $\bar{A}$ *is*

$$\bar{A} = \begin{bmatrix} 1-i & -2 & -5-3i \\ 7 & 6 & 1+2i \end{bmatrix}.$$

Proposition 4.3 *Let A be a square matrix of order n and let p be a positive integer. Then, the following hold.*

(i) The characteristic polynomial of A coincides with the characteristic polynomial of A^T.

(ii) If $\lambda \in \sigma(A)$ and $\boldsymbol{x}$ is an eigenvector associated with λ, then $\lambda^p \in \sigma(A^p)$ and $\boldsymbol{x}$ is a eigenvector associated with λ^p.

(iii) If $A \in \mathrm{M}_{n\times n}(\mathbb{C})$ is a matrix with real entries, then $\lambda \in \sigma(A)$ if and only if $\bar{\lambda} \in \sigma(A)$. Moreover,

$$E(\bar{\lambda}) = \{\bar{\boldsymbol{x}} : \boldsymbol{x} \in E(\lambda)\}.$$

Proof *(i) Let $p_{A^T}(\lambda)$ and $p_A(\lambda)$ the be characteristic polynomials of A^T and A, respectively. Since the determinant remains unchanged by transposition (cf. Proposition 2.4), we have*

$$p_{A^T}(\lambda) = |A^T - \lambda I| = |(A - \lambda I)^T| = |A - \lambda I| = p_A(\lambda),$$

as required.

(ii) Let λ be a eigenvalue of A and let $\boldsymbol{x}$ be a vector in the eigenspace $E(\lambda)$. The result is trivially true for $p = 1$. We shall prove the remaining cases by induction.

Suppose then that (ii) holds for $p-1$, with $p \geq 2$, that is, $A^{p-1}\mathbf{x} = \lambda^{p-1}\mathbf{x}$. Then,

$$A^p\mathbf{x} = A(A^{p-1}\mathbf{x}) = A(\lambda^{p-1}\mathbf{x}) = \lambda^{p-1}A\mathbf{x} = \lambda^p\mathbf{x}.$$

(iii) Given $\lambda \in \mathbb{K}, \boldsymbol{x} \in \mathbb{K}^n$, we have $A = \lambda\mathbf{x}$ if and only if

$$\overline{A\mathbf{x}} = \overline{\lambda\mathbf{x}}.$$

But $\overline{\lambda\mathbf{x}} = \bar{\lambda}\bar{\mathbf{x}}$ and, since the entries of A are real numbers,

$$\overline{A\mathbf{x}} = \bar{A}\bar{\mathbf{x}} = A\bar{\mathbf{x}}.$$

It follows that $A = \lambda\mathbf{x}$ if and only if

$$A\bar{\mathbf{x}} = \bar{\lambda}\bar{\mathbf{x}}.$$

Moreover, for each $\lambda \in \sigma(A)$,

$$E(\bar{\lambda}) = \{\bar{\boldsymbol{x}} \colon \boldsymbol{x} \in E(\lambda)\}.$$

Assertion (iii) above can be expressed in an informal (but catchy) way as 'complex eigenvalues of matrices with real entries always come in pairs: a number and its conjugate'.

Definition 36 *A complex square matrix $A = [a_{ij}]$ is said to be a* ***hermitian matrix*** *if $A = \bar{A}^T$.*

The diagonal entries of a hermitian matrix are real numbers.

Proposition 4.4 *Let A be a hermitian matrix. Then, the spectrum $\sigma(A)$ is a subset of $\mathbb{R}$.*

Proof *Let λ be an eigenvalue of A and let $\boldsymbol{x} = (x_1, x_2, \ldots, x_n)$ be an eigenvector of A associated with λ, that is, $A\mathbf{x} = \lambda\mathbf{x}$. Then*

$$\begin{aligned}(\overline{A\mathbf{x}})^T &= (\overline{\lambda\mathbf{x}})^T \Rightarrow \\ (\overline{A}\overline{\mathbf{x}})^T &= \overline{\lambda}\overline{\mathbf{x}}^T \Rightarrow \\ \overline{\mathbf{x}}^T\overline{A}^T &= \overline{\lambda}\overline{\mathbf{x}}^T.\end{aligned}$$

Since A is hermitian, we have

$$\overline{\mathbf{x}}^T A = \overline{\lambda}\overline{\mathbf{x}}^T.$$

Multiplying both members of this equality by $\mathbf{x}$ on the right,

$$\begin{aligned}\overline{\mathbf{x}}^T A\mathbf{x} &= \overline{\lambda}\overline{\mathbf{x}}^T\mathbf{x} \Rightarrow \\ \overline{\mathbf{x}}^T(\lambda\mathbf{x}) &= \overline{\lambda}\overline{\mathbf{x}}^T\mathbf{x} \Rightarrow \\ \lambda\overline{\mathbf{x}}^T\mathbf{x} &= \overline{\lambda}\overline{\mathbf{x}}^T\mathbf{x} \Rightarrow \\ \lambda(x_1^2 + x_2^2 + \cdots + x_n^2) &= \overline{\lambda}(x_1^2 + x_2^2 + \cdots + x_n^2).\end{aligned}$$

Hence

$$(\lambda - \overline{\lambda})\underbrace{(x_1^2 + x_2^2 + \cdots + x_n^2)}_{\neq 0} = 0,$$

from which follows that $\lambda = \overline{\lambda}$, that is, $\lambda \in \mathbb{R}$.

An immediate consequence of this proposition is the following result.

Corollary 4.1 *The spectrum of a real symmetric matrix is a non-empty subset of $\mathbb{R}$.*

Example 4.3 *When A is a real matrix, there is no guarantee that the characteristic polynomial has a factorisation over $\mathbb{R}$ as that of (4.5). For example, the characteristic polynomial of*

$$A = \begin{bmatrix} 0 & -1 \\ 1 & 0 \end{bmatrix}$$

is $p(\lambda) = \lambda^2 + 1$ which has no real roots. ***This matrix has no real eigenvalues*** *(see EX 5.8.13).*

How to find the spectrum and eigenspaces of a matrix

Let A be an $n \times n$ matrix. To obtain the spectrum $\sigma(A)$ and the eigenspaces proceed as indicated below.

1. Find the roots of the characteristic polynomial $|A - \lambda I|$. In other words, determine a factorisation
$$|A - \lambda I| = (\lambda_1 - \lambda)^{r_1}(\lambda_2 - \lambda)^{r_2} \cdots (\lambda_k - \lambda)^{r_k}.$$
The spectrum $\sigma(A) = \{\lambda_1, \lambda_2, \ldots, \lambda_k\}$. The algebraic multiplicity of each λ_i, for $i = 1, \ldots, k$, is r_i.

2. For each λ_i, with $i = 1, \ldots, k$, find the null space $N(A - \lambda_i I)$, that is, solve the homogeneous system $A - \lambda_i I = 0$. The eigenspace $E(\lambda_i)$ is this solution set.

3. For each λ_i, find a basis $\mathcal{B}_{E(\lambda_i)}$ of $E(\lambda_i)$. The geometric multiplicity $m_g(\lambda_i)$ of the eigenvalue λ_i is $\dim E(\lambda_i)$, that is, $m_g(\lambda_i)$ is the number of vectors in $\mathcal{B}_{E(\lambda_i)}$.

The invertibility of matrices has been characterised in every chapter so far and this chapter will not be an exception. The next theorem provides a necessary and sufficient condition for a matrix to be non-singular in terms of its spectrum.

Theorem 4.1 (Necessary and sufficient conditions of invertibility (IV)) *Let A be a square matrix over $\mathbb{K}$. The following assertions are equivalent.*

(i) A is invertible.

(ii) $0 \notin \sigma(A)$.

Proof *By Theorem 3.7 (ii), A is non-singular if and only if $N(A) = \{\mathbf{0}\}$. Since $N(A) = N(A - 0I)$, it is immediate that A is non-singular if and only if $0 \notin \sigma(A)$.*

It is worth noticing that, whenever 0 is an eigenvalue of a matrix A, that is, whenever $0 \in \sigma(A)$, the null space of A coincides with the eigenspace $E(0)$:

$$E(0) = N(A - 0I) = N(A).$$

This simple (albeit important) fact, already implicit in the proof above, is strangely often overlooked. The reader ought to ***keep it in mind****.*

4.3 Similarity and Diagonalisation

We begin here the discussion of similarity mentioned at the beginnig of this chapter.

Definition 37 *Let A and B be real (resp., complex) $n \times n$ matrices. B is said to be* ***similar*** *to A if there exists an invertible matrix S such that*

$$B = S^{-1}AS$$

or, equivalently, if

$$SB = AS.$$

It is easy to see that B is similar to A if and only if A is similar to B. Consequently, to simplify one says simply that A and B are ***similar matrices****.*

Theorem 4.2 *Let A and B be $n \times n$ similar matrices.*

(i) $|A| = |B|$.

(ii) A is invertible if and only if B is invertible.

(iii) The characteristic polynomial $p_A(\lambda)$ of A coincides with the characteristic polynomial $p_B(\lambda)$ of B.

(iv) $\operatorname{tr} A = \operatorname{tr} B$.

(v) $\sigma(A) = \sigma(B)$ and the corresponding algebraic multiplicities (respectively, geometric multiplicities) of each eigenvalue coincide.

(vi) $\dim N(A) = \dim N(B)$.

(vii) $\operatorname{rank} A = \operatorname{rank} B$.

Proof *(i) Since A and B are similar, we have $|B| = |S^{-1}AS|$. Hence, by Propositon 2.3 and Corollary 2.1,*

$$|B| = |S^{-1}||A||S| = |A||S^{-1}||S| = |A|.$$

(ii) By (i), we know that $|A| \neq 0$ if and only if $|B| \neq 0$. Hence, A is invertible if and only if B is invertible.

(iii) We have

$$p_B(\lambda) = |B - \lambda I| = |S^{-1}AS - \lambda I| = |S^{-1}AS - \lambda S^{-1}IS|.$$

Hence

$$p_B(\lambda) = |S^{-1}(A - \lambda I)S| = |A - \lambda I| = p_A(\lambda).$$

(iv) This is a direct consequence of (iii) and Proposition 4.2 (ii).

(v) By (iii), it is clear that $\sigma(A) = \sigma(B)$. *It is also clear that the algebraic multiplicities coincide.*

As to the geometric multiplicities of $\lambda \in \sigma(A) = \sigma(B)$, *notice that, if* $\mathbf{x}$ *is a vector such that* $B\mathbf{x} = \lambda\mathbf{x}$, *i.e.,* $\mathbf{x}$ *is a eigenvector of* B *associated with the eigenvalue* λ, *then*

$$\lambda\mathbf{x} = B\mathbf{x} = S^{-1}AS\mathbf{x}.$$

Hence,

$$\lambda S\mathbf{x} = AS\mathbf{x},$$

which shows that $S\mathbf{x}$ *is an eigenvector of* A *associated with* λ. *In fact,* $\mathbf{x}$ *is eigenvector of* B *associated with* λ *if and only if* $S\mathbf{x}$ *is an eigenvector of* A *associated with the eigenvalue* λ.

It follows that

$$E_A(\lambda) = \{\mathbf{y} \in \mathbb{K}^n : \mathbf{y} = S\mathbf{x}, \mathbf{x} \in E_B(\lambda)\}.$$

Here $E_A(\lambda)$ *and* $E_B(\lambda)$ *are the eigenspaces of* A *and* B, *respectively, corresponding to the eigenvalue* λ.

It is easy to see that a subset $\mathcal{B} = \{\mathbf{x}_1, \mathbf{x}_2, \ldots, \mathbf{x}_k\}$ *of* $E_B(\lambda)$ *is linearly independent if and only if the subset* $\mathcal{B}' = \{S\mathbf{x}_1, S\mathbf{x}_2, \ldots, S\mathbf{x}_k\}$ *of* $E_A(\lambda)$ *is linearly independent. It now follows that* $\mathcal{B}$ *is a basis of* $E_B(\lambda)$ *if and only if* $\mathcal{B}'$ *is a basis of* $E_A(\lambda)$, *which proves the equality of the geometric multiplicities, as required.*

(vi) By (v), $0 \in \sigma(A)$ *if and only if* $0 \in \sigma(B)$. *Hence, if both matrices are invertible, then* $N(A) = \{0\} = N(B)$.

The only other possibility, is 0 being an eigenvalue of both A *and* B. *In this case, letting* $E_A(0)$ *and* $E_B(0)$ *denote, respectively, the eigenspace of* A *and* B *corresponding to 0, by (v), we have*

$$\dim N(A) = \dim E_A(0) = \dim E_B(0) = \dim N(B).$$

(vii) This follows immediately from (vi) observing that, by Theorem 3.6 and Proposition 3.11,

$$\operatorname{rank}(A) = n - \dim N(A) = n - \dim N(B) = \operatorname{rank}(B).$$

Exercise 4.1 *Let* D *be the diagonal matrix over* $\mathbb{K}$

$$D = \begin{bmatrix} d_1 & 0 & 0 \\ 0 & d_2 & 0 \\ 0 & 0 & d_3 \end{bmatrix}.$$

Find the eigenvalues and eigenspaces of D. *(Take into account that the diagonal entries in the matrix might not be all distinct.)*

As it can be extrapolated from the exercise above, the spectrum of a diagonal matrix D can be easily determined: it consists of the diagonal entries of D. Although finding the spectrum of a general matrix A is not usually easy, it sometimes helps knowing when A is 'comparable' with a diagonal matrix.

Definition 38 *An $n \times n$ matrix A over $\mathbb{K}$ is said to be* ***diagonalisable*** *if there exist a diagonal matrix D and an invertible matrix S such that*

$$D = S^{-1}AS$$

or, equivalently, if

$$SD = AS.$$

Under these conditions, S is said to be a ***diagonalising matrix*** *for A.*

As one can see from the definition, A shares many features with D, for example, the characteristic polynomial and, hence, the spectrum (see Theorem 4.2).

Theorem 4.3 *Let A be a square matrix of order n over $\mathbb{K}$. The following are equivalent.*

(i) A is diagonalisable.

(ii) A has n linearly independent eigenvectors.

(iii) There exists a basis of $\mathbb{K}^n$ consisting entirely of eigenvectors of A.

Before embarking in the proof of the theorem, it is worth taking some time to ponder its statement. Assertions (i) and (iii) are clear enough, however assertion (ii) might cause some misunderstanding.

A common mistake. Assertion (ii) does not mean that A has exactly n eigenvectors which are linearly independent. It says that, amongst the eigenvectors of A, one can extract a linearly independent subset having n eigenvectors.

An $n \times n$ matrix over $\mathbb{K}$ has either no eigenvectors in $\mathbb{K}^n$ (e.g., Example 4.3) or infinitely many (Why?).

Example 4.4 *The spectrum of the matrix*

$$A = \begin{bmatrix} i & 1 \\ 0 & i \end{bmatrix}$$

is $\sigma(A) = \{i\}$ and the corresponding eigenspace is $E(i) = \{(z, 0) \colon z \in \mathbb{C}\}$.

Obviously, this matrix has infinitely many eigenvectors but any two eigenvectors are linearly dependent. In other words, this matrix does not satisfy the conditions of Theorem 4.3.

Exercise 4.2 *Find two linearly independent eigenvectors for*

$$\begin{bmatrix} -i & 1 \\ 0 & i \end{bmatrix}.$$

We prove Theorem 4.3 next.

Proof *We show firstly that (ii) implies (i). Suppose that A has n linearly independent eigenvectors $\mathbf{v}_1, \mathbf{v}_2, \ldots, \mathbf{v}_n$ in $\mathbb{K}^n$. Let S be the non-singular matrix*

$$S = \left[\mathbf{v}_1 \mid \mathbf{v}_2 \mid \cdots \mid \mathbf{v}_n\right].$$

Then

$$AS = A\left[\mathbf{v}_1 \mid \mathbf{v}_2 \mid \cdots \mid \mathbf{v}_n\right] = \left[A\mathbf{v}_1 \mid A\mathbf{v}_2 \mid \cdots \mid A\mathbf{v}_n\right] =$$

$$= \left[\mathbf{v}_1 \mid \mathbf{v}_2 \mid \cdots \mid \mathbf{v}_n\right] \begin{bmatrix} \lambda_1 & 0 & 0 & \cdots & 0 & 0 \\ 0 & \lambda_2 & 0 & \cdots & 0 & 0 \\ \\ \vdots & & & \vdots & & \vdots \\ 0 & 0 & 0 & \cdots & 0 & \lambda_n \end{bmatrix} = SD,$$

as required.

We show now that (i) implies (ii). Suppose then that A is diagonalisable. It follows that $AS = SD$, where

$$S = \left[\mathbf{v}_1 \mid \mathbf{v}_2 \mid \cdots \mid \mathbf{v}_n\right].$$

Observe that, since S is invertible, its columns are linearly independent vectors in $\mathbb{K}^n$. We have

$$AS = \left[\mathbf{v}_1 \mid \mathbf{v}_2 \mid \cdots \mid \mathbf{v}_n\right] \begin{bmatrix} \lambda_1 & 0 & 0 & \cdots & 0 & 0 \\ 0 & \lambda_2 & 0 & \cdots & 0 & 0 \\ \vdots & & & \vdots & & \vdots \\ 0 & 0 & 0 & \cdots & 0 & \lambda_n \end{bmatrix}$$

$\iff$

$$A\left[\mathbf{v}_1 \mid \mathbf{v}_2 \mid \cdots \mid \mathbf{v}_n\right] = \left[\lambda_1\mathbf{v}_1 \mid \lambda_2\mathbf{v}_2 \mid \cdots \mid \lambda_n\mathbf{v}_n\right]$$

$\iff$

$$\left[A\mathbf{v}_1 \mid A\mathbf{v}_2 \mid \cdots \mid A\mathbf{v}_n\right] = \left[\lambda_1\mathbf{v}_1 \mid \lambda_2\mathbf{v}_2 \mid \cdots \mid \lambda_n\mathbf{v}_n\right].$$

Hence,

$$A\mathbf{v}_1 = \lambda_1 \mathbf{v}_1, \quad A\mathbf{v}_2 = \lambda_2 \mathbf{v}_2, \quad \ldots, \quad A\mathbf{v}_n = \lambda_n \mathbf{v}_n,$$

from which follows that A has n linearly independent eigenvectors.

The equivalence between (ii) and (iii) is obvious (see Theorem 3.5 (i)).

We are now able to identify which matrices are diagonalisable and which are not. For example, Theorem 4.3 gives us a clear answer: the matrix in Example 4.4 is not diagonalisable whilst that in Example 4.2 is a diagonalisable matrix.

Example 4.5 *We shall show that*

$$A = \begin{bmatrix} 0 & -1 \\ -1 & 0 \end{bmatrix}$$

is diagonalisable and shall see how this helps calculate the power A^{2020}.

A simple calculation yields that the eigenvalues of A are ± 1:

eigenvalue	$m_a(\lambda)$	$m_g(\lambda)$
$\lambda_1 = 1$	1	1
$\lambda_2 = -1$	1	1

The eigenspace $E(1)$ is the straight line $x = -y$ and the eigenspace $E(-1)$ is the straight line $x = y$. Choosing two linearly independent vectors, say, $\boldsymbol{v}_1 = (-1, 1) \in E(1)$ e $\boldsymbol{v}_2 = (1, 1) \in E(-1)$, Theorem 4.3 guarantees that A is diagonalisable and that

$$\underbrace{\begin{bmatrix} 1 & 0 \\ 0 & -1 \end{bmatrix}}_{D} = \underbrace{\begin{bmatrix} -1 & 1 \\ 1 & 1 \end{bmatrix}^{-1}}_{S^{-1}} \underbrace{\begin{bmatrix} 0 & -1 \\ -1 & 0 \end{bmatrix}}_{A} \underbrace{\begin{bmatrix} -1 & 1 \\ 1 & 1 \end{bmatrix}}_{S}.$$

The proof of Theorem 4.3 is constructive and, for that reason, provides a way of constructing S. One has just to find the required number of linearly independent eigenvectors and build a matrix whose columns are those vectors, in no particular order. However, having done that, one has to be sure that, in each column of D, the diagonal entry is the eigenvalue corresponding to the eigenvector in the same column of S.

The diagonalising matrix S is a possible solution amongst many others, since S depends on the chosen eigenvectors.

To find A^{2020}, we begin by calculating A^2:

$$A^2 = (SDS^{-1})^2 = SD\underbrace{S^{-1}S}_{I}DS^{-1} = SD^2S^{-1}.$$

Generalising this process, it is easily seen that $A^{2020} = SD^{2020}S^{-1}$. *Hence,*

$$\begin{aligned} A^{2020} &= \begin{bmatrix} -1 & 1 \\ 1 & 1 \end{bmatrix} \begin{bmatrix} 1 & 0 \\ 0 & -1 \end{bmatrix}^{2020} \begin{bmatrix} -1 & 1 \\ 1 & 1 \end{bmatrix}^{-1} \\ &= \underbrace{\begin{bmatrix} -1 & 1 \\ 1 & 1 \end{bmatrix}}_{S} \underbrace{\begin{bmatrix} 1^{2020} & 0 \\ 0 & (-1)^{2020} \end{bmatrix}}_{D} \underbrace{\begin{bmatrix} -1/2 & 1/2 \\ 1/2 & 1/2 \end{bmatrix}}_{S^{-1}} \\ &= \begin{bmatrix} 1 & 0 \\ 0 & 1 \end{bmatrix}. \end{aligned}$$

This is a very simple case for which would be very easy to find A^{2020} *without using diagonalisation, since* $A = -P_{12}$. *However, this serves the purpose of illustrating a general process that is very useful when calculating powers of (more complicated) matrices.*

Collecting the insight given by this example:

Proposition 4.5 *Let* A, B *be matrices over* $\mathbb{K}$ *such that* $A = SBS^{-1}$, *for some non-singular matrix* S, *and let* $n \in \mathbb{N}$. *Then,*

$$A^n = SB^nS^{-1}.$$

Proof *Exercise.*

A difficulty that may occur when diagonalising a matrix is how to choose the eigenvectors in order for them to be linearly independent, as required in Theorem 4.3. The next proposition helps overcoming this difficulty.

Proposition 4.6 *Let* $\boldsymbol{v}_1, \boldsymbol{v}_2, \dots, \boldsymbol{v}_n$ *be eigenvectors of a square matrix* A *of order* k *and let* $\lambda_1, \lambda_2, \dots, \lambda_n$ *be the corresponding eigenvalues, all distinct. Then* $\boldsymbol{v}_1, \boldsymbol{v}_2, \dots, \boldsymbol{v}_n$ *are linearly independent.*

Proof *The assertion is trivially true for* $n = 1$. *Let* $1 < n$ *and consider the eigenvectors* $\boldsymbol{v}_1, \boldsymbol{v}_2, \dots, \boldsymbol{v}_n$ *and the corresponding eigenvalues* $\lambda_1, \lambda_2, \dots, \lambda_n$, *all distinct.*

Suppose that the assertion holds for $\boldsymbol{v}_1, \boldsymbol{v}_2, \dots, \boldsymbol{v}_{n-1}$, *i.e., the eigenvectors* $\boldsymbol{v}_1, \boldsymbol{v}_2, \dots, \boldsymbol{v}_{n-1}$ *are linearly independent. We want to show that* $\boldsymbol{v}_1, \boldsymbol{v}_2, \dots, \boldsymbol{v}_{n-1}, \boldsymbol{v}_n$ *are also linearly independent.*

Suppose that, on the contrary, the vector $\boldsymbol{v}_n$ *is a linear combination of the remaining eigenvectors. Observe that there is no loss of generality with this assumption. We have then that there exist scalars* $\alpha_1, \alpha_2, \dots, \alpha_{n-1}$ *such that*

$$\boldsymbol{v}_n = \alpha_1 \boldsymbol{v}_1 + \alpha_2 \boldsymbol{v}_2 + \cdots + \alpha_{n-1} \boldsymbol{v}_{n-1}.$$

It follows that

$$A\boldsymbol{v}_n = \lambda_n \boldsymbol{v}_n$$
$$\Longleftrightarrow$$
$$A(\alpha_1\boldsymbol{v}_1 + \alpha_2\boldsymbol{v}_2 + \cdots + \alpha_{n-1}\boldsymbol{v}_{n-1}) = \lambda_n(\alpha_1\boldsymbol{v}_1 + \alpha_2\boldsymbol{v}_2 + \cdots + \alpha_{n-1}\boldsymbol{v}_{n-1})$$
$$\Longleftrightarrow$$
$$\alpha_1\lambda_1\boldsymbol{v}_1 + \alpha_2\lambda_2\boldsymbol{v}_2 + \cdots + \alpha_{n-1}\lambda_{n-1}\boldsymbol{v}_{n-1} = \lambda_n(\alpha_1\boldsymbol{v}_1 + \alpha_2\boldsymbol{v}_2 + \cdots + \alpha_{n-1}\boldsymbol{v}_{n-1})$$

Hence,

$$\alpha_1(\lambda_n - \lambda_1)\boldsymbol{v}_1 + \alpha_2(\lambda_n - \lambda_2)\boldsymbol{v}_2 + \cdots + \alpha_{n-1}(\lambda_n - \lambda_{n-1})\boldsymbol{v}_{n-1} = \mathbf{0}$$

Since the eigenvectors $\boldsymbol{v}_1, \boldsymbol{v}_2, \ldots, \boldsymbol{v}_{n-1}$ *are linearly independent, we have*

$$\begin{cases} \alpha_1(\lambda_n - \lambda_1) = 0 \\ \alpha_2(\lambda_n - \lambda_2) = 0 \\ \vdots \\ \alpha_{n-1}(\lambda_n - \lambda_{n-1}) = 0\ . \end{cases}$$

Since there exists at least one scalar $\alpha_i \neq 0$*, it follows that* $\lambda_n = \lambda_i$*, contradicting the initial assumption of all the eigenvalues being distinct.*

Corollary 4.2 *Let A be an $n \times n$ matrix over $\mathbb{K}$ having n distinct eigenvalues. Then A is diagonalisable.*

Proof *This is an immediate consequence of Theorem 4.3 and Proposition 4.6.*

Proposition 4.7 *Let $\lambda_1, \lambda_2, \ldots, \lambda_p$ be eigenvalues, all distinct, of a square matrix A of order k, and let $\mathcal{B}_{E(\lambda_1)}, \mathcal{B}_{E(\lambda_2)}, \ldots, \mathcal{B}_{E(\lambda_p)}$ be bases of the corresponding eigenspaces. Then $\mathcal{B}_{E(\lambda_1)} \cup \mathcal{B}_{E(\lambda_2)} \cup \cdots \cup \mathcal{B}_{E(\lambda_p)}$ is a linearly independent set.*

Proof *For $i = 1, 2, \ldots, p$, let $\mathcal{B}_{E(\lambda_i)} = \{\boldsymbol{v}_1^{(i)}, \boldsymbol{v}_2^{(i)}, \ldots \boldsymbol{v}_{r_i}^{(i)}\}$ be a basis of the eigenspace $E(\lambda_i)$. We want to show that the set*

$$\mathcal{B} = \{\boldsymbol{v}_1^{(1)}, \boldsymbol{v}_2^{(1)}, \ldots \boldsymbol{v}_{r_1}^{(1)}\} \cup \{\boldsymbol{v}_1^{(2)}, \boldsymbol{v}_2^{(2)}, \ldots \boldsymbol{v}_{r_2}^{(2)}\} \cup \cdots \cup \{\boldsymbol{v}_1^{(p)}, \boldsymbol{v}_2^{(p)}, \ldots \boldsymbol{v}_{r_p}^{(p)}\}$$

is linearly independent.

Suppose, without loss of generality, that $\boldsymbol{v}_1^{(1)}$ is a linear combination of the remaining vectors:

$$\boldsymbol{v}_1^{(1)} = \underbrace{\alpha_2^{(1)}\boldsymbol{v}_2^{(1)} + \cdots + \alpha_{r_1}^{(1)}\boldsymbol{v}_{r_1}^{(1)}}_{\boldsymbol{u}_1} + \underbrace{\alpha_1^{(2)}\boldsymbol{v}_1^{(2)} + \alpha_2^{(2)}\boldsymbol{v}_2^{(2)} + \cdots + \alpha_{r_2}^{(2)}\boldsymbol{v}_{r_2}^{(2)}}_{\boldsymbol{u}_2} +$$
$$+ \cdots + \underbrace{\alpha_1^{(p)}\boldsymbol{v}_1^{(p)} + \alpha_2^{(p)}\boldsymbol{v}_2^{(p)} + \cdots + \alpha_{r_p}^{(p)}\boldsymbol{v}_{r_p}^{(p)}}_{\boldsymbol{u}_p}\ .$$

where, for some $i = 2, \dots, p$, *the vector* $\boldsymbol{u}_i$ *is non-zero. Observe that, otherwise,* $\boldsymbol{v}_1^{(1)}$ *would lie in the span of* $\{\boldsymbol{v}_2^{(1)}, \dots \boldsymbol{v}_{r_1}^{(1)}\}$, *which is impossible.*

We have

$$A\boldsymbol{v}_1^{(1)} = A\boldsymbol{u}_1 + A\boldsymbol{u}_2 + \dots + A\boldsymbol{u}_p,$$

from which follows that

$$\lambda_1(\boldsymbol{u}_1 + \boldsymbol{u}_2 + \dots + \boldsymbol{u}_p) = \lambda_1\boldsymbol{u}_1 + \lambda_2\boldsymbol{u}_2 + \dots + \lambda_p\boldsymbol{u}_p.$$

Hence,

$$(\lambda_1 - \lambda_2)\boldsymbol{u}_2 + \dots + (\lambda_1 - \lambda_p)\boldsymbol{u}_p = 0. \tag{4.6}$$

The non-zero vectors in $\{\boldsymbol{u}_1, \boldsymbol{u}_2, \dots, \boldsymbol{u}_p\}$ *are eigenvectors corresponding to distinct eigenvalues. It follows from Proposition 4.6 that they are linearly independent.*

Since there exists in (4.6) *at least one* $i \in \{2, \dots, p\}$ *such that* $\boldsymbol{u}_i \neq 0$, *the eigenvalue* λ_1 *coincides with the eigenvalue* λ_i, *which contradicts the initial assumption of all eingenvalues being distinct.*

Proposition 4.8 *Let* A *be a square matrix of order* n *over* $\mathbb{C}$. *The following are equivalent.*

(i) A *is a diagonalisable matrix.*

(ii) For all eigenvalues λ *of* A, *the algebraic multiplicity of* λ *coincides with the geometric multiplicity of* λ.

Proof *We begin by showing that (i)⇒(ii). Suppose that* A *is diagonalisable and that* $\lambda_1, \lambda_2, \dots, \lambda_p$ *are the eigenvalues of* A. *Then, the matrix* A *is similar to a diagonal matrix* D *having in the diagonal the eigenvalues, repeated according to their algebraic multiplicity (cf. the proof of Theorem 4.3). Hence, for all* $i = 1, 2, \dots, p$, *the geometric multiplicity* $m_g(\lambda_i)$ *satisfies*

$$\begin{aligned} m_g(\lambda_i) &= \dim N(A - \lambda_i I) \\ &= \dim N(D - \lambda_i I) && \textit{(by Theorem 4.2)} \\ &= n - \operatorname{rank}(D - \lambda_i I) && \textit{(by Theorem 3.6)} \\ &= n - (n - m_a(\lambda_i)) \\ &= m_a(\lambda_i), \end{aligned}$$

where $m_a(\lambda_i)$ *is the algebraic multiplicity of* λ_i.

We show now that (ii)⇒(i). Let $\lambda_1, \lambda_2, \dots, \lambda_p$ *be the eigenvalues of* A *and suppose that, for all* $i = 1, 2, \dots, p$, *we have* $m_g(\lambda_i) = m_a(\lambda_i) = r_i$. *Notice that*

$$r_1 + r_2 + \dots + r_p = n.$$

For $i = 1, 2, \ldots, p$, let $\{\boldsymbol{v}_1^{(i)}, \boldsymbol{v}_2^{(i)}, \ldots \boldsymbol{v}_{r_i}^{(i)}\}$, a basis of the eigenspace $E(\lambda_i)$. We shall show that the set defined by

$$\mathcal{B} = \{\boldsymbol{v}_1^{(1)}, \boldsymbol{v}_2^{(1)}, \ldots \boldsymbol{v}_{r_1}^{(1)}\} \cup \{\boldsymbol{v}_1^{(2)}, \boldsymbol{v}_2^{(2)}, \ldots \boldsymbol{v}_{r_2}^{(2)}\} \cup \cdots \cup \{\boldsymbol{v}_1^{(p)}, \boldsymbol{v}_2^{(p)}, \ldots \boldsymbol{v}_{r_p}^{(p)}\}$$

is a basis of $\mathbb{K}^n$ (cf. Teorema 4.3). Observing that $\mathcal{B}$ has n elements, it suffices to show that $\mathcal{B}$ is linearly independent. But this follows immediately from Proposition 4.7.

Given a matrix, as pointed out previously, the algebraic and geometric multiplicities of an eigenvalue might not coincide. In fact, if this happens for each of the eigenvalues, by Proposition 4.8, the matrix is necessarily diagonalisable.

How to diagonalise a matrix

Let A be an $n \times n$ matrix over $\mathbb{K}$.

1. Find the eigenvalues and bases for the eigenspaces of A. If the sum of the dimensions of the eigenspaces is n (meaning that A has n linearly independent eigenvectors), then A is diagonalisable. Otherwise, it is not.

2. If A is diagonalisable, then let $\lambda_1, \lambda_2, \ldots, \lambda_p$ be the distinct eigenvalues of A.

   For all $i = 1, 2, \ldots, p$, let $\{\boldsymbol{v}_1^{(i)}, \boldsymbol{v}_2^{(i)}, \ldots \boldsymbol{v}_{r_i}^{(i)}\}$ be a basis of the eigenspace $E(\lambda_i)$. Build the $n \times n$ matrix S whose columns consist of the vectors in these bases arranged by juxtaposition, in no particular order.

3. Build the diagonal matrix D whose diagonal entry in any column j coincides with the eigenvalue corresponding to the eingenvector in the column j of S.

4. Now we have $A = SDS^{-1}$.

We end this section with an important result about similarity which is a weaker version of a theorem known as Schur's Triangularisation Theorem[1]. We will prove next that any complex square matrix is similar to an upper triangular matrix.

Theorem 4.4 (Schur's triangularisation theorem) *Let A be a $k \times k$ complex matrix. Then, A is similar to a $k \times k$ upper triangular matrix whose diagonal consists of the k (not necessarily distinct) eigenvalues of A.*

[1] Schur's Triangularisation Theorem states that any $n \times n$ complex matrix is similar to an upper triangular matrix and the similarity matrix might be chosen to be a unitary matrix. The definition of unitary matrix is given in Chapter 6.

Proof *We will prove the result by induction on the size of the matrix. If* $n = 1$*, then the result holds trivially. Suppose now that the result holds for* $n \times n$ *matrices. Let* A *be an* $n+1 \times n+1$ *matrix with eigenvalues* $\lambda_1, \dots, \lambda_n, \lambda_{n+1}$.

Let $\mathbf{u}$ *be an eigenvector in* $\mathbb{C}^{n+1}$ *corresponding to the eigenvalue* λ_{n+1} *and let* $\mathcal{B} = \{\mathbf{u}, \mathbf{b}_1, \dots, \mathbf{b}_n\}$ *be a basis of* $\mathbb{C}^{n+1}$ *including* $\mathbf{u}$*. Let* N *be the matrix whose columns are the vectors of* $\mathcal{B}$*, i.e.,*

$$N = \left[\mathbf{u} \mid \mathbf{b}_1 \mid \dots \mid \mathbf{b}_n\right].$$

It follows that

$$AN = \left[A\mathbf{u} \mid A\mathbf{b}_1 \mid \dots \mid A\mathbf{b}_n\right] = \left[\lambda_{n+1}\mathbf{u} \mid A\mathbf{b}_1 \mid \dots \mid A\mathbf{b}_n\right],$$

that is

$$N^{-1}AN = \left[\begin{array}{c|ccc} \lambda_{n+1} & * & \dots & * \\ \hline 0 & * & \dots & * \\ 0 & * & \dots & * \\ \vdots & \vdots & \vdots & \vdots \\ 0 & * & \dots & * \end{array}\right],$$

where the lower right hand corner B *of* A *is an* $n \times n$ *matrix such that* $\sigma(B) = \{\lambda_1, \dots, \lambda_n\}$.

By the induction hypothesis, there exists an $n+1 \times n+1$ *invertible matrix* M *such that*

$$\left[\begin{array}{c|c} 1 & 0 \\ \hline 0 & B \end{array}\right] M = M \left[\begin{array}{c|cccc} 1 & 0 & & & \\ \hline 0 & \lambda_1 & * & \dots & * \\ 0 & 0 & \lambda_2 & \dots & * \\ \vdots & \vdots & \vdots & \ddots & * \\ 0 & 0 & 0 & \dots & \lambda_n \end{array}\right].$$

Hence,

$$M^{-1}N^{-1}ANM = \underbrace{\left[\begin{array}{c|cccc} \lambda_{n+1} & * & * & \dots & * \\ \hline 0 & \lambda_1 & * & \dots & * \\ 0 & 0 & \lambda_2 & \dots & * \\ \vdots & \vdots & \vdots & \ddots & * \\ 0 & 0 & 0 & \dots & \lambda_n \end{array}\right]}_{U}.$$

Finally, $A = (NM)U(NM)^{-1}$*, where* U *is an upper triangular matrix having the* n *eigenvalues of* A *in the diagonal.*

Summing up: Theorem 4.4 states that, given a complex square matrix A, there exist an upper triangular matrix U and an invertible matrix S such that

$$S^{-1}AS = U.$$

It is then obvious that both A and U must have the same eigenvalues which appear in the diagonal of U. Notice also that the proof above can be done in a way that equal eigenvalues in the diagonal of U are grouped together.

4.4 Jordan Canonical Form

Diagonalising matrices is a fine thing. It simplifies life in many ways, most notably when calculating matrix powers. Unfortunately, not all matrices are diagonalisable. However, we have a next best thing. The aim of this section is to show how any given complex matrix A is similar to an upper triangular matrix with a 'lot' of zero entries, said a Jordan canonical form of A, which is 'almost' a diagonal matrix. To do this, we need firstly to collect some background and will do just that in the following two subsections. We will digress through nilpotent matrices and generalised eigenvectors.

This section develops a considerable amount of theory to justify the existence of a Jordan canonical form for any given complex matrix. Notwithstanding the possible 'dryness' of these proofs, Example 4.10 is a rather pedestrian walk through the construction of a Jordan canonical form and gives a hands on guide to the problem.

4.4.1 Nilpotent matrices

We want to consider here matrices which have a particular behaviour: those for which there exists a power, depending on the matrix in hand, which is the zero matrix. For example, if

$$A = \begin{bmatrix} 0 & 1 & 0 \\ 0 & 0 & 1 \\ 0 & 0 & 0 \end{bmatrix},$$

then you can see for yourself that $A^2 \neq 0$ but $A^3 = 0$. Obviously, for any $p \geq 3$, we have $A^p = 0$. This matrix A has many zero entries but this is not a requirement to display this behaviour. In fact,

$$A = \begin{bmatrix} 5 & 15 & 10 \\ -3 & -9 & -6 \\ 2 & 6 & 4 \end{bmatrix}$$

is such that $A^2 = 0$.

Definition 39 *A square matrix A over $\mathbb{K}$ is said to be* **nilpotent** *if there exists a positive integer p such that $A^p = 0$.*

As we see, this is all about the null space of some power of a matrix being the whole space. It is worth making a note of a simple fact about the null spaces of matrix powers:

> Given an $n \times n$ matrix B,
>
> $$\{\mathbf{0}\} = N(B^0) \subseteq N(B) \subseteq N(B^2) \subseteq \cdots \subseteq N(B^p) \subseteq N(B^{p+1}) \subseteq \ldots$$

In fact, if, for some non-negative integer p, we take any vector $\mathbf{x} \in N(B^p)$, then

$$B^{p+1}\mathbf{x} = B(B^p\mathbf{x}) = B\mathbf{0} = \mathbf{0}.$$

We have then that the null spaces of the powers of an $n \times n$ matrix form an ascending chain of subspaces of $\mathbb{K}^n$ which grows with the exponent of the power.

Does it ever stop? Well, it must, since none of (the dimensions of) these null spaces can go 'past' (the dimension of) $\mathbb{K}^n$. More precisely,

Proposition 4.9 *Let B be an $n \times n$ matrix over $\mathbb{K}$. Then, there exists an integer p with $0 \leq p \leq n$ such that, for all $j \in \mathbb{N}$,*

$$N(B^p) = N(B^{p+j}) = N(B^n).$$

Proof *Suppose that for all p with $0 \leq p \leq n$, $N(B^p) \neq N(B^{p+1})$. Then,*

$$\dim N(B^{p+1}) \geq \dim N(B^p) + 1.$$

It follows that, under these circumstances,

$$\dim N(B^{n+1}) \geq n + 1 > \dim \mathbb{K}^n$$

which is impossible.

Hence, at this point, we have an integer p, with $0 \leq p \leq n$, for which $N(B^p) = N(B^{p+1})$. We will show now that

$$N(B^p) = N(B^{p+1}) = N(B^{p+2}) = N(B^{p+3}) = \ldots,$$

that is, after the power B^p the null spaces of the larger powers stabilise. Contrarily, suppose that there exists a positive integer r such that

$$N(B^{p+r}) \subsetneq N(B^{p+r+1}).$$

Hence, there exists a vector $\mathbf{x}$ *such that*

$$B^{p+r}\mathbf{x} \neq \mathbf{0} \quad \textit{and} \quad B^{p+r+1}\mathbf{x} = \mathbf{0}.$$

But then

$$B^p(B^r\mathbf{x}) \neq \mathbf{0} \quad \textit{and} \quad B^{p+1}(B^r\mathbf{x}) = \mathbf{0},$$

which shows that $B^r\mathbf{x}$ *lies in the null space of* B^{p+1} *but not in the null space of* B^p. *This however yields a contradiction, since* $N(B^p) = N(B^{p+1})$.

The proposition we just proved has the following immediate consequence.

Corollary 4.3 *Let* A *be an* $n \times n$ *nilpotent matrix. Then* $A^n = 0$.

Corollary 4.4 *The only eigenvalue of a (real or complex) nilpotent matrix is zero.*

Proof *Let* A *be an* $n \times n$ *nilpotent matrix over* $\mathbb{K}$. *If* $A = 0$, *the assertion holds trivially.*

Suppose now that $A \neq 0$. *We must show that* $\sigma(A) = \{0\}$. *We begin by showing that, if* λ *is an eigenvalue of* A, *then* $\lambda = 0$.

Let λ *be an eigenvalue of* A. *Then, there exists a non-zero* $\mathbf{x} \in \mathbb{K}^n$ *such that* $A\mathbf{x} = \lambda\mathbf{x}$. *By Corollary 4.3, it follows that*

$$\mathbf{0} = A^n\mathbf{x} = \lambda^n\mathbf{x},$$

yielding that $\lambda = 0$.

It remains to show that $\sigma(A) \neq \emptyset$, *which might not happen only for real matrices (see Example 4.3). Since* $A \neq 0$ *and* $A^n = 0$, *by Proposition 4.9, there exist a non-zero vector* $\mathbf{x} \notin N(A)$ *and a positive integer* $1 < j \leq n$ *such that*

$$\mathbf{0} = A^n\mathbf{x} = A^j\mathbf{x} \quad \textit{and} \quad A^{j-1}\mathbf{x} \neq \mathbf{0}.$$

But, in this case,

$$A(A^{j-1}\mathbf{x}) = \mathbf{0},$$

showing that $A^{j-1}\mathbf{x}$ *is an eigenvector of* A *with the associated eigenvalue* 0.

Exercise 4.3 *Let* A, B, C *be the block upper triangular matrices*

$$A = \begin{bmatrix} 0 & A_{12} & A_{13} \\ 0 & A_{22} & A_{23} \\ 0 & 0 & A_{33} \end{bmatrix}, \quad B = \begin{bmatrix} B_{11} & B_{12} & B_{13} \\ 0 & 0 & B_{23} \\ 0 & 0 & B_{33} \end{bmatrix}, \quad C = \begin{bmatrix} C_{11} & C_{12} & C_{13} \\ 0 & C_{22} & C_{23} \\ 0 & 0 & 0 \end{bmatrix}.$$

Show that $ABC = 0$. *Here we suppose that the sizes of the blocks are compatible for the multiplication purpose. (Solving this exercise will help you to better understand the proof of the next theorem.)*

We prove next the Cayley–Hamilton Theorem as a consequence of Theorem 4.4. Firstly, we make a definition: for a polynomial $q(\lambda) = a_0 + a_1\lambda + a_2\lambda^2 + \cdots + a_n\lambda^n$, given a square matrix A, define $q(A)$ by

$$q(A) = a_0 I + a_1 A + a_2 A^2 + \cdots + a_n A^n.$$

Corollary 4.5 (Cayley–Hamilton Theorem) *Let A be a complex square matrix and let $p(\lambda)$ be its characteristic polynomial. Then $p(A) = 0$.*

Proof *In this proof, we write I for the identity matrices of any order. Let $p(\lambda) = (\lambda - \lambda_1)^{n_1}(\lambda - \lambda_2)^{n_2} \dots (\lambda - \lambda_p)^{n_p}$, where $\lambda_1, \lambda_2, \dots, \lambda_p$ are all distinct, be the characteristic polynomial of A. By Theorem 4.4, there exists an upper triangular matrix U and an invertible matrix S such that $A = S^{-1}US$. Moreover, U can be chosen to be*

$$U = \begin{bmatrix} U_1 & * & \cdots & \cdots & \cdots & \cdots & * \\ & U_2 & * & \cdots & \cdots & \cdots & * \\ & & \ddots & & & & \\ & & & U_j & * & \cdots & * \\ & & & & \ddots & & \\ & & & & & & U_p \end{bmatrix}, \tag{4.7}$$

where, for $j = 1, \dots, p$, the block U_j is an $n_j \times n_j$ upper triangular matrix whose diagonal entries coincide with λ_j. It follows from Corollary 4.3 that $(U_j - \lambda_j I)^{n_j} = 0$. Hence,

$$(U - \lambda_j I)^{n_j} = \begin{bmatrix} * & * & \cdots & \cdots & \cdots & \cdots & * \\ & * & * & \cdots & \cdots & \cdots & * \\ & & \ddots & & & & \\ & & & 0 & * & \cdots & * \\ & & & & \ddots & & \\ & & & & & & * \end{bmatrix},$$

where the 0 in the diagonal corresponds to the block U_j. Consequently, we have

$$p(U) = (U - \lambda_1 I)^{n_1}(U - \lambda_2 I)^{n_2} \dots (U - \lambda_p I)^{n_p} = 0$$

(see Exercise 4.3). Since,

$$\begin{aligned}(A - \lambda_1 I)^{n_1}&(A - \lambda_2 I)^{n_2} \dots (A - \lambda_p I)^{n_p} \\ &= S^{-1}(U - \lambda_1 I)^{n_1}(U - \lambda_2 I)^{n_2} \dots (U - \lambda_p I)^{n_p} S,\end{aligned}$$

we have that

$$p(A) = S^{-1}p(U)S = 0.$$

Definition 40 *A square matrix is said to be* ***strictly upper triangular*** *if it is upper triangular and has a null diagonal.*

A strictly upper triangular $n \times n$ matrix is nilpotent (notice that $p(\lambda) = (-1)^n \lambda^n$) but we can say more.

Lemma 4.1 *A complex nilpotent matrix is similar to a strictly upper triangular matrix.*

Proof *This is an immediate consequence of Theorem 4.4, since the only eigenvalue of a nilpotent matrix is* 0.

Proposition 4.10 *A complex square matrix is nilpotent if and only if its only eigenvalue is* 0.

Proof *We already know that if an $n \times n$ matrix A is nilpotent, then its spectrum coincides with $\{0\}$ (see Corollary 4.4).*

Conversely, suppose that $\sigma(A) = \{0\}$. In other words, $p(\lambda) = (-1)^n \lambda^n$. By the Cayley–Hamilton Theorem (see Corollary 4.5),

$$0 = p(A) = (-1)^n A^n,$$

showing that A is nilpotent.

This result cannot be extended to real matrices.

A real matrix with the single eigenvalue 0 which is not nilpotent

The spectrum over the reals of

$$A = \left[\begin{array}{c|cc} 0 & 0 & 0 \\ \hline 0 & 0 & 1 \\ 0 & -1 & 0 \end{array}\right]$$

is $\{0\}$ and this matrix is not nilpotent. In fact, $A^p \mathbf{e}_3 \neq \mathbf{0}$ whichever the positive integer p (check it yourself). Here $\mathbf{e}_3 = (0, 0, 1)$ is the third vector of the standard basis of $\mathbb{R}^3$, as usual.

Definition 41 *Let A be an $n \times n$ matrix over $\mathbb{K}$. A subspace S of $\mathbb{K}^n$ is said to be* ***A-invariant*** *or* ***invariant under A*** *if, for all $\mathbf{x} \in S$, $A\mathbf{x} \in S$.*

Proposition 4.11 *Let A be a square matrix over $\mathbb{K}$ and let $p(\lambda)$ be a polynomial over $\mathbb{K}$. Then, the null space and the column space of $p(A)$ are A-invariant.*

Proof *Let* $\mathbf{x}$ *lie in* $N(p(A))$*. Then,*

$$p(A)A\mathbf{x} = A(p(A)\mathbf{x}) = A\mathbf{0} = \mathbf{0}$$

and, consequently, $A\mathbf{x} \in N(p(A))$*.*

Suppose now that $\mathbf{x} \in C(p(A))$*. Then, for some vector* $\mathbf{y}$ *in* $\mathbb{K}^n$*,*

$$A\mathbf{x} = A(p(A)\mathbf{y}) = p(A)(A\mathbf{y}) \in C(p(A)).$$

We turn now to ponder over the column space of an $n \times n$ matrix A.

> Given an $n \times n$ matrix A,
>
> $$\mathbb{K}^n = C(A^0) \supseteq C(A) \supseteq C(A^2) \supseteq \cdots \supseteq C(A^p) \supseteq C(A^{p+1}) \supseteq \dots$$

Observe that, fixing some non-negative integer p and any given vector $\mathbf{y} \in C(A^{p+1})$, there exists $\mathbf{x} \in \mathbb{K}^n$ such that

$$\mathbf{y} = A^{p+1}\mathbf{x} = A^p(\underbrace{A\mathbf{x}}_{\mathbf{z}}).$$

That is, $\mathbf{y} = A^p\mathbf{z}$ and, therefore, lies in $C(A^p)$.

Does this descending chain of subspaces stabilise? Yes, of course. Why? Well, $\{\mathbf{0}\}$ is a lower bound, obviously! But, we can say more. On one hand we have, by the Rank-nullity Theorem (see Theorem 3.6),

$$n = \dim N(A^p) + \dim C(A^p),$$

on the other hand we have a smallest non-negative integer p such that $N(A^p) = N(A^n)$. That is, p is the smallest integer stabilising the null spaces of the powers of A and, consequently, the formula above 'freezes' the dimension of $C(A^p)$ and of the column spaces of powers beyond. Hence, we have established a counterpart of Proposition 4.9 for column spaces.

Proposition 4.12 *Let* A *be an* $n \times n$ *matrix over* $\mathbb{K}$*. Then, there exists an integer* p *with* $0 \leq p \leq n$ *such that, for all* $j \in \mathbb{N}$*,*

$$C(A^p) = C(A^{p+j}) = C(A^n).$$

Notice that the smallest numbers p *for which Proposition 4.9 and Proposition 4.12 hold are the same.*

Suppose that A is such that $N(A) \cap C(A) = \{\mathbf{0}\}$, then, by Theorem 3.8,

$$\dim(N(A)+C(A)) = \dim N(A)+\dim C(A) - \dim(N(A)\cap C(A)) = n-0 = n.$$

Under this circumstance, it follows that

$$\mathbb{K}^n = N(A) \oplus C(A).$$

However, it is not always the case that $N(A)\cap C(A) = \{\mathbf{0}\}$. A simple counter-example is

$$A = \begin{bmatrix} 1 & 1 \\ -1 & -1 \end{bmatrix},$$

for which

$$N(A) = C(A) = \{(a, -a) \colon a \in \mathbb{R}\}.$$

There is however a way of bypassing this.

Proposition 4.13 *Let A be an $n \times n$ matrix. The following hold.*

(i) $N(A^n) \cap C(A^n) = \{\mathbf{0}\}$.

(ii) $N(A^n) \oplus C(A^n) = \mathbb{K}^n$.

Proof *(i) Suppose that on the contrary there exists $\mathbf{y} \neq \mathbf{0}$ such that*

$$A^n\mathbf{y} = \mathbf{0} \quad \textit{and} \quad \mathbf{y} = A^n\mathbf{x},$$

where $\mathbf{x}$ is some vector in $\mathbb{K}^n$. Hence,

$$\mathbf{0} = A^n\mathbf{y} = A^{2n}\mathbf{x}$$

from which follows that $\mathbf{x}$ lies in the null space of A^{2n} but not in $N(A^n)$. But this is impossible, since it contradicts Proposition 4.9.

(ii) This is a consequence of (i), as discussed above.

The splitting of $\mathbb{K}^n$ by the null space and the column space of a power of a matrix

Given any $n \times n$ matrix A,

$$N(A^n) \oplus C(A^n) = \mathbb{K}^n.$$

4.4.2 Generalised eigenvectors

Suppose that B is an $n \times n$ matrix. Did you notice that Proposition 4.9 forces the existence of a subspace of vectors $\mathbf{x} \in \mathbb{K}^n$ to satisfy $B^r\mathbf{x} = \mathbf{0}$, for some non-negative integer r which depends on $\mathbf{x}$? (We are referring to the subspace $N(B^n)$.) Surely it suffices to impose $r = n$. But we may find that this holds for $r < n$, depending on the vectors $\mathbf{x}$ in $N(B^n)$. Hence, given a non-zero $\mathbf{x} \in N(B^n)$, should it exist, there is a smallest r such that

$$B^r\mathbf{x} = \mathbf{0} \quad \text{and} \quad B^{r-1}\mathbf{x} \neq \mathbf{0}.$$

Hold this thought when reading the next definition.

Definition 42 *Let A be an $n \times n$ matrix over $\mathbb{K}$ and let $\lambda \in \mathbb{K}$ be an eigenvalue of A. A non-zero vector $\mathbf{x} \in \mathbb{K}^n$ is said to be a* **generalised eigenvector** *of A associated with λ if there exists $k \in \mathbb{N}$ such that*

$$(A - \lambda I)^k\mathbf{x} = \mathbf{0}. \tag{4.8}$$

The **order** *of the generalised eigenvector $\mathbf{x}$ is the smallest $k \in \mathbb{N}$ such that $\mathbf{x} \in N(A - \lambda I)^k$.*

Notice that (4.8) is the same as requiring that $\mathbf{x}$ lie in the null space $N(A - \lambda I)^k$ of the matrix $(A - \lambda I)^k$ and, consequently, in all the null spaces of higher powers of $A - \lambda I$. Hence the definition of the order of the generalised eigenvector as the smallest k for which (4.8) holds. In other words, $\mathbf{x}$ is a generalised eigenvector of order k of A, associated with λ, if

$$(A - \lambda I)^k\mathbf{x} = \mathbf{0} \qquad \text{and} \qquad (A - \lambda I)^{k-1}\mathbf{x} \neq \mathbf{0}. \tag{4.9}$$

It follows that, in these conditions,

- $\mathbf{x}$ is a vector in $N((A - \lambda I)^k)$ but not in $N((A - \lambda I)^{k-1})$.
- $(A - \lambda I)^{k-1}\mathbf{x}$ is a eigenvector of A associated with the eigenvalue λ.

The vector sequence

$$(A - \lambda I)^{k-1}\mathbf{x}, (A - \lambda I)^{k-2}\mathbf{x}, \dots, (A - \lambda I)\mathbf{x}, \mathbf{x} \tag{4.10}$$

is called a **Jordan chain of length** k associated with λ. Denote by $\mathbf{C}(\mathbf{x}, \lambda)$ the set consisting of the k vectors of the Jordan chain (4.10), i.e.,

$$C(\mathbf{x}, \lambda) = \{\underbrace{(A - \lambda I)^{k-1}\mathbf{x}}_{\mathbf{u}_1}, \underbrace{(A - \lambda I)^{k-2}\mathbf{x}}_{\mathbf{u}_2}, \dots, \underbrace{(A - \lambda I)\mathbf{x}}_{\mathbf{u}_{k-1}}, \underbrace{\mathbf{x}}_{\mathbf{u}_k}\}.$$

Example 4.6 *We are going to determine a generalised eigenvector* $\mathbf{x}$ *of order* 2 *of the matrix*

$$A = \begin{bmatrix} 2 & 1 & 1 \\ 0 & 2 & 1 \\ 0 & 0 & 1 \end{bmatrix},$$

and the corresponding Jordan chain. Calculations similar to those done previously in this chapter yield that the characteristic polynomial of A *is*

$$p(\lambda) = (2-\lambda)^2(1-\lambda).$$

Consequently, the spectrum of A *is* $\sigma(A) = \{1, 2\}$. *Both eigenvalues have geometric multiplicity equal to* 1, *and it is easy to see that bases for the eigenspaces are* $B_{E(2)} = \{(1,0,0)\}$ *and* $B_{E(-1)} = \{(0,-1,1)\}$.

We want to find a generalised eigenvector $\mathbf{x}$ *of order* 2, *i.e., to find an eigenvalue* λ *(which will be* 1 *or* 2*) such that*

$$\mathbf{u}_1 = (A - \lambda I)\mathbf{x} \qquad\qquad \mathbf{u}_2 = \mathbf{x}.$$

Hence, $\mathbf{u}_1$ *must be an eigenvector but* $\mathbf{u}_2$ *cannot be an eigenvector. The only possibility corresponds to the eigenvalue* 2. *In other words, we have to solve the equation*

$$(A - 2I)\mathbf{x} = \begin{bmatrix} 1 \\ 0 \\ 0 \end{bmatrix}.$$

(What would have happened if you were to choose $\mathbf{u}_1 = (0,-1,1)$*? Try it.)*

The solution set of this non-homogeneous system consists of the vectors

$$\mathbf{x} = t\begin{bmatrix} 1 \\ 0 \\ 0 \end{bmatrix} + \begin{bmatrix} 0 \\ 1 \\ 0 \end{bmatrix} \qquad (t \in \mathbb{R}). \tag{4.11}$$

Choosing, for example, $t = 0$, *we have that* $(0,1,0)$ *is a generalised eigenvector of order* 2.

We have now that the Jordan chain is

$$(1,0,0),\ (0,1,0)$$

and that, therefore, $C((0,1,0),2) = \{(1,0,0),(0,1,0)\}$.

Notice that the set of generalised eigenvectors $\{(1,0,0),(0,1,0),(0,-1,1)\}$ *is a basis of* $\mathbb{R}^3$.

Given a matrix $A \in M_n(\mathbb{K})$, let $G(\lambda)$ be the set consisting of $\mathbf{0}$ and all the generalised eigenvectors (of any order) of A associated with λ. This set is in fact a subspace of $\mathbb{K}^n$ called the **generalised eigenspace** of A associated with the eigenvalue λ. (It is an easy exercise to show that $G(\lambda)$ is closed under vector addition and scalar multiplication. See EX 4.5.9.)

A clarifying reminder:

> *An eigenvector is a generalised eigenvector of order 1.*
>
> *The eigenspace $E(\lambda)$ and the generalised eigenspace $G(\lambda)$ corresponding to a given eigenvalue λ satisfy $E(\lambda) \subseteq G(\lambda)$.*

Hence, we have here a slight abuse of notation: when the generalised eigenvectors consist only of eigenvectors proper (as was the case for $\lambda = 1$ in Example 4.6), then $G(\lambda)$ is what we called $E(\lambda)$ in Section 4.1. But once we keep this in mind, there will be no source of confusion.

Another question is how are we to find a method to determine the generalised eigenspaces? Or Jordan chains? In Example 4.6, we were advised from the start about the order of the generalised eigenvector. What if we were not? Do two distinct generalised eigenvectors in the same generalised eigenspace have necessarily the same order?

These are important questions that need to be answered. Proposition 4.14 below is a first but crucial step towards the answers.

Proposition 4.14 *Let A be a square matrix of order n over $\mathbb{K}$ and let $\lambda \in \mathbb{K}$ be an eigenvalue of A. Then*

$$G(\lambda) = N((A - \lambda I)^n).$$

Moreover, if $\mathbf{x} \in G(\lambda)$ is a generalised eigenvector of order k, then $1 \leq k \leq n$.

Proof *Let $\mathbf{x} \neq \mathbf{0}$ lie in $N((A-\lambda I)^n)$. Then, there exists a positive integer k for which $(A - \lambda I)^k \mathbf{x} = \mathbf{0}$. Indeed, it suffices to take $k = n$. Hence, $\mathbf{x}$ is a generalised eigenvector in $G(\lambda)$.*

Conversely, suppose that $\mathbf{x} \in G(\lambda)$. Then, $\mathbf{x}$ lies in the null space of some power of $A - \lambda I$. Hence, by Proposition 4.9, $\mathbf{x} \in N(A - \lambda I)^n$.

By the definition of order of a generalised eigenvector, we have finally that this order must be at most equal to n.

Observe that, by Propositions 4.11 and 4.14,

> Any generalised eigenspace of a matrix A is A-invariant.

Example 4.7 *Let us revisit Example 4.6 under the new light of this proposition. We have that*

$$(A - 2I)^3 = \begin{bmatrix} 0 & 1 & 1 \\ 0 & 0 & 1 \\ 0 & 0 & -1 \end{bmatrix}^3 = \begin{bmatrix} 0 & 0 & 0 \\ 0 & 0 & 1 \\ 0 & 0 & -1 \end{bmatrix}$$

and

$$(A-I)^3 = \begin{bmatrix} 1 & 1 & 1 \\ 0 & 1 & 1 \\ 0 & 0 & 0 \end{bmatrix}^3 = \begin{bmatrix} 1 & 3 & 3 \\ 0 & 1 & 1 \\ 0 & 0 & 0 \end{bmatrix}.$$

Hence,

$$G(2) = N(A-2I)^3 = \{(x,y,0)\colon x,y \in \mathbb{R}\}$$

and

$$G(1) = N(A-I)^3 = \{(0,-z,z)\colon z \in \mathbb{R}\}.$$

Observe that $G(1)$ must consist of eigenvectors only because it has dimension 1 and that the generalised eigenvectors (of order higher than 1), therefore, lie all of them in $G(2)$. Compare with what we found in Example 4.6.

Notice also that

$$G(1) \cap G(2) = \{\mathbf{0}\}$$

and

$$G(1) + G(2) = \mathbb{R}^3.$$

In other words,

$$G(1) \oplus G(2) = \mathbb{R}^3.$$

A way of determining the generalised eigenspaces

Let A be an $n \times n$ matrix over $\mathbb{K}$. To obtain its generalised eigenspaces:

1. Find the spectrum $\sigma(A)$ of A;
2. For each $\lambda \in \sigma(A)$, determine the null space $N(A-\lambda I)^n$;
3. The corresponding generalised eigenspace is $G(\lambda) = N(A-\lambda I)^n$.

Now we have sorted out a way to find the generalised eigenspaces. We have then answered the question 'How can we find the generalised eigenvectors systematically?' Answer: proceed as indicated above. It so happens that we do not want to find the generalised eigenvectors *per se*, we want rather use a well-chosen few Jordan chains to complete our program of 'almost' diagonalising every complex matrix.

One of the questions still pending about the order is 'Do two distinct generalised eigenvectors in the same generalised eigenspace have necessarily the same order?' This is a **No**.

For example, consider the matrix

$$\begin{bmatrix} 1 & 1 & 0 & 0 & 0 \\ 0 & 1 & 0 & 0 & 0 \\ 0 & 0 & 1 & 1 & 0 \\ 0 & 0 & 0 & 1 & 1 \\ 0 & 0 & 0 & 0 & 1 \end{bmatrix}. \tag{4.12}$$

Its only eigenvalue is 1. A basis for the eigenspace is

$$\mathcal{B}_{E(1)} = \{(1,0,0,0,0),(0,0,1,0,0)\}.$$

We can verify that $(1,1,1,0,0)$ is a generalised eigenvector of order 2 and that $(1,1,1,1,1)$ is a generalised eigenvector of order 3.

But even more obvious than this example is that

all vectors in a Jordan chain are generalised eigenvectors, each one having a different order.

Exercise 4.4 *Find the Jordan chains of the generalised eigenvectors above. Verify that the set consisting of these Jordan chains is a basis of* $\mathbb{R}^5$.

4.4.3 Jordan canonical form

As seen in Example 4.6, we found a basis of $\mathbb{R}^3$ consisting only of generalised eigenvectors. In particular, the vectors in each of the Jordan chains in the example were linearly independent. In turns out that this is a general result.

Proposition 4.15 *Let A be a matrix in $\mathrm{M}_n(\mathbb{K})$, let $\lambda \in \mathbb{K}$ be an eigenvalue of A and let $\mathbf{x}$ be a generalised eigenvector of order k of A associated with the eigenvalue λ. Then, the vectors in the Jordan chain*

$$(A-\lambda I)^{k-1}\mathbf{x}, \quad (A-\lambda I)^{k-2}\mathbf{x}, \quad \dots, \quad (A-\lambda I)\mathbf{x}, \quad \mathbf{x}$$

are linearly independent.

Proof *Let k and $\mathbf{x}$ be as above and let*

$$\underbrace{(A-\lambda I)^{k-1}\mathbf{x},}_{\mathbf{u}_1} \quad \underbrace{(A-\lambda I)^{k-2}\mathbf{x},}_{\mathbf{u}_2} \quad \dots \quad \underbrace{(A-\lambda I)\mathbf{x},}_{\mathbf{u}_{k-1}} \quad \underbrace{\mathbf{x}}_{\mathbf{u}_k}.$$

If $k=1$, then the set $\{\mathbf{u}_1\}$ is linearly independent, since $\mathbf{u}_1 \neq 0$. Hence, the proposition is proved for $k=1$.

Let now $k>1$, let p be an integer such that $1 \leq p < k$ and consider the set $S_p = \{\mathbf{u}_1, \dots, \mathbf{u}_p\}$. Observe that, if $p=1$, it can be shown similarly to the above paragraph that S_1 is linearly independent. We wish to show that, fixing p and assuming that S_p is linearly independent, then

$$S_{p+1} = \{\mathbf{u}_1, \dots, \mathbf{u}_p, \mathbf{u}_{p+1}\}$$

is also linearly independent.

If $\{\mathbf{u}_1, \ldots, \mathbf{u}_p, \mathbf{u}_{p+1}\}$ *were linearly dependent, then there would exist scalars* $\alpha_1, \ldots, \alpha_p, \alpha_{p+1}$, *not all equal to zero, such that*

$$\alpha_1 \mathbf{u}_1 + \cdots + \alpha_p \mathbf{u}_p + \alpha_{p+1} \mathbf{u}_{p+1} = \mathbf{0}.$$

Notice that, under these conditions, $\alpha_{p+1} \neq 0$, *since otherwise* $\{\mathbf{u}_1, \ldots, \mathbf{u}_p\}$ *would be linearly dependent, contradicting the hypothesis.*

Hence, we have

$$\alpha_{p+1}(A - \lambda I)^p \mathbf{u}_{p+1} = -\alpha_1 (A - \lambda I)^p \mathbf{u}_1 - \cdots - \alpha_p (A - \lambda I)^p \mathbf{u}_p.$$

Observe that, for $1 \leq j \leq p$,

$$(A - \lambda I)^p \mathbf{u}_j = (A - \lambda I)^p (A - \lambda I)^{k-j} \mathbf{x} = (A - \lambda I)^{k+(p-j)} \mathbf{x} = 0,$$

since $k + (p - j) \geq k$ *(compare with* (4.9)*). Moreover,*

$$(A - \lambda I)^p \mathbf{u}_{p+1} = (A - \lambda I)^{p+k-(p+1)} \mathbf{x} = (A - \lambda I)^{k-1} \mathbf{x} \neq 0$$

(see (4.9)*). It follows that*

$$\alpha_{p+1} \underbrace{(A - \lambda I)^p \mathbf{u}_{p+1}}_{\neq 0} = -\alpha_1 \underbrace{(A - \lambda I)^p \mathbf{u}_1}_{=0} - \cdots - \alpha_p \underbrace{(A - \lambda I)^p \mathbf{u}_p}_{=0},$$

which cannot be.

In fact, generalised eigenvectors corresponding to different eigenvalues are also linearly independent. To be precise:

Lemma 4.2 *Let* $\mathbf{x}_1, \mathbf{x}_2, \ldots, \mathbf{x}_p$ *be generalised eigenvectors of a square matrix* A *of order* n *and let* $\lambda_1, \lambda_2, \ldots, \lambda_p$ *be the corresponding eigenvalues, all distinct. Then* $\mathbf{x}_1, \mathbf{x}_2, \ldots, \mathbf{x}_p$ *are linearly independent.*

We had already a similar result for eigenvectors. Compare this lemma with Proposition 4.6.

Proof *Let* $\alpha_1, \alpha_2, \ldots, \alpha_p$ *be scalars such that*

$$\alpha_1 \mathbf{x}_1 + \alpha_2 \mathbf{x}_2 + \cdots + \alpha_p \mathbf{x}_p = \mathbf{0}.$$

Let k_1 *be the order of the generalised eigenvector* $\mathbf{x}_1$. *Then, we have*

$$\begin{aligned}\mathbf{0} = {} & \alpha_1 (A - \lambda_1 I)^{k_1 - 1} (A - \lambda_2 I)^n \ldots (A - \lambda_p I)^n \mathbf{x}_1 \\ & + \alpha_2 (A - \lambda_1 I)^{k_1 - 1} (A - \lambda_2 I)^n \ldots (A - \lambda_p I)^n \mathbf{x}_2 \\ & + \cdots + \alpha_p (A - \lambda_1 I)^{k_1 - 1} (A - \lambda_2 I)^n \ldots (A - \lambda_p I)^n \mathbf{x}_p.\end{aligned}$$

Observing that all the matrix powers commute, it follows from Proposition 4.14 that

$$\mathbf{0} = \alpha_1 (A - \lambda_2 I)^n \ldots (A - \lambda_p I)^n (A - \lambda_1 I)^{k_1 - 1} \mathbf{x}_1 + \mathbf{0} + \cdots + \mathbf{0}. \tag{4.13}$$

Notice that

$$(A-\lambda_1 I)(A-\lambda_1 I)^{k_1-1}\mathbf{x}_1=\mathbf{0},$$

from which follows that

$$A(A-\lambda_1 I)^{k_1-1}\mathbf{x}_1=\lambda_1(A-\lambda_1 I)^{k_1-1}\mathbf{x}_1.$$

Consequently, given any non-negative integer r *and* $\lambda\in\mathbb{K}$,

$$(A-\lambda I)^r(A-\lambda_1 I)^{k_1-1}\mathbf{x}_1=(\lambda_1-\lambda)^r(A-\lambda_1 I)^{k_1-1}\mathbf{x}_1.$$

Using this in (4.13),

$$\alpha_1(\lambda_1-\lambda_2)^n\dots(\lambda_1-\lambda_p)^n\underbrace{(A-\lambda_1 I)^{k_1-1}\mathbf{x}_1}_{\neq\mathbf{0}}=\mathbf{0}.$$

Hence, $\alpha_1=0$. *Repeating this reasoning sufficiently many times, we will have that* $\alpha_1,\alpha_2,\dots,\alpha_p=0$, *as required.*

We begin here a series of results which will lead us to reach our aim of showing that any given complex matrix is similar to some Jordan canonical form.

Proposition 4.16 *Let* A *be a matrix in* $M_n(\mathbb{C})$, *let* $\lambda_1,\lambda_2,\dots,\lambda_p\in\mathbb{C}$ *be all the distinct eigenvalues of* A *and let*

$$p(\lambda)=(\lambda_1-\lambda)^{n_1}(\lambda_2-\lambda)^{n_2}\dots(\lambda_p-\lambda)^{n_p}$$

be its characteristic polynomial. Then, for all $j=1,\dots,p$, $\dim G(\lambda_j)=n_j$ *and, for any generalised eigenvector* $\mathbf{x}\in G(\lambda_j)$, *the order of* $\mathbf{x}$ *is less than or equal to* n_j. *Moreover,*

$$\mathbb{C}^n=G(\lambda_1)\oplus G(\lambda_2)\oplus\dots\oplus G(\lambda_p) \tag{4.14}$$

In (4.14), we are taking a direct sum of finitely many subspaces. This is a generalisation of our definition in Chapter 3 of the direct sum of two subspaces. For details, see Section 8.3 of the Appendix.

Proof *It is clear that, once we prove that* $\dim G(\lambda_j)=n_j$, *any generalised eigenvector* $\mathbf{x}\in G(\lambda_j)$ *must have order at most* n_j, *since its Jordan chain consists of linearly independent vectors. We prove now that* $\dim G(\lambda_j)=n_j$.

We know, by Proposition 4.14, that $G(\lambda_j)=N(A-\lambda_j I)^n$. *On the other hand, as in* (4.7), *we know that* $(A-\lambda_j I)^n$ *is similar to the upper triangular matrix*

$$(U-\lambda_j I)^n=\begin{bmatrix}(U_1-\lambda_j I)^n & * & \dots & & \dots & \dots & *\\ & \ddots & & & & & \\ & & & (U_j-\lambda_j I)^n & * & \dots & *\\ & & & & \ddots & & \\ & & & & & & (U_p-\lambda_j I)^n\end{bmatrix},$$

Notice that $(U - \lambda_j I)^n$ *has all diagonal entries different from zero, except for the* n_j *diagonal entries of* $(U_j - \lambda_j I)^n$*, which are equal to* 0*. Since the strictly upper triangular matrix* $U_j - \lambda_j I$ *satisfies*

$$0 = (U_j - \lambda_j I)^{n_j} = (U_j - \lambda_j I)^n,$$

it follows that the null space of $(U - \lambda_j I)^n$ *has dimension* n_j*. Hence, by Theorem 4.2 (vi),*

$$\dim G(\lambda_j) = \dim N(A - \lambda_j I)^n = n_j.$$

Since $\sum_{j=1}^{p} \dim G(\lambda_j) = n$*, by Proposition 8.3 (ii), it suffices to show that with* $n > 1$ *the only vector lying in any given generalised eigenspace* $G(\lambda_j)$ *and also in the sum of the other generalised eigenspaces is* $\mathbf{0}$*. That is, we must show that, given any* $j = 1, \ldots, p$*,*

$$G(\lambda_j) \bigcap \sum_{l \in \{1,2,\ldots,p\} \setminus \{j\}} G(\lambda_j) = \{\mathbf{0}\}.$$

To simplify the notation, we prove only for $j = 1$*, but the proof is easily generalised for any* j*. Let* $\mathbf{x}_1 \in G(\lambda_1)$ *be such that*

$$\mathbf{x}_1 = \mathbf{x}_2 + \cdots + \mathbf{x}_p,$$

where $\mathbf{x}_2 + \cdots + \mathbf{x}_p \in \sum_{l \in \{1,2,\ldots,p\} \setminus \{j\}} G(\lambda_j)$*. Re-writing the above equality, we have*

$$\mathbf{x}_1 - \mathbf{x}_2 - \cdots - \mathbf{x}_p = \mathbf{0}.$$

It now follows from Lemma 4.2 that all these vectors must coincide with $\mathbf{0}$*.*

We can now make the following note:

Algebraic multiplicity versus geometric multiplicity

The algebraic multiplicity of an eigenvalue λ coincides with the dimension of the corresponding generalised eigenspace $G(\lambda)$.

The geometric multiplicity of an eigenvalue λ coincides with the dimension of the corresponding eigenspace $E(\lambda)$.

Now we are finally able to prove Proposition 4.1 for complex matrices.

Corollary 4.6 *Let* A *be a complex matrix with an eigenvalue* λ*. Then the algebraic multiplicity of* λ *is greater than or equal to its geometric multiplicity.*

Proof *Let* A *be a complex matrix. By Proposition 4.16,*

$$m_a(\lambda) = \dim G(\lambda) \geq \dim E(\lambda) = m_g(\lambda).$$

Example 4.8 *Let $\mathbf{x} \in \mathbb{K}^n$ be a generalised eigenvector of order k of an $n \times n$ matrix A associated with the eigenvalue λ and let*

$$\underbrace{(A-\lambda I)^{k-1}\mathbf{x},}_{\mathbf{u}_1} \quad \underbrace{(A-\lambda I)^{k-2}\mathbf{x},}_{\mathbf{u}_2} \quad \ldots, \quad \underbrace{(A-\lambda I)\mathbf{x},}_{\mathbf{u}_{k-1}} \quad \underbrace{\mathbf{x}}_{\mathbf{u}_k}.$$

be the corresponding Jordan chain. Since, for all $j = 2, \ldots k$,

$$\begin{aligned} \mathbf{u}_{j-1} &= (A-\lambda I)\mathbf{u}_j \\ &= A\mathbf{u}_j - \lambda\mathbf{u}_j, \end{aligned}$$

we have

$$A\mathbf{u}_j = \mathbf{u}_{j-1} + \lambda\mathbf{u}_j. \tag{4.15}$$

If, for example, $\mathbf{x}$ were a generalised eigenvector of order n, then

$$A\left[\mathbf{u}_1|\mathbf{u}_2|\ldots|\mathbf{u}_n\right] = \left[A\mathbf{u}_1|A\mathbf{u}_2| \quad \ldots|A\mathbf{u}_n\right] \tag{4.16}$$

$$A\left[\mathbf{u}_1|\mathbf{u}_2|\ldots|\mathbf{u}_n\right] = \left[\mathbf{u}_1|\mathbf{u}_2|\ldots|\mathbf{u}_n\right] \underbrace{\begin{bmatrix} \lambda & 1 & 0 & \ldots & 0 \\ 0 & \lambda & 1 & \ddots & 0 \\ 0 & 0 & \lambda & \ddots & 0 \\ \vdots & \vdots & \vdots & \ddots & 1 \\ 0 & 0 & \ldots & 0 & \lambda \end{bmatrix}}_{J_n(\lambda)}, \tag{4.17}$$

where $J_n(\lambda)$ is an $n \times n$ matrix. The matrix $J_n(\lambda)$ is said to be a ***Jordan block of degree n****. (If $n = 1$, $J_1(\lambda) = [\lambda]$.)*

If $S = \left[\mathbf{u}_1|\mathbf{u}_2|\ldots|\mathbf{u}_n\right]$, then (4.17) becomes

$$J_n(\lambda) = S^{-1}AS, \tag{4.18}$$

since, by Proposition 4.15, S is invertible.

An $n \times n$ matrix J is said to be a **Jordan canonical form** if

$$J = \begin{bmatrix} J_{n_1}(\lambda_1) & & & \\ & J_{n_2}(\lambda_2) & & \\ & & \ddots & \\ & & & J_{n_p}(\lambda_p) \end{bmatrix}, \qquad n_1+n_2+\cdots+n_p = n,$$

where the positive integers $n_1, n_2, \ldots, n_p$ may not be all distinct and the scalars $\lambda_1, \lambda_2, \ldots, \lambda_p$ may also be repeated.

Example 4.9 *The matrix*

$$\begin{bmatrix} 1 & 1 & 0 & 0 & 0 \\ 0 & 1 & 0 & 0 & 0 \\ 0 & 0 & 1 & 1 & 0 \\ 0 & 0 & 0 & 1 & 1 \\ 0 & 0 & 0 & 0 & 1 \end{bmatrix}$$

of (4.12) *is a* 5×5 *Jordan canonical form consisting of two Jordan blocks*

$$J_2(1) = \begin{bmatrix} 1 & 1 \\ 0 & 1 \end{bmatrix}$$

and

$$J_3(1) = \begin{bmatrix} 1 & 1 & 0 \\ 0 & 1 & 1 \\ 0 & 0 & 1 \end{bmatrix}$$

of degree 2 *and degree* 3*, respectively.*

It is worth making a note of the following facts.

> The spectrum of a Jordan block $J_n(\lambda)$ consists of the single eigenvalue λ whose algebraic multiplicity is n and geometric multiplicity is 1.

Exercise 4.5 *Consider the Jordan block*

$$J_3(\lambda) = \begin{bmatrix} \lambda & 1 & 0 \\ 0 & \lambda & 1 \\ 0 & 0 & \lambda \end{bmatrix}.$$

Which vectors in $\mathbb{C}^3$ *are generalised eigenvectors of order* 3*? Can a single Jordan chain be a basis of* $\mathbb{C}^3$*? If yes, find such a basis.*

Solution. We are looking for a vector $\boldsymbol{x} = (a, b, c)$ *for which* $(J_3 - \lambda I)^2\mathbf{x} \neq \mathbf{0}$*. (Notice that* $(J_3 - \lambda I)^3 = 0$*.) Calculations yield that the generalised eigen vectors of order* 3 *are* $\{(a, b, c) \in \mathbb{C}^3 : c \neq 0\}$*. Consider, for example, the vector* $(1, 1, 1)$*. Its Jordan chain is*

$$(1, 0, 0), (1, 1, 0), (1, 1, 1),$$

which is a basis of $\mathbb{C}^3$*. In fact, the Jordan chain of any generalised eigenvector of order* 3 *consists of three linearly independent vectors and, consequently, forms a basis of* $\mathbb{C}^3$*.*

Equality (4.18) is reminiscent of the diagonalisation of matrices. In fact, it amounts to saying that, in this particular case, A is similar to the Jordan block $J_n(\lambda)$. It is not always true that a given matrix A is similar to a single

Jordan block. However, we will see that any complex matrix A is similar to a Jordan canonical form.

In Proposition 4.16, we saw that, given an $n \times n$ complex matrix, $\mathbb{C}^n$ is the direct sum of its generalised eigenspaces. This hints (at the very least!) at the possibility of $\mathbb{C}^n$ having a basis consisting entirely of generalised eigenvectors. But even better, we can have a basis consisting of Jordan chains only. We are going to prove this, starting with nilpotent matrices.

The proofs of the next two results are inspired by those in [4].

Lemma 4.3 *Let A be an $m \times m$ nilpotent complex matrix. Then there exist generalised eigenvectors $\mathbf{x}_1, \mathbf{x}_2, \dots, \mathbf{x}_p$ such that*

$$\mathcal{B} = C(\mathbf{x}_1, 0) \cup C(\mathbf{x}_2, 0) \cup \dots \cup C(\mathbf{x}_p, 0)$$

is a basis of $\mathbb{C}^m$.

Proof *We prove this by induction on the size $n \times n$ of the matrix. If $n = 1$, then A is the 1×1 zero matrix and, therefore, the result is obvious since $E(0) = \mathbb{C}$.*

Assume now that the lemma holds for all sizes less than or equal to n. Let A be a square matrix of order $n+1$, and suppose that $A \neq 0$. (If $A = 0$, then $E(0) = \mathbb{C}^{n+1}$ and, hence, the result would hold trivially.)

Let $\mathcal{B}_1$ be a basis of the column space of A and let $\mathcal{B}$ be a basis of $\mathbb{C}^{n+1}$ containing $\mathcal{B}_1$ (see Theorem 3.5 (iii)).

Let S be a matrix whose first columns are the vectors of $\mathcal{B}_1$ followed by the remaining vectors of $\mathcal{B}$. Since $C(A)$ is A-invariant,

$$AS = S \begin{bmatrix} B & C \\ 0 & D \end{bmatrix},$$

where B is a $r \times r$ nilpotent matrix such that $1 \leq r \leq n$. Notice that r is the number of vectors in $\mathcal{B}_1$.

By the induction hypothesis, there exist vectors $\mathbf{y}_1, \mathbf{y}_2, \dots, \mathbf{y}_p \in \mathbb{C}^r$ such that

$$C_B(\mathbf{y}_1, 0) \cup C_B(\mathbf{y}_2, 0) \cup \dots \cup C_B(\mathbf{y}_p, 0) \tag{4.19}$$

is a basis of $\mathbb{C}^r$. Here the subscript B emphasises the fact that these Jordan chains are taken with respect to matrix B. Consider the vectors $\mathbf{w}_1, \mathbf{w}_2, \dots, \mathbf{w}_p$ in $C(A)$ such that

$$[\mathbf{w}_1]_{\mathcal{B}_1} = \mathbf{y}_1, \quad [\mathbf{w}_2]_{\mathcal{B}_1} = \mathbf{y}_2, \quad \dots, \quad [\mathbf{w}_p]_{\mathcal{B}_1} = \mathbf{y}_p.$$

In other words, for each $i = 1, 2, \dots, p$, we have

$$\mathbf{w}_i = S \begin{bmatrix} \mathbf{y}_i \\ \mathbf{0} \end{bmatrix}. \tag{4.20}$$

Notice that, given a positive integer k,

$$A^k\mathbf{w}_i = A^k S\begin{bmatrix}\mathbf{y}_i\\ \mathbf{0}\end{bmatrix} = S\begin{bmatrix}B & C\\ 0 & D\end{bmatrix}^k\begin{bmatrix}\mathbf{y}_i\\ \mathbf{0}\end{bmatrix} = S\begin{bmatrix}B^k & 0\\ 0 & 0\end{bmatrix}\begin{bmatrix}\mathbf{y}_i\\ \mathbf{0}\end{bmatrix}.$$

Consequently, for all $i = 1, 2, \dots, p$, we have that $\mathbf{w_i}$ is a generalised eigenvector whose Jordan chain $C(\mathbf{w}_i, 0)$ has the same length of $C_B(\mathbf{y}_i, 0)$.

Since, for all $i = 1, 2, \dots, p$, the vectors $\mathbf{w}_i$ lie in the column space of A, there exists $\mathbf{x}_i \in \mathbb{C}^{n+1}$ such that $\mathbf{w}_i = A\mathbf{x}_i$. Moreover, since the set (4.19) is linearly independent, it is easily seen using (4.20) that

$$C(\mathbf{w}_1, 0) \cup C(\mathbf{w}_2, 0) \cup \cdots \cup C(\mathbf{w}_p, 0) \tag{4.21}$$

is also linearly independent. Consider, for all $i = 1, 2, \dots, p$, the Jordan chain $C(\mathbf{x}_i, 0)$, whose length is that of $C(\mathbf{w}_i, 0)$ plus 1.

We prove next that the union of the Jordan chains

$$C(\mathbf{x}_1, 0) \cup C(\mathbf{x}_2, 0) \cup \cdots \cup C(\mathbf{x}_p, 0) \tag{4.22}$$

is a linearly independent set. Consider a linear combination of these vectors equal to $\mathbf{0}$. Multiplying this linear combination by A on the left, one gets a linear combination of (4.21) , which forces all the coefficients in the former linear combination to be equal to 0, except possibly the coefficients of

$$A^{k_1}\mathbf{x}_1, A^{k_2}\mathbf{x}_2, \dots, A^{k_p}\mathbf{x}_p,$$

where, for for all $i = 1, 2, \dots, p$, the number k_i is the order of the chain $C(\mathbf{w}_i, 0)$. However these coefficients correspond to the set consisting of

$$A^{k_1-1}\mathbf{w}_1, A^{k_2-1}\mathbf{w}_2, \dots, A^{k_p-1}\mathbf{w}_p,$$

which is linearly independent, since (4.21) is linearly independent.

Let

$$C(\mathbf{x}_1, 0) \cup C(\mathbf{x}_2, 0) \cup \cdots \cup C(\mathbf{x}_p, 0) \cup \{\mathbf{z}_1, \mathbf{z}_2, \dots, \mathbf{z}_l\} \tag{4.23}$$

be a basis of $\mathbb{C}^{n+1}$ (see Theorem 3.5 (iii)). Notice that there exist vectors $\mathbf{v}_i$ in the span of (4.22) such that, for each $j = 1, 2, \dots, l$,

$$A\mathbf{z}_j = A\mathbf{v}_j,$$

since the set (4.21) spans $C(A)$.

For all $j = 1, 2, \dots, l$, define $\mathbf{u}_{p+j}$ by $\mathbf{u}_{p+j} = \mathbf{z}_j - \mathbf{v}_j$, and observe that

$$A\mathbf{u}_{p+j} = \mathbf{0}.$$

In other words, each of the vectors $\mathbf{u}_{p+j}$ is an eigenvector of A. Moreover,

$$C(\mathbf{x}_1, 0) \cup C(\mathbf{x}_2, 0) \cup \cdots \cup C(\mathbf{x}_p, 0) \cup \{\mathbf{u}_{p+1}, \mathbf{u}_{p+2}, \dots, \mathbf{u}_{p+l}\}$$

is a linearly independent set which spans $\mathbb{C}^{n+1}$, since its cardinality is that of (4.23). This ends the proof.

Theorem 4.5 (Jordan canonical form) *Let A be a matrix in $M_n(\mathbb{C})$. Then, A is similar to a Jordan canonical form*

$$J = \begin{bmatrix} J_{n_1}(\mu_1) & & & \\ & J_{n_2}(\mu_2) & & \\ & & \ddots & \\ & & & J_{n_l}(\mu_l) \end{bmatrix}, \qquad n_1+n_2+\cdots+n_l = n,$$

where $\{\mu_1, \mu_2, \ldots, \mu_l\} = \sigma(A)$.

Notice that the positive integers $n_1, n_2, \ldots, n_l$ might not be all distinct and the scalars $\mu_1, \mu_2, \ldots, \mu_l$ might also be repeated.

Proof *Let $\lambda_1, \lambda_2, \ldots, \lambda_p$ be the (all distinct) eigenvalues of A. By* (4.14),

$$\mathbb{C}^n = G(\lambda_1) \oplus G(\lambda_2) \oplus \cdots \oplus G(\lambda_p),$$

and, by Proposition 4.11, each of these summands is A-invariant. Consequently, if one constructs a matrix S whose columns are formed by the juxtaposition of the bases of these summands, then

$$AS = SM,$$

where M is a block upper triangular matrix where the size of each block in M equals the dimension of the respective generalised eigenspace. It follows that

$$(A - \lambda_1 I)S = S(M - \lambda_1 I).$$

If $\mathbf{x} \in G(\lambda_1)$, then

$$S^{-1}(A - \lambda_1 I)S \begin{bmatrix} \mathbf{x}_{\mathcal{B}_1} \\ \mathbf{0} \end{bmatrix} = (M - \lambda_1 I) \begin{bmatrix} \mathbf{x}_{\mathcal{B}_1} \\ \mathbf{0} \end{bmatrix} = \begin{bmatrix} M_1 - \lambda_1 I & 0 \\ 0 & 0 \end{bmatrix} \begin{bmatrix} \mathbf{x}_{\mathcal{B}_1} \\ \mathbf{0} \end{bmatrix},$$

where $\mathcal{B}_1$ is the basis of $G(\lambda_1)$ formed by the columns of S lying in $G(\lambda_1)$, and M_1 is the block of M corresponding to this generalised eigenspace. Here, to simplify the notation, we denote by I the identity irrespective of the size of the matrix in hand.

It follows that $M_1 - \lambda_1 I$ is a $r_1 \times r_1$ nilpotent matrix. By Lemma 4.3, there exists a basis of $\mathbb{C}^{r_1}$ consisting of Jordan chains associated with $M_1 - \lambda_1 I$.

Notice that, if

$$(M_1 - \lambda_1 I)^{k_1 - 1}\mathbf{u}, (M_1 - \lambda_1 I)^{k_1 - 1}\mathbf{u}, \ldots, (M_1 - \lambda_1 I)\mathbf{u}, \mathbf{u}, \qquad (\mathbf{u} \in \mathbb{C}^{r_1})$$

is a Jordan chain associated with $M_1 - \lambda_1 I$, then

$$(A - \lambda_1 I)^{k_1 - 1}\mathbf{x}, (A - \lambda_1 I)^{k_1 - 2}\mathbf{x}, \ldots, (A - \lambda_1 I)\mathbf{x}, \mathbf{x}$$

is a Jordan chain for A, where

$$\mathbf{x} = S \begin{bmatrix} \mathbf{u} \\ \mathbf{0} \end{bmatrix}.$$

A reasoning similar to this can be applied to each of the generalised eigenspaces, leading to the construction of a basis of $\mathbb{C}^n$ consisting of Jordan chains. Let S be the matrix whose columns are formed by the juxtaposition of these chains associated with the matrix A. As in Example 4.8, we have that $AS = SJ$, where

$$J = \begin{bmatrix} J_{n_1}(\mu_1) & & & \\ & J_{n_2}(\mu_2) & & \\ & & \ddots & \\ & & & J_{n_l}(\mu_l) \end{bmatrix},$$

$\{\mu_1, \mu_2, \dots, \mu_l\} = \sigma(A)$, and $n_1 + n_2 + \cdots + n_l = n$. The proof is complete.

Notice that an eigenvalue appears in the diagonal of the Jordan canonical form as many times as its algebraic multiplicity.

Exercise 4.6 *Find a Jordan canonical form of the matrix in Example 4.6 and the corresponding similarity matrix.*

Solution. We already know that $(0, 1, 0)$ is a generalised eigenvector of order 2 whose Jordan chain is $\boldsymbol{u}_1 = (1, 0, 0), \boldsymbol{u}_2 = (0, 1, 0)$. Hence, $A = SJS^{-1}$ with

$$J = \begin{bmatrix} 2 & 1 & 0 \\ 0 & 2 & 0 \\ 0 & 0 & 1 \end{bmatrix}, \qquad S = \begin{bmatrix} 1 & 0 & 0 \\ 0 & 1 & -1 \\ 0 & 0 & 1 \end{bmatrix}.$$

Example 4.10 *Find a Jordan canonical form and a corresponding similarity matrix for*

$$A = \begin{bmatrix} 1 & 0 & 0 & 0 & 0 \\ 1 & 1 & 0 & 0 & 0 \\ 1 & 0 & 0 & -1 & 0 \\ -1 & 0 & 1 & 2 & 0 \\ 0 & 1 & 0 & 1 & 1 \end{bmatrix}.$$

It is easily obtained that the characteristic polynomial of A is $p(\lambda) = (1 - \lambda)^5$. Calculations will yield that a basis for the eigenspace $E(1) = N(A - I)$ is

$$\mathcal{B}_{E(1)} = \{(0, -1, -1, 1, 0), (0, 0, 0, 0, 1)\}.$$

Since the the geometric multiplicity is 2 and the dimension of the eigenspace of any Jordan block is 1, we have that the Jordan canonical form must have two Jordan blocks. Hence, the only two possibilities are

$$J = \begin{bmatrix} 1 & 0 & 0 & 0 & 0 \\ 0 & 1 & 1 & 0 & 0 \\ 0 & 0 & 1 & 1 & 0 \\ 0 & 0 & 0 & 1 & 1 \\ 0 & 0 & 0 & 0 & 1 \end{bmatrix} \quad or \quad J = \begin{bmatrix} 1 & 1 & 0 & 0 & 0 \\ 0 & 1 & 0 & 0 & 0 \\ 0 & 0 & 1 & 1 & 0 \\ 0 & 0 & 0 & 1 & 1 \\ 0 & 0 & 0 & 0 & 1 \end{bmatrix},$$

apart from the relative position of the Jordan blocks. Since we have

$$(A - I)^2 = \begin{bmatrix} 0 & 0 & 0 & 0 & 0 \\ 0 & 0 & 0 & 0 & 0 \\ 0 & 0 & 0 & 0 & 0 \\ 0 & 0 & 0 & 0 & 0 \\ 0 & 0 & 1 & 1 & 0 \end{bmatrix}$$

and $(A - I)^3 = 0$, *it follows that we have a Jordan chain of length* 3. *Consequently,*

$$J = \begin{bmatrix} 1 & 1 & 0 & 0 & 0 \\ 0 & 1 & 0 & 0 & 0 \\ 0 & 0 & 1 & 1 & 0 \\ 0 & 0 & 0 & 1 & 1 \\ 0 & 0 & 0 & 0 & 1 \end{bmatrix}$$

is the only possibility.

To construct a Jordan chain of length 3, *we need a vector in the null space of* $(A - I)^3$ *that does not lie in the null space of* $(A - I)^2$. *On the other hand, to get a Jordan chain of length* 2, *we need a vector in* $N((A - I)^2)$ *which is not an eigenvector.*

We start by finding a basis of $N((A - I)^2)$. *We choose a basis containing* $\mathcal{B}_{E(1)} = \{(0, -1, -1, 1, 0), (0, 0, 0, 0, 1)\}$ *to make things easier. Hence, a possible basis is*

$$\mathcal{B}_{N((A-I)^2)} = \{(1, 0, 0, 0, 0), (0, 1, 0, 0, 0), (0, -1, -1, 1, 0), (0, 0, 0, 0, 1)\}.$$

Picking a(ny) vector from the basis which is not an eigenvector, for example, $\boldsymbol{x} = (1, 0, 0, 0, 0)$, *we have the Jordan chain of length* 2

$$\boldsymbol{u}_1 = (A - I)\mathbf{x} = (0, 1, 1, -1, 0), \quad \boldsymbol{u}_2 = (1, 0, 0, 0, 0).$$

Choosing now a vector which is not a linear combination of the vectors of $\mathcal{B}_{N((A-I)^2)}$, *for example,* $\boldsymbol{y} = (0, 0, 0, 1, 0) = \boldsymbol{v}_3$, *we have*

$$\boldsymbol{v}_2 = (A - I)\mathbf{y} = (0, 0, -1, 1, 1), \quad \boldsymbol{v}_1 = (A - I)\mathbf{v}_2 = (0, 0, 0, 0, 1).$$

Hence,

$$S = \begin{bmatrix} 0 & 1 & 0 & 0 & 0 \\ 1 & 0 & 0 & 0 & 0 \\ 1 & 0 & 0 & -1 & 0 \\ -1 & 0 & 0 & 1 & 1 \\ 0 & 0 & 1 & 1 & 0 \end{bmatrix}.$$

Finally,

$$A = SJS^{-1} = \begin{bmatrix} 0 & 1 & 0 & 0 & 0 \\ 1 & 0 & 0 & 0 & 0 \\ 1 & 0 & 0 & -1 & 0 \\ -1 & 0 & 0 & 1 & 1 \\ 0 & 0 & 1 & 1 & 0 \end{bmatrix} \begin{bmatrix} 1 & 1 & 0 & 0 & 0 \\ 0 & 1 & 0 & 0 & 0 \\ 0 & 0 & 1 & 1 & 0 \\ 0 & 0 & 0 & 1 & 1 \\ 0 & 0 & 0 & 0 & 1 \end{bmatrix} \begin{bmatrix} 0 & 1 & 0 & 0 & 0 \\ 1 & 0 & 0 & 0 & 0 \\ 1 & 0 & 0 & -1 & 0 \\ -1 & 0 & 0 & 1 & 1 \\ 0 & 0 & 1 & 1 & 0 \end{bmatrix}^{-1}.$$

4.5 Exercises

EX 4.5.1. Consider the matrix

$$A = \begin{bmatrix} 1 & 2 & -3 \\ 0 & -1 & 1 \\ -1 & -1 & 2 \end{bmatrix}$$

Find which of the vectors below are eigenvectors of A. For those that are, find the corresponding eigenvalue.

a) $(5, -1, -4)$ b) $(1, -1, -1)$ c) $(0, 0, 0)$ d) $(1, -1, 0)$ e) $(-1, -1, -1)$

EX 4.5.2. Verify if $\lambda = 12$ is an eigenvalue of

$$\begin{bmatrix} 10 & 4 & 16 \\ -6 & 0 & 14 \\ 0 & 0 & 6 \end{bmatrix}.$$

If it is, find a corresponding eigenvector.

EX 4.5.3. Determine the characteristic polynomial, the spectrum, and a basis for the eigenspace for each of the following matrices.

a) $\begin{bmatrix} -3i & 0 \\ -8i & i \end{bmatrix}$ b) $\begin{bmatrix} 10 & 0 & 2 \\ 2 & 2 & 0 \\ -14 & 2 & 0 \end{bmatrix}$

EX 4.5.4. Which matrices are hermitian? For those which are, determine their spectrum (use technology).

a) $\begin{bmatrix} i & -8 \\ -8 & i \end{bmatrix}$ b) $\begin{bmatrix} 10 & 0 & 1+i \\ 0 & 2 & 3 \\ 1-i & 3 & 7 \end{bmatrix}$ c) $A = \begin{bmatrix} 2 & -3+2i & 1 \\ -3-2i & -1 & 1 \\ 1 & 1 & -5 \end{bmatrix}.$

EX 4.5.5. Diagonalise

$$A = \begin{bmatrix} 1 & 2 & 2 \\ -1 & -2 & -1 \\ 1 & 1 & 0 \end{bmatrix}$$

and calculate A^{21}.

EX 4.5.6. Let $\mathbf{v}$ be an eigenvector of an invertible matriz A corresponding to an eigenvalue λ. Show that the same vector $\mathbf{v}$ is an eigenvector of A^{-1} corresponding to the eigenvalue λ^{-1}.

EX 4.5.7. Let A be a matrix with an eigenvector $\boldsymbol{v}$ corresponding to an eigenvector λ. Show that λ^3 is an eigenvalue of A^3 and that $\boldsymbol{v}$ is an eigenvector of A^3 corresponding to λ^3.

EX 4.5.8. Consider the matrix

$$A = \begin{bmatrix} 1 & b \\ -b & -1 \end{bmatrix},$$

where b is a real number. Suppose that $0 \in \sigma(A)$ and that B is a 2×2 matrix. Consider the following assertions:

I) $\dim N(A) = 1$;

II) $0 \in \sigma(BA)$;

III) A is diagonalisable;

IV) A is not invertible.

Select all the correct assertions.

A) II, III B) I, IV C) I, III, IV D) I, II, IV

EX 4.5.9. Let A be a square matrix and let λ be an eigenvector of A. Show that $G(\lambda)$ is closed under vector addition and scalar multiplication.

EX 4.5.10. Prove directly that $(J_n(\lambda) - \lambda I)^n = 0$. Do not use any properties presented in Section 4.4.

EX 4.5.11. Determine which matrices are
a) a Jordan block b) a Jordan canonical form.

(i)

$$\begin{bmatrix} 5 & 1 & 0 & 0 & 0 & 0 \\ 0 & 5 & 1 & 0 & 0 & 0 \\ 0 & 0 & 5 & 0 & 0 & 0 \\ 0 & 0 & 0 & -2 & 1 & 0 \\ 0 & 0 & 0 & 0 & 4 & 0 \\ 0 & 0 & 0 & 0 & 0 & 2 \end{bmatrix}$$

(ii)

$$\begin{bmatrix} 5 & 1 & 0 & 0 & 0 & 0 \\ 0 & 5 & 1 & 0 & 0 & 0 \\ 0 & 0 & 5 & 0 & 0 & 0 \\ 0 & 0 & 0 & -2 & 0 & 0 \\ 0 & 0 & 0 & 0 & -2 & 0 \\ 0 & 0 & 0 & 0 & 0 & 4 \end{bmatrix}$$

(iii)

$$\begin{bmatrix} 5 & 1 & 0 & 0 & 0 & 0 \\ 0 & 5 & 1 & 0 & 0 & 0 \\ 0 & 0 & 5 & 0 & 0 & 0 \\ 0 & 0 & 0 & -2 & 0 & 0 \\ 0 & 0 & 0 & 0 & 4 & 0 \\ 0 & 0 & 0 & 0 & 0 & -2 \end{bmatrix}$$

(iv)

$$\begin{bmatrix} 3 & 1 & 0 & 0 \\ 0 & 3 & 1 & 0 \\ 0 & 0 & 3 & 0 \\ 0 & 0 & 0 & 1 \end{bmatrix}$$

EX 4.5.12. Determine all Jordan canonical forms whose characteristic polynomial is

$$p(\lambda) = (-2 - \lambda)^2(1 - \lambda).$$

EX 4.5.13. Find a Jordan canonical form J and an invertible matrix S such that $J = S^{-1}AS$ for each of the matrices A below.

$$\text{a) } A = \begin{bmatrix} 2 & 2 \\ -2 & -2 \end{bmatrix} \quad \text{b) } A = \begin{bmatrix} 3 & 1 & 1 \\ 0 & 3 & 2 \\ 0 & 0 & 1 \end{bmatrix} \quad \text{c) } A = \begin{bmatrix} 3 & -3 & 1 & 1 \\ 0 & 1 & 1 & 1 \\ 0 & 0 & 3 & 1 \\ 0 & 0 & 0 & 3 \end{bmatrix}.$$

4.6 At a Glance

The spectrum $\sigma(A)$ of a square matrix A consists of the roots of its characteristic polynomial $p(\lambda) = |A - \lambda I|$. These roots are called the eigenvalues of A.

The eigenspace $E(\lambda)$ corresponding to the eigenvalue λ is the solution set of the homogeneous system of linear equations $(A - \lambda I)\mathbf{x} = \mathbf{0}$. The non-zero vectors in $E(\lambda)$ are eigenvectors of A.

The algebraic multiplicity of an eigenvalue λ is, by definition, the multiplicity of that eigenvalue as a root of the characteristic polynomial of A. The dimension of $E(\lambda)$ is called the geometric multiplicity of λ.

A matrix A is diagonalisable if it is similar to a diagonal matrix, that is, $A = SDS^{-1}$, for some diagonal matrix D and invertible matrix S. A complex matrix is diagonalisable if and only if the geometric and algebraic multiplicities coincide for each eigenvalue. If this is the case, then it is very easy to calculate any power of A, for in these circumstances, $A^n = SD^nS^{-1}$.

Not all matrices are diagonalisable but, given a complex matrix A, it is always possible to find an upper triangular matrix J similar to A. That is, $A = SJS^{-1}$, where J is a Jordan canonical form of A.

The columns of the similarity matrix S are juxtapositions of corresponding Jordan chains. A Jordan chain corresponding to some eigenvalue consists of the so-called generalised eigenvectors. The dimension of a generalised eigenspace equals the algebraic multiplicity of the corresponding eigenvalue.

Chapter 5

Linear Transformations

5.1	Linear Transformations	176
5.2	Matrix Representations	179
5.3	Null Space and Image	185
	5.3.1 Linear transformations $T : \mathbb{K}^n \to \mathbb{K}^k$	185
	5.3.2 Linear transformations $T : U \to V$	187
5.4	Isomorphisms and Rank-nullity Theorem	189
5.5	Composition and Invertibility	191
5.6	Change of Basis	195
5.7	Spectrum and Diagonalisation	198
5.8	Exercises	201
5.9	At a Glance	203

The most significant functions between vector spaces are those which respect the linear structure inasmuch as they map sums of vectors to sums of their images and scalar multiplication of a vector to scalar multiplication of its image. These are the so-called linear transformations.

Linear transformations are intrinsically bound together with matrices. As will be shown in the sequel, once we fix bases both in the domain and the codomain of a linear transformation, there exists a one-to-one correspondence between linear transformations and a space of matrices. This runs so deep that one can think of linear transformations as matrices and vice-versa. This interchanging might come in very handy: linear transformations will benefit from our accumulated knowledge of matrices and, conversely, the theory of matrices might also gain from perceiving them as linear transformations.

The most relevant numbers here are the dimensions of the null space and the image of a linear transformation. The formula linking these dimensions is a very important result called the Rank-nullity Theorem (Theorem 5.2).

DOI: 10.1201/9781351243452-5

5.1 Linear Transformations

Definition 43 *Let U and V be vector spaces over $\mathbb{K}$. A function $T : U \to V$ is called a* **linear transformation** *if, for all $\boldsymbol{x}, \boldsymbol{y} \in U$ and $\alpha \in \mathbb{K}$,*

$$T(\boldsymbol{x}+\boldsymbol{y}) = T(\boldsymbol{x}) + T(\boldsymbol{y}), \tag{5.1}$$
$$T(\alpha\boldsymbol{x}) = \alpha T(\boldsymbol{x}). \tag{5.2}$$

In other words, a function $T : U \to V$ is a linear transformation if it is additive (5.1) and homogeneous (5.2). As usual, U is the domain of T and V is its codomain.

Proposition 5.1 *Let U and V be vector spaces over $\mathbb{K}$ and let $T : U \to V$ be a linear transformation. Then*

$$T(\mathbf{0}_U) = \mathbf{0}_V,$$

where $\mathbf{0}_U$ is the zero vector in U and $\mathbf{0}_V$ is the zero vector in V.

Proof *Let $\boldsymbol{x}$ be a vector in U. Hence, by* (5.1)*,*

$$T(\boldsymbol{x}) = T(\boldsymbol{x} + \mathbf{0}_U) = T(\boldsymbol{x}) + T(\mathbf{0}_U),$$

from which follows that $T(\mathbf{0}_U) = \mathbf{0}_V$.

Example 5.1 *Find which of the following functions are linear transformations.*

a) $T : \mathbb{R}^2 \to \mathbb{R}^2$ is a reflection relative to the x-axis.

b) $T : \mathbb{R}^3 \to \mathbb{R}^3$ is an orthogonal projection on the xy-plane.

c) $T : \mathbb{R}^2 \to \mathbb{R}^2$ is a translation by the vector $\boldsymbol{u} = (1, 0)$.

The function T in a) is defined, for all $\boldsymbol{x} = (x_1, x_2)$ in $\mathbb{R}^2$, by

$$T(x_1, x_2) = (x_1, -x_2).$$

Checking whether this is a linear transformation consists of verifying if both equalities (5.1) *and* (5.2) *hold. Beginning with* (5.2)*, let α be a real number and let (x_1, x_2) be a vector in $\mathbb{R}^2$. Then,*

$$\begin{aligned} T(\alpha(x_1, x_2)) &= T(\alpha x_1, \alpha x_2) \\ &= (\alpha x_1, -\alpha x_2) \\ &= \alpha(x_1, -x_2) \\ &= \alpha T(x_1, x_2), \end{aligned}$$

which shows that (5.2) *holds.*

Now let $\boldsymbol{x} = (x_1, x_2), \boldsymbol{y} = (y_1, y_2) \in \mathbb{R}^2$. *We have*

$$\begin{aligned} T((x_1, x_2) + (y_1, y_2)) &= T(x_1 + y_1, x_2 + y_2) \\ &= (x_1 + y_1, -(x_2 + y_2)) \\ &= (x_1, -x_2) + (y_1, -y_2) \\ &= T(x_1, x_2) + T(y_1, y_2). \end{aligned}$$

Hence,(5.1) *holds, yielding finally that* T *is a linear transformation on* $\mathbb{R}^2$.

Observing that the function T *in b) is defined, for all* $\boldsymbol{x} = (x_1, x_2, x_3) \in \mathbb{R}^3$, *by*

$$T(x_1, x_2, x_3) = (x_1, x_2, 0)$$

and proceeding analogously to a) above, it is easy to see that T *is a linear transformation.*

In case c) the analytic expression of T *is* $T(x_1, x_2) = (x_1 + 1, x_2)$, *from which we see that* $T(0, 0) = (1, 0)$. *Hence,* $T(0, 0) \neq (0, 0)$, *contradicting Proposition 3.2. Consequently, the function* T *is not a linear transformation.*

Example 5.2 *The image of a line segment.*
Consider the triangle whose vertices are $\boldsymbol{a} = (0, 0), \boldsymbol{b} = (1, 1)$, *and* $\mathbf{c} = (2, 0)$. *We are going to see how this triangle is mapped through the linear transformation* $T : \mathbb{R}^2 \to \mathbb{R}^2$ *satisfying* $T(\boldsymbol{b}) = (2, 1)$ *e* $T(\mathbf{c}) = (1, 0)$ *(see Figure 5.1).*

Let $\boldsymbol{x}$ *be a point lying in the line segment with end points* $\boldsymbol{b}$ *and* $\mathbf{c}$. *Then,*

$$\boldsymbol{x} = \mathbf{c} + \alpha(\boldsymbol{b} - \mathbf{c}), \qquad \alpha \in [0, 1], \tag{5.3}$$

from which follows that

$$T(\boldsymbol{x}) = T(\mathbf{c}) + \alpha(T(\boldsymbol{b}) - T(\mathbf{c})), \qquad \alpha \in [0, 1].$$

Hence, every point in this line segment is mapped by T *onto the line segment connecting the images* $T(\boldsymbol{b})$ *and* $T(\mathbf{c})$. *Similarly, it is easily seen that the same holds for the two remaining line segments connecting* $\boldsymbol{a}$ *and* $\boldsymbol{b}$ *and* $\boldsymbol{a}$ *and* $\mathbf{c}$, *respectively. It now follows that the image of the triangle is a triangle whose vertices are* $(0, 0), (2, 1), (1, 0)$. *Observe that, by Proposition 3.2,* $T(\boldsymbol{a}) = (0, 0)$.

A comment is in order: at this point is not clear why fixing the image of the two given points $\boldsymbol{b} = (1, 1)$ and $\mathbf{c} = (2, 0)$, we have a linear transformation satisfying these data, or indeed whether it is unique. However, there exists a unique linear transformation which satisfies these requirements as we will see later (see §5.2).

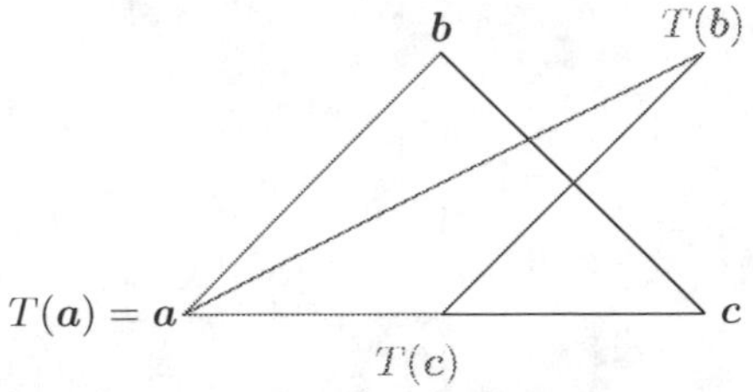

FIGURE 5.1: The image of a triangle.

What we saw in (5.3), allows for us to conclude that:

A line segment connecting two points is mapped through a linear transformation onto the line segment which connects the images of those points.

Example 5.3 *Show that the function* $T\colon \mathrm{M}_2(\mathbb{C}) \to \mathbb{C}$ *defined, for all* $A \in \mathrm{M}_2(\mathbb{C})$, *by* $T(A) = \mathrm{tr}(A)$, *is a linear transformation.*

As before, we must verify that (5.1) *and* (5.2) *hold. Let* A, B *be matrices in* $\mathrm{M}_2(\mathbb{C})$. *Then, by Proposition 1.13 (i),*

$$T(A + B) = \mathrm{tr}(A + B) = \mathrm{tr}\, A + \mathrm{tr}\, B = T(A) + T(B).$$

Hence,(5.1) *holds. Considering now* (5.2), *let* $\alpha \in \mathbb{C}$ *and let* A *be a matrix in* $\mathrm{M}_2(\mathbb{C})$. *Then, by Proposition 1.13 (ii),*

$$T(\alpha A) = \mathrm{tr}(\alpha A) = \alpha\, \mathrm{tr}\, A = \alpha T(A),$$

which concludes the proof that T *is a linear transformation.*

How to find if a function is a linear transformation

Let U and V be vector spaces over $\mathbb{K}$ and let $T\colon U \to V$ be a function. To see whether T is a linear transformation take the following two steps.

1. Check whether $T(\mathbf{0}_U) = \mathbf{0}_V$. If this is not true, then T is not a linear transformation and you can stop here. On the other hand, if $T(\mathbf{0}_U) = \mathbf{0}_V$, then one cannot conclude anything: T might or might not be a linear transformation. Proceed to step 2 below.

2. Verify, as in Examples 5.1, 5.3, if (5.1), (5.2) hold. If both hold, then T is a linear transformation. If at least one of them fails, then T is not a linear transformation.

Exercise 5.1 *Let A be a $k \times n$ matrix over $\mathbb{K}$. Show that the function $T\colon \mathbb{K}^n \to \mathbb{K}^k$ defined, for all $\boldsymbol{x} \in \mathbb{K}^n$ by $T\mathbf{x} = A\mathbf{x}$ is a linear transformation.*

5.2 Matrix Representations

Let $T\colon \mathbb{K}^n \to \mathbb{K}^k$ be a linear transformation and let $\mathcal{E}_n = (\boldsymbol{e}_1, \boldsymbol{e}_2, \dots, \boldsymbol{e}_n)$ be the ordered standard basis of $\mathbb{K}^n$ and $\mathcal{E}_k$ the ordered standard basis of $\mathbb{K}^k$. Then, for $\boldsymbol{x} = (x_1, x_2, \dots, x_n)$, we have

$$T(\boldsymbol{x}) = T(x_1\boldsymbol{e}_1 + x_2\boldsymbol{e}_2 + \cdots + x_n\boldsymbol{e}_n) = x_1T(\boldsymbol{e}_1) + x_2T(\boldsymbol{e}_2) + \cdots + x_nT(\boldsymbol{e}_n).$$

Denoting by $[\boldsymbol{x}]$ the vector column version of a vector $\boldsymbol{x}$,

$$[T(\boldsymbol{x})] = \underbrace{\big[[T(\boldsymbol{e}_1)] \;\mid\; [T(\boldsymbol{e}_2)] \;\mid\; \dots \;\mid\; [T(\boldsymbol{e}_n)]\big]}_{\text{matrix associated with } T} \begin{bmatrix} x_1 \\ x_2 \\ \vdots \\ x_n \end{bmatrix}. \tag{5.4}$$

Observe that, given a vector $\boldsymbol{u} \in \mathbb{R}^m$, we have that $[\boldsymbol{u}] = [\boldsymbol{u}]_{\mathcal{E}_m}$, where $[\boldsymbol{u}]_{\mathcal{E}_m}$ is the coordinate vector of $\boldsymbol{u}$ relative to the basis $\mathcal{E}_m$. Hence, we can now rewrite (5.4) as

$$[T(\boldsymbol{x})]_{\mathcal{E}_k} = \underbrace{\big[[T(\boldsymbol{e}_1)]_{\mathcal{E}_k} \;\mid\; [T(\boldsymbol{e}_2)]_{\mathcal{E}_k} \;\mid\; \dots \;\mid\; [T(\boldsymbol{e}_n)]_{\mathcal{E}_k}\big]}_{[T]_{\mathcal{E}_k,\mathcal{E}_n}}[\boldsymbol{x}]_{\mathcal{E}_n} \tag{5.5}$$

That is,

$$[T(\boldsymbol{x})]_{\mathcal{E}_k} = [T]_{\mathcal{E}_k,\mathcal{E}_n}[\boldsymbol{x}]_{\mathcal{E}_n}, \tag{5.6}$$

where $[T]_{\mathcal{E}_k,\mathcal{E}_n}$ is a $k \times n$ matrix called the **matrix of T relative to the standard bases of the domain $\mathbb{K}^n$ and codomain $\mathbb{K}^k$**. In what follows this matrix might be denoted simply by $[T]$.

Example 5.4 *Find the matrix which represents each of the linear transformations below relative to the relevant standard basis.*

a) The reflection relative to the x-axis in $\mathbb{R}^2$ (see Figure 5.2).

b) The orthogonal projection on the xy-plane in $\mathbb{R}^3$ (see Figure 5.3).

c) The counter-clockwise rotation in $\mathbb{R}^2$ around $(0,0)$ by an angle θ (see Figure 5.4). Find also an analytic expression for this linear transformation.

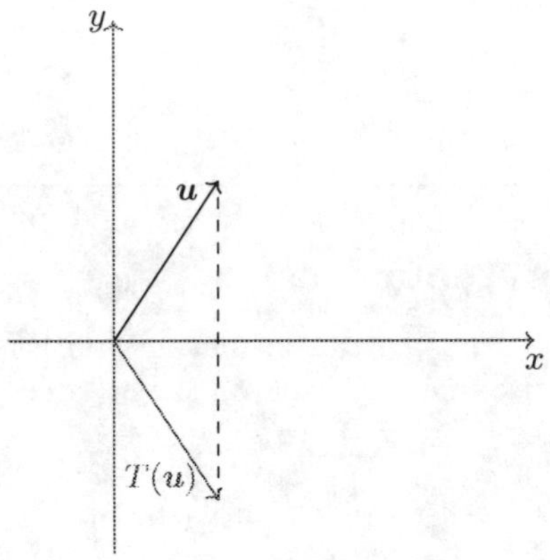

FIGURE 5.2: Reflection relative to the x-axis.

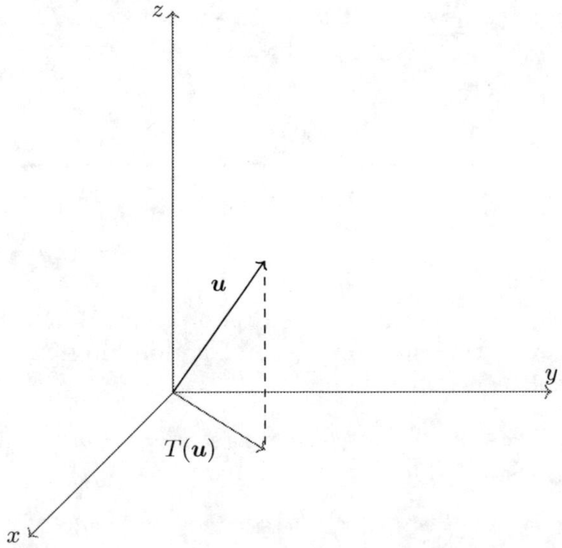

FIGURE 5.3: Orthogonal projection on the xy-plane.

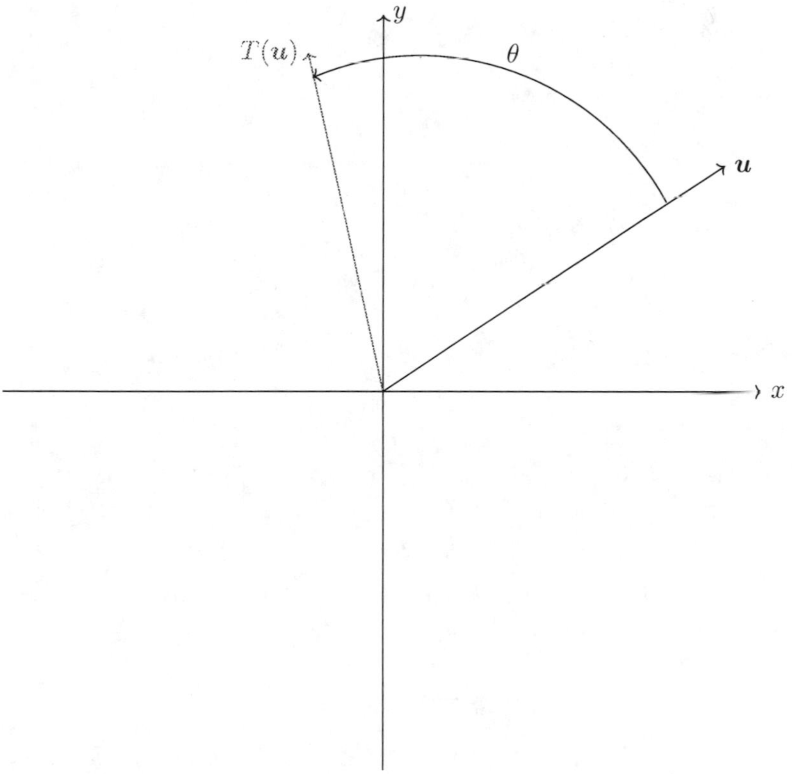

FIGURE 5.4: Counter-clockwise rotation around $(0, 0)$ by an angle θ.

The linear transformation in a) is defined on $\mathbb{R}^2$ *and, therefore, we are looking for the matrix* $[T]_{\mathcal{E}_2,\mathcal{E}_2}$ *which represents* T *relative to the standard basis of* $\mathbb{R}^2$. *By* (5.5),

$$[T]_{\mathcal{E}_2,\mathcal{E}_2} = \left[[T(\boldsymbol{e}_1)]_{\mathcal{E}_2} \quad | \quad [T(\boldsymbol{e}_2)]_{\mathcal{E}_2} \right] = \left[[T(\boldsymbol{e}_1)] \quad | \quad [T(\boldsymbol{e}_2)] \right].$$

We need then calculate $T(1,0)$ *and* $T(0,1)$. *The vector* $(1,0)$ *is fixed by* T *and* $T(0,1) = (0,-1)$. *Hence*

$$[T] = \begin{bmatrix} 1 & 0 \\ 0 & -1 \end{bmatrix}.$$

Consequently, given a vector $(x,y) \in \mathbb{R}^2$,

$$T(x,y) = \begin{bmatrix} 1 & 0 \\ 0 & -1 \end{bmatrix} \begin{bmatrix} x \\ y \end{bmatrix} = \begin{bmatrix} x \\ -y \end{bmatrix}. \tag{5.7}$$

In (5.7) *we made a slight abuse of notation as to be completely precise one should have written* $[T(x,y)]$ *instead of* $T(x,y)$. *However we shall adopt this throughout to simplify the notation.*

Analogously, we have in b)

$$[T]_{\mathcal{E}_3,\mathcal{E}_3} = \left[[T(\boldsymbol{e}_1)]_{\mathcal{E}_3} \quad | \quad [T(\boldsymbol{e}_2)]_{\mathcal{E}_3} \quad | \quad [T(\boldsymbol{e}_3)]_{\mathcal{E}_3} \right] = \begin{bmatrix} 1 & 0 & 0 \\ 0 & 1 & 0 \\ 0 & 0 & 0 \end{bmatrix}.$$

Hence, for $(x,y,z) \in \mathbb{R}^3$,

$$T(x,y,z) = \begin{bmatrix} 1 & 0 & 0 \\ 0 & 1 & 0 \\ 0 & 0 & 0 \end{bmatrix} \begin{bmatrix} x \\ y \\ z \end{bmatrix} = \begin{bmatrix} x \\ y \\ 0 \end{bmatrix}.$$

The columns of the matrix $[T]$ *associated with the rotation* T *in c) are the images of the vectors of the standard basis* $\mathcal{E}_2$. *Since* $T(1,0) = (\cos\theta, \sin\theta)$ *and* $T(0,1) = (-\sin\theta, \cos\theta)$, *we have*

$$[T] = \begin{bmatrix} \cos\theta & -\sin\theta \\ \sin\theta & \cos\theta \end{bmatrix}.$$

Hence, for all $\boldsymbol{x} = (x_1, x_2) \in \mathbb{R}^2$,

$$\begin{aligned} T(\mathbf{x}) &= [T] \begin{bmatrix} x_1 \\ x_2 \end{bmatrix} \\ &= \begin{bmatrix} \cos\theta & -\sin\theta \\ \sin\theta & \cos\theta \end{bmatrix} \begin{bmatrix} x_1 \\ x_2 \end{bmatrix} \\ &= \begin{bmatrix} x_1\cos\theta - x_2\sin\theta \\ x_1\sin\theta + x_2\cos\theta \end{bmatrix}. \end{aligned}$$

An analytic expression is

$$T(x_1, x_2) = (x_1\cos\theta - x_2\sin\theta, x_1\sin\theta + x_2\cos\theta).$$

We are now ready to make a general statement about the matrix representation of a linear transformation relative to fixed bases in its domain and codomain.

Theorem 5.1 *Let U and V be vector spaces over $\mathbb{K}$ with* $\dim U = n, \dim V = k$, *let $\mathcal{B}_1 = (\boldsymbol{b}_1, \boldsymbol{b}_2, \dots, \boldsymbol{b}_n)$ be a basis of U and $\mathcal{B}_2$ be a basis of V and let $T : U \to V$ be a linear transformation. Then, there exists uniquely a $k \times n$ matrix $[T]_{\mathcal{B}_2,\mathcal{B}_1}$ such that, for all $\boldsymbol{x} \in U$,*

$$[T(\boldsymbol{x})]_{\mathcal{B}_2} = [T]_{\mathcal{B}_2,\mathcal{B}_1}\mathbf{x}_{\mathcal{B}_1},$$

where $\mathbf{x}_{\mathcal{B}_1}$ is the coordinate vector of $\boldsymbol{x}$ relative to the basis $\mathcal{B}_1$ and $[T(\boldsymbol{x})]_{\mathcal{B}_2}$ is the coordinate vector of $T(\boldsymbol{x})$ relative to the basis $\mathcal{B}_2$. Moreover,

$$[T]_{\mathcal{B}_2,\mathcal{B}_1} = \left[[T(\boldsymbol{b}_1)]_{\mathcal{B}_2} \mid [T(\boldsymbol{b}_2)]_{\mathcal{B}_2} \mid \dots \mid [T(\boldsymbol{b}_n)]_{\mathcal{B}_2}\right]. \tag{5.8}$$

The matrix $[T]_{\mathcal{B}_2,\mathcal{B}_1}$ is called the **matrix of T relative to the basis $\mathcal{B}_1$ of the domain and the basis $\mathcal{B}_2$ of the codomain.**

Proof *Let $\boldsymbol{x}$ be a vector in U such that its coordinate vector $(\boldsymbol{x})_{\mathcal{B}_1} = (\alpha_1, \alpha_2, \dots, \alpha_n)$. Then*

$$T(\boldsymbol{x}) = T(\alpha_1\boldsymbol{b}_1 + \alpha_2\boldsymbol{b}_2 + \dots + \alpha_n\boldsymbol{b}_n)$$

and, by Proposition 3.5,

$$\begin{aligned}(T\boldsymbol{x})_{\mathcal{B}_2} &= T((\alpha_1\boldsymbol{b}_1 + \alpha_2\boldsymbol{b}_2 + \dots + \alpha_n\boldsymbol{b}_n))_{\mathcal{B}_2} \\ &= \alpha_1(T(\boldsymbol{b}_1))_{\mathcal{B}_2} + \alpha_2(T(\boldsymbol{b}_2))_{\mathcal{B}_2} + \dots + \alpha_n(T(\boldsymbol{b}_n))_{\mathcal{B}_2}.\end{aligned}$$

It follows that

$$[T\boldsymbol{x}]_{\mathcal{B}_2} = \left[[T(\boldsymbol{b}_1)]_{\mathcal{B}_2} \mid [T(\boldsymbol{b}_2)]_{\mathcal{B}_2} \mid \dots \mid [T(\boldsymbol{b}_n)]_{\mathcal{B}_2}\right]\begin{bmatrix}\alpha_1\\ \alpha_2\\ \vdots\\ \alpha_n\end{bmatrix},$$

which shows that

$$[T(\boldsymbol{x})]_{\mathcal{B}_2} = [T]_{\mathcal{B}_2,\mathcal{B}_1}\mathbf{x}_{\mathcal{B}_1}.$$

Suppose now that A is a $k \times n$ matrix such that $[T(\boldsymbol{x})]_{\mathcal{B}_2} = A\mathbf{x}_{\mathcal{B}_1}$. Then, for all $i = 1, \dots, n$,

$$\mathbf{0} = [T(\boldsymbol{b}_i)]_{\mathcal{B}_2} - [T(\boldsymbol{b}_i)]_{\mathcal{B}_2} = [T]_{\mathcal{B}_2,\mathcal{B}_1}(\mathbf{b}_i)_{\mathcal{B}_1} - A(\mathbf{b}_i)_{\mathcal{B}_1} = ([T]_{\mathcal{B}_2,\mathcal{B}_1} - A)\begin{bmatrix}0\\ \vdots\\ 0\\ 1\\ 0\\ \vdots\\ 0\end{bmatrix},$$

where the entry equal to 1 in the column vector lies in row i. Hence, for all $i = 1, \dots, n$, the columns i of both matrices $[T]_{\mathcal{B}_2,\mathcal{B}_1}$ and A coincide, yielding that $[T]_{\mathcal{B}_2,\mathcal{B}_1} = A$.

Observe that by means of Theorem 5.1, to the linear transformation

$$\begin{aligned} T\colon U &\to V \\ \boldsymbol{x} &\mapsto T(\boldsymbol{x}) \end{aligned}$$

corresponds another linear transformation

$$\begin{aligned} S\colon \mathbb{K}^n &\to \mathbb{K}^k \\ \mathbf{y} &\mapsto A\mathbf{y}, \end{aligned}$$

where $A = [T]_{\mathcal{B}_2,\mathcal{B}_1}$ (cf. Exercise 5.1). This linear transformation maps the coordinate vectors of the vectors in U relative to the basis $\mathcal{B}_1$ to the coordinate vectors of their images relative to the basis $\mathcal{B}_2$.

Example 5.5 *Let $T : \mathbb{P}_2 \to \mathbb{P}_1$ be the linear transformation $p \mapsto Dp$, where Dp is the derivative of the polynomial p. Find the matrix associated with T relative to the standard bases of $\mathbb{P}_2$ and $\mathbb{P}_1$. Use this matrix to calculate the image of the polynomial $p(t) = 1 - 2t + 3t^2$.*

Using (5.8),

$$[T]_{\mathcal{P}_1,\mathcal{P}_2} = \begin{bmatrix}(D1)_{\mathcal{P}_1} & (Dt)_{\mathcal{P}_1} & (Dt^2)_{\mathcal{P}_1}\end{bmatrix} = \begin{bmatrix}(0)_{\mathcal{P}_1} & (1)_{\mathcal{P}_1} & (2t)_{\mathcal{P}_1}\end{bmatrix} = \begin{bmatrix} 0 & 1 & 0 \\ 0 & 0 & 2 \end{bmatrix}.$$

To obtain the image of $p(t) = 1 - 2t + 3t^2$ using this matrix, we have

$$[T(1 - 2t + 3t^2)]_{\mathcal{P}_1} = \begin{bmatrix} 0 & 1 & 0 \\ 0 & 0 & 2 \end{bmatrix} \begin{bmatrix} 1 \\ -2 \\ 3 \end{bmatrix} = \begin{bmatrix} -2 \\ 6 \end{bmatrix},$$

from which follows that $T(1 - 2t + 3t^2) = -2 + 6t$.

How to find the matrix of a linear transformation

Let $T\colon U \to V$ be a linear transformation and let $\mathcal{B}_1$ and $\mathcal{B}_2$ be bases of U and V, respectively.

1. Find the images of the vectors in the ordered basis $\mathcal{B}_1$ of U.
2. Find the coordinate vectors of these images relative to the ordered basis $\mathcal{B}_2$.
3. Construct a matrix whose columns are these vectors respecting the order of $\mathcal{B}_1$. This matrix is $[T]_{\mathcal{B}_2,\mathcal{B}_1}$.

A comment: Now it is clear how we can guarantee the existence and uniqueness of a linear transformation in Example 5.2 just by giving the image of two vectors. Indeed we can because these two vectors form a basis of $\mathbb{R}^2$.

5.3 Null Space and Image

Definition 44 *Let U and V be vector spaces over $\mathbb{K}$ and let $T\colon U \to V$ be a linear transformation. The* **null space** *or* **kernel** *$N(T)$ of the linear transformation T is the subspace of U defined by*

$$N(T) = \{\boldsymbol{x} \in U \, : \, T(\boldsymbol{x}) = \mathbf{0}_V\}.$$

The **image** *$I(T)$ of the linear transformation T is the subspace of V defined by*

$$I(T) = \{T(\boldsymbol{x}) \in V \, : \, \boldsymbol{x} \in U\}.$$

In other words, the null space is the subset of U consisting of the vectors in the domain mapped by T to the zero vector of V, and the image of T is the subset of V consisting of the images of all the vectors in U.

Exercise 5.2 *Verify that, given a linear transformation $T\colon U \to V$, the sets $N(T)$ and $I(T)$ are subspaces of U and V, respectively.*

5.3.1 Linear transformations $T : \mathbb{K}^n \to \mathbb{K}^k$

We begin by finding a way of calculating the null space and image of a linear transformation $T : \mathbb{K}^n \to \mathbb{K}^k$. We have

$$N(T) = \{\boldsymbol{x} \in \mathbb{K}^n \, : \, T(\boldsymbol{x}) = \mathbf{0}_{\mathbb{K}^k}\}.$$

Let $A = [T]_{\mathcal{E}_k,\mathcal{E}_n}$ be the matrix of T relative to the standard bases of $\mathbb{K}^n$ and $\mathbb{K}^k$. Then,

$$T(\boldsymbol{x}) = \mathbf{0} \qquad \text{if and only if} \qquad A\mathbf{x} = \mathbf{0}.$$

Hence,

$$N(T) = N(A).$$

That is, the null space of T is the null space of the matrix A which represents T when one considers the standard bases in the domain and codomain.

As to the image $I(T)$, we have by definition that

$$I(T) = \{T(\boldsymbol{x}) \in \mathbb{K}^k \, : \, \boldsymbol{x} \in \mathbb{K}^n\}.$$

Observing that

$$[T(\boldsymbol{x})] = A\mathbf{x},$$

the image $I(T)$ is found by obtaining all the linear combinations $A\mathbf{x}$ of the columns of A, i.e.,

$$I(T) = C(A).$$

Hence,

$$I(T) = \text{span}(\{T(\boldsymbol{e}_1), T(\boldsymbol{e}_2), \ldots, T(\boldsymbol{e}_n)\}),$$

which shows that $\{T(\boldsymbol{e}_1), T(\boldsymbol{e}_2), \ldots, T(\boldsymbol{e}_n)\}$ is a spanning set for $I(T)$, although not necessarily a basis of $I(T)$.

Example 5.6 *Find the null spaces and the images of the the linear transformations of Example 5.4.*

In a), the null space of the matrix $[T]$ *is* $\{(0,0)\}$ *and, hence* $N(T) = \{(0,0)\}$.

The image $I(T)$ *is the subspace generated by the columns of* $[T]$. *Since this corresponds to the vectors* $(1,0), (0,-1)$, *it follows that* $I(T) = \mathbb{R}^2$. *The image of this linear transformation coincides with the codomain* $\mathbb{R}^2$, *that is,* T *is a surjective function. Indeed, it is also an injective function.*

Similarly, in b) we have that $N(T)$ *coincides with the* z*-axis and the image* $I(T)$ *is the* xy*-plane, which shows that* T *is neither surjective nor injective.*

As to c), observe that

$$\det[T] = \cos^2\theta + \sin^2\theta = 1$$

and, consequently, the matrix is invertible. Hence, $N(T) = N([T]) = \{(0,0)\}$. *Maybe it is not so easy to justify that* $I(T) = \mathbb{R}^2$. *The next theorem will help.*

Let T be a linear transformation. The **nullity** of T, $\text{nul}(T)$, is the dimension of its null space and the **rank** of T, $\text{rank}\,(T)$, is the dimension of its image.

Theorem 5.2 (Rank-nullity theorem) *Let* $T\colon \mathbb{K}^n \to \mathbb{K}^k$ *be a linear transformation. Then*

$$n = \text{nul}(T) + \text{rank}\,(T).$$

Before proving the result, we apply it to c) in Example 5.6. We have that

$$2 = \dim N(T) + \dim I(T) = 0 + \dim I(T),$$

yielding that $I(T)$ is a subspace having dimension 2 within $\mathbb{R}^2$. Hence, the only possibility is $I(T) = \mathbb{R}^2$, that is, T is surjective.

Proof *Let A be the $k \times n$ matrix of T relative to the bases $\mathcal{E}_n$ and $\mathcal{E}_k$. Then, by Proposition 3.11 and Theorem 3.6,*

$$\begin{aligned} n &= \dim N(A) + \operatorname{rank}(A) \\ &= \dim N(A) + \dim C(A) \\ &= \dim N(T) + \dim I(T) \ , \end{aligned}$$

which concludes the proof.

5.3.2 Linear transformations $T : U \to V$

Let U and V be vector spaces over $\mathbb{K}$, let $\mathcal{B}_1 = (\boldsymbol{u}_1, \boldsymbol{u}_2, \dots, \boldsymbol{u}_n)$ be a basis of U and let $\mathcal{B}_2 = (\boldsymbol{v}_1, \boldsymbol{v}_2, \dots, \boldsymbol{v}_k)$ be a basis of V. Let $T\colon U \to V$ be a linear transformation. We are now interested in devising a way to determine the null space and the image of such a general linear transformation by means of its representing matrix relative to the bases of the domain and the codomain, as we did in §5.3.1, for the particular kind of linear transformations under scrutiny in that part of the book.

Tackling firstly the null space of $T\colon U \to V$, we are then interested in determining the vectors $\boldsymbol{x} \in U$ such that $T(\boldsymbol{x}) = \mathbf{0}$. If $A = [T]_{\mathcal{B}_2,\mathcal{B}_1}$ is the matrix of T relative to the bases of the domain and codomain, we have

$$T(\boldsymbol{x}) = \mathbf{0} \qquad \text{if and only if} \qquad [T(\boldsymbol{x})]_{\mathcal{B}_2} = \mathbf{0}.$$

It follows that $T(\boldsymbol{x}) = \mathbf{0}$ if and only if

$$A[\boldsymbol{x}]_{\mathcal{B}_1} = \mathbf{0},$$

where this equality corresponds to determining the null space of A. Hence, once $N(A) \subseteq \mathbb{K}^n$ is determined, we have the coordinate vectors relative to the basis $\mathcal{B}_1$ of the vectors in the null space $N(T)$ of the linear transformation T, i.e.,

$$N(T) = \{\alpha_1 \boldsymbol{u}_1 + \alpha_2 \boldsymbol{u}_2 + \cdots + \alpha_n \boldsymbol{u}_n \ : \ (\alpha_1, \alpha_2, \dots, \alpha_n) \in N(A)\} \subseteq U. \quad (5.9)$$

Example 5.7 *Find the null space of the linear transformation of Example 5.5.*

The null space of

$$[T] = \begin{bmatrix} 0 & 1 & 0 \\ 0 & 0 & 2 \end{bmatrix}$$

is

$$N([T]) = \{(\alpha, 0, 0) \colon \alpha \in \mathbb{R}\}.$$

Hence, the polynomials in $N(T)$ are those of the form

$$p(t) = \alpha + 0t + 0t^2,$$

i.e., $N(T)$ consists of the polynomials of degree 0.

As defined before, the image $I(T)$ of $T\colon U \to V$ is the subspace of V defined by

$$I(T) = \{T(\boldsymbol{x}) \,:\, \boldsymbol{x} \in U\}.$$

Consequently, given the matrix

$$[T(\boldsymbol{x})]_{\mathcal{B}_2} = A[\boldsymbol{x}]_{\mathcal{B}_1}$$

of T, we want to obtain all the linear combinations $A[\boldsymbol{x}]_{B_1}$ of the columns of A. In other words, the set of the coordinate vectors relative to the basis $\mathcal{B}_2$ of the images of the vectors in U coincides with the column space $C(A)$. Observe that $C(A) \subseteq \mathbb{K}^k$. Then

$$I(T) = \{\alpha_1\boldsymbol{v}_1 + \alpha_2\boldsymbol{v}_2 + \cdots + \alpha_k\boldsymbol{v}_n \,:\, (\alpha_1, \alpha_2, \ldots, \alpha_k) \in C(A)\}.$$

Example 5.8 *Find the image $I(T)$ of the linear transformation of Example 5.5.*

Now we have

$$C([T]) = C\left(\begin{bmatrix} 0 & 1 & 0 \\ 0 & 0 & 2 \end{bmatrix}\right) = \mathbb{R}^2.$$

It follows that

$$I(T) = \{\alpha_1 1 + \alpha_2 t\colon (\alpha_1, \alpha_2) \in \mathbb{R}^2\} = \mathbb{P}_1.$$

How to find the null space and the image of a linear transformation

Let $T\colon U \to V$ be a linear transformation, let $\mathcal{B}_1, \mathcal{B}_2$ be bases of U and V, respectively, and let A be the matrix representing T relative to these bases.

1. Find the null space and the column space of A.
2. Find all the vectors in U whose coordinate vectors lie in $N(A)$: this is the null space of T.
3. Find all the vectors in V whose coordinate vectors lie in $C(A)$: this is the image of T.

5.4 Isomorphisms and Rank-nullity Theorem

An isomorphism is a particularly important type of linear transformation, the word coming from the ancient Greek: *equal (iso) form (morph)*. An isomorphism between two spaces means that as linear spaces they can be thought of as being the 'same'. Without explicitly saying it, we have been dealing with isomorphisms since Chapter 3. Indeed, the linear transformation which maps each vector in a space of dimension n to its coordinate vector in $\mathbb{K}^n$ is an isomorphism. And how convenient it might be replacing general vectors by its coordinates in $\mathbb{K}^n$, as we already experienced!

We have seen linear transformations that are injective or surjective as functions. Now we define formally the concepts of injective and surjective linear transformations, which amount exactly to being so as functions.

Definition 45 *A linear transformation $T\colon U \to V$ between vector spaces U, V over $\mathbb{K}$ is said to be* **injective** *if, for all $\boldsymbol{x}, \boldsymbol{y} \in U$,*

$$\boldsymbol{x} \neq \boldsymbol{y} \Rightarrow T(\boldsymbol{x}) \neq T(\boldsymbol{y})$$

or, equivalently, if

$$T(\boldsymbol{x}) = T(\boldsymbol{y}) \Rightarrow \boldsymbol{x} = \boldsymbol{y}.$$

Notice that

$$T(\boldsymbol{x}) = T(\boldsymbol{y})$$

if and only if

$$T(\boldsymbol{x} - \boldsymbol{y}) = 0 \Leftrightarrow \boldsymbol{x} - \boldsymbol{y} \in N(T).$$

Hence,we see that

$$T(\boldsymbol{x} + N(T)) = \{T(\boldsymbol{x})\},$$

where we define $\boldsymbol{x} + N(T)$ by

$$\boldsymbol{x} + N(T) = \{\boldsymbol{x} + \boldsymbol{z} \colon \boldsymbol{z} \in N(T)\}.$$

That is to say, any vector in U which differs from $\boldsymbol{x}$ by a vector in $N(T)$ is mapped under T to the image of $\boldsymbol{x}$.

We have proved the following proposition.

Proposition 5.2 *Let U, V be a vector spaces over $\mathbb{K}$ and let $T\colon U \to V$ be a linear transformation. Then T is injective if and only if $N(T) = \{0\}$.*

Definition 46 *A linear transformation $T\colon U \to V$ between vector spaces U, V over $\mathbb{K}$ is said to be* **surjective** *if $I(T) = V$. If T is injective and surjective, then T is called a* **bijective** *linear transformation or an* **isomorphism** *between the space U and V. In this case, U and V are said to be* **isomorphic** *vector spaces.*

Example 5.9 *The reflection a) in Example 5.4 is an isomorphism, the projection b) is not, and the rotation c) is an isomorphism.*

Proposition 5.3 *Let U be a vector space over $\mathbb{K}$ and let $\mathcal{B} = (\boldsymbol{b}_1, \boldsymbol{b}_2, \ldots, \boldsymbol{b}_n)$ be a basis of U. The linear transformation $T : U \to \mathbb{K}^n$ defined by*

$$\boldsymbol{x} \mapsto (\boldsymbol{x})_{\mathcal{B}}$$

is an isomorphism.

Proof *See Proposition 3.5.*

We shall now state a more general version of Theorem 5.2 valid for any linear transformation between vector spaces.

Theorem 5.3 (Rank-nullity theorem) *Let U, V be vector spaces over $\mathbb{K}$, let $\dim U = n$ and let $T \colon U \to V$ a linear transformation. Then.*

$$n = \mathrm{nul}(T) + \mathrm{rank}\,(T).$$

Proof *Let $\mathcal{B}_1, \mathcal{B}_2$ be bases of U and V, respectively, and let $A = [T]_{\mathcal{B}_2,\mathcal{B}_1}$. By Proposition 5.3, $N(T)$ and $N(A)$ are isomorphic and, hence, have the same dimension. Similarly, $I(T)$ and $C(A)$ are isomorphic and, therefore, have the same dimension.*

It follows from Theorem 5.2 that

$$\dim N(T) + \dim I(T) = \dim N(A) + \dim C(A) = n.$$

Proposition 5.4 *Let $T : U \to V$ be a linear transformation between vector spaces U, V over $\mathbb{K}$ such that $\dim U = n = \dim V$. The following are equivalent.*

(i) T is injective.

(ii) T is surjective.

(iii) T is an isomorphism.

Proof *We show firstly that (i) $\Rightarrow$ (ii). Suppose that T is injective. Then, by Theorem 5.3,*

$$n = \dim N(T) + \dim I(T)$$

from which follows that $n = 0 + \dim I(T)$. Hence, $I(T) = V$, i.e., T is surjective.

Suppose now that T is surjective. Then,

$$n = \dim N(T) + n,$$

yielding that $\dim N(T) = 0$. Hence, (ii) $\Rightarrow$ (iii).

The implication (iii) $\Rightarrow$ (i) is clear. The proof is concluded.

This proposition tells us that for a linear transformation T between vector spaces of the same dimension n, either T is both surjective and injective or is neither. Hence, when checking if such a transformation is an isomorphism, it suffices to determine if it is injective or surjective. It is not necessary to go the extra length of verifying both properties.

This is not so for linear transformations between spaces of different dimensions.

Exercise 5.3 *Give examples of linear transformations which are (i) injective but not surjective, (ii) surjective but not injective.*

5.5 Composition and Invertibility

Let U, V, and W be vector spaces over $\mathbb{K}$ and let $T\colon U \to V$ and $S\colon V \to W$ be linear transformations. Let ST be the composite function defined by

$$\begin{aligned} ST\colon U &\to W \\ \boldsymbol{x} &\mapsto S(T(\boldsymbol{x})) \,. \end{aligned}$$

That is,

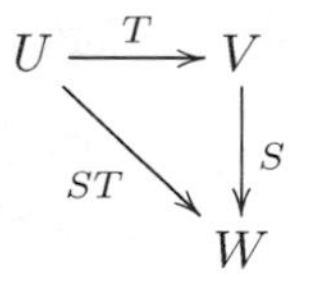

Proposition 5.5 *Let U, V, and W be vector spaces over $\mathbb{K}$ and let $T\colon U \to V$ and $S\colon V \to W$ be linear transformations. Then the function $ST\colon U \to W$ is a linear transformation.*

Proof *Exercise.*

Suppose that U, V, and W are vector spaces over $\mathbb{K}$ whose dimensions are

$$\dim U = n, \qquad \dim V = p, \text{ and } \qquad \dim W = k$$

and let $\mathcal{B}_U$, $\mathcal{B}_V$, and $\mathcal{B}_W$ be bases of U, V, and W, respectively. Let $A = [T]_{\mathcal{B}_V,\mathcal{B}_U}$ and $B = [S]_{\mathcal{B}_W,\mathcal{B}_V}$ be the matrices of the linear transformations T, S relative to the fixed bases in U, V, W. We have, for all $\boldsymbol{x} \in U$,

$$\begin{aligned} [(ST)(\boldsymbol{x})]_{\mathcal{B}_W} &= [S(T(\boldsymbol{x}))]_{\mathcal{B}_W} \\ &= B[(T(\boldsymbol{x}))]_{\mathcal{B}_V} \\ &= BA[\boldsymbol{x}]_{\mathcal{B}_U} \,. \end{aligned}$$

Hence, the matrix $[ST]_{\mathcal{B}_W,\mathcal{B}_U}$ of the linear transformation ST relative to the basis $\mathcal{B}_U$ in the domain and the basis $\mathcal{B}_W$ in the codomain is

$$[ST]_{\mathcal{B}_W,\mathcal{B}_U} = BA.$$

We have just proved the following proposition.

Proposition 5.6 *Let U, V, and W be vector spaces over $\mathbb{K}$ and let $T\colon U \to V$ and $S\colon V \to W$ be linear transformations. Then, the matrix $[ST]_{\mathcal{B}_W,\mathcal{B}_U}$ of the linear transformation ST relative to the basis $\mathcal{B}_U$ in the domain and the basis $\mathcal{B}_W$ in the codomain is*

$$[ST]_{\mathcal{B}_W,\mathcal{B}_U} = [S]_{\mathcal{B}_W,\mathcal{B}_V}\,[T]_{\mathcal{B}_V,\mathcal{B}_U}.$$

In other words, given two linear transformations T and S and bases in their domains and codomains, the matrix of the composite ST relative to the relevant bases is the product of the matrices of T and S in the same order. That is, if $A = [T]_{\mathcal{B}_V,\mathcal{B}_U}$ and $B = [S]_{\mathcal{B}_W,\mathcal{B}_V}$, then

$$[ST]_{\mathcal{B}_W,\mathcal{B}_U} = BA.$$

Composition of linear transformations corresponds to matrix multiplication (in an appropriate order) of the matrices associated with those linear transformations.

In terms of the coordinate vectors, we have the scheme

$$
\begin{array}{ccc}
\mathbb{K}^n & \xrightarrow{\ A\ } & \mathbb{K}^p \\
 & \searrow & \downarrow B \\
BA=[ST]_{\mathcal{B}_W,\mathcal{B}_U} & & \mathbb{K}^k
\end{array}
\qquad
\begin{array}{ccc}
[\boldsymbol{x}]_{B_U} & \xmapsto{\ A\ } & [T(\boldsymbol{x})]_{B_V} \\
 & \searrow & \downarrow B \\
BA=[ST]_{B_W,B_U} & & [S(T(\boldsymbol{x}))]_{B_W}
\end{array}
$$

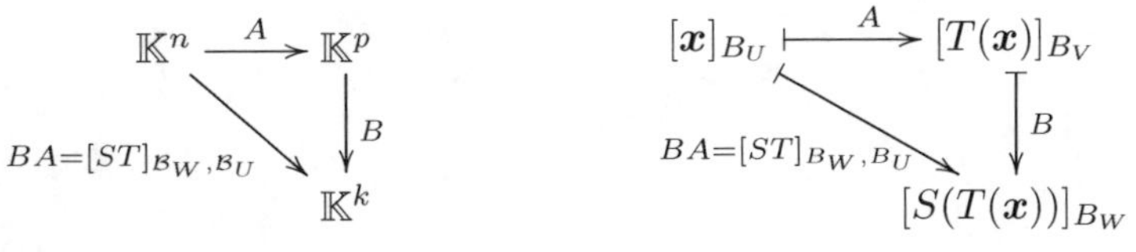

Example 5.10 *Let T be a reflection relative to the x-axis in $\mathbb{R}^2$ and let S be a counter-clockwise rotation around the origin in $\mathbb{R}^2$ by an angle $\theta = \frac{\pi}{2}$. Find:*

a) the matrix of ST relative to the standard basis $\mathcal{E}_2$;

b) an analytic expression for ST.

a) The matrices of T and S relative to the standard basis of $\mathbb{R}^2$ are

$$[T] = \begin{bmatrix} 1 & 0 \\ 0 & -1 \end{bmatrix}, \qquad [S] = \begin{bmatrix} 0 & -1 \\ 1 & 0 \end{bmatrix}.$$

Hence,

$$[ST] = \begin{bmatrix} 0 & -1 \\ 1 & 0 \end{bmatrix} \begin{bmatrix} 1 & 0 \\ 0 & -1 \end{bmatrix} = \begin{bmatrix} 0 & 1 \\ 1 & 0 \end{bmatrix}$$

b) As to the analytic expression of the composite transformation, we have

$$ST(x, y) = \begin{bmatrix} 0 & 1 \\ 1 & 0 \end{bmatrix} \begin{bmatrix} x \\ y \end{bmatrix} = \begin{bmatrix} y \\ x \end{bmatrix},$$

that is, $ST(x, y) = (y, x)$.

Let U, V be vector spaces over $\mathbb{K}$ and let $\mathcal{B}_U, \mathcal{B}_V$ be the corresponding bases. Let $T\colon U \to V$ be an isomorphism, i.e., T is a bijective linear transformation. Observe that, by Theorem 5.3, $\dim U = \dim V$. Consequently, the matrix $[T]_{\mathcal{B}_V, \mathcal{B}_U}$ is a square matrix. Under this circumstances, it is possible to define the inverse function T^{-1} of T by

$$\begin{aligned} T^{-1}\colon V &\to U \\ \boldsymbol{y} &\mapsto \boldsymbol{x}, \end{aligned}$$

where $\boldsymbol{y} = T(\boldsymbol{x})$.

Proposition 5.7 *Let* U, V *be vector spaces over* $\mathbb{K}$ *and let* $T\colon U \to V$ *be an isomorphism. Then* T *is invertible and* T^{-1} *is a linear transformation.*

Proof *Since* T *is bijective, the function* T^{-1} *is well-defined. In fact,* T^{-1} *is additive and homogeneous because*

$$\begin{aligned} T^{-1}(\boldsymbol{y}_1 + \boldsymbol{y}_2) &= T^{-1}(T(\boldsymbol{x}_1) + T(\boldsymbol{x}_2)) \\ &= T^{-1}(T(\boldsymbol{x}_1 + \boldsymbol{x}_2)) \\ &= \boldsymbol{x}_1 + \boldsymbol{x}_2 = T^{-1}(\boldsymbol{y}_1) + T^{-1}(\boldsymbol{y}_2), \end{aligned}$$

and

$$T^{-1}(\alpha \boldsymbol{y}) = T^{-1}(\alpha T(\boldsymbol{x})) = T^{-1}(T(\alpha \boldsymbol{x})) = \alpha \boldsymbol{x} = \alpha T^{-1}(\boldsymbol{y}).$$

Here $\boldsymbol{y}, \boldsymbol{y}_1, \boldsymbol{y}_2 \in V$ *are arbitrary and such that* $\boldsymbol{y} = T(\boldsymbol{x}), \boldsymbol{y}_1 = T(\boldsymbol{x}_1), \boldsymbol{y}_2 = T(\boldsymbol{x}_2)$, *for some* $\boldsymbol{x}, \boldsymbol{x}_1, \boldsymbol{x}_2 \in X$, *and* $\alpha \in \mathbb{K}$.

Let $B = [T^{-1}]_{\mathcal{B}_U, \mathcal{B}_V}$ be the matrix of T^{-1} relative to the bases $\mathcal{B}_U$ of the codomain U and $\mathcal{B}_V$ of the domain V and let $A = [T]_{\mathcal{B}_V, \mathcal{B}_U}$ the matrix of T relative to the same bases. We have, for all $\boldsymbol{x} \in U$, that

$$\begin{aligned} [(T^{-1}T)(\boldsymbol{x})]_{\mathcal{B}_U} &= [T^{-1}(T\boldsymbol{x})]_{\mathcal{B}_U} \\ &= [T^{-1}]_{\mathcal{B}_U, \mathcal{B}_V} [(T\boldsymbol{x})]_{\mathcal{B}_V} \\ &= [T^{-1}]_{\mathcal{B}_U, \mathcal{B}_V} [T]_{\mathcal{B}_V, \mathcal{B}_U} [\boldsymbol{x}]_{\mathcal{B}_U} \\ &= BA[\boldsymbol{x}]_{\mathcal{B}_U} . \end{aligned}$$

Observing that $T^{-1}T$ is the **identity** I_U on U, i.e, the linear transformation which assigns to each $\boldsymbol{x} \in U$ the image $I_U(\boldsymbol{x}) = \boldsymbol{x}$, it follows that $BA = I$. Hence, we have the following.

Proposition 5.8 *Let U, V be vector spaces over $\mathbb{K}$ and let $\mathcal{B}_U$, and $\mathcal{B}_V$ be bases of U and V, respectively. Let $T: U \to V$ be an isomorphism. Then, the matrix of T^{-1} relative to the bases $\mathcal{B}_U$ and $\mathcal{B}_V$ is*

$$[T^{-1}]_{\mathcal{B}_U,\mathcal{B}_V} = \left([T]_{\mathcal{B}_V,\mathcal{B}_U}\right)^{-1}. \tag{5.10}$$

The matrix associated with the inverse T^{-1} of a linear transformation T is the inverse of the matrix associated with T.

Example 5.11 *Let U be the subspace of the real polynomials $\mathbb{P}_2$ defined by*

$$U = \{a_1 t + a_2 t^2 \colon a_1, a_2 \in \mathbb{R}\},$$

and let $T: U \to \mathbb{P}_1$ be the linear transformation which assigns to each polynomial its derivative. Find the matrix of T^{-1} relative to the basis $\mathcal{B}_U = (t, t^2)$ of U and the standard basis of $\mathbb{P}_1$.

Solution: The matrix of T relative to this bases is

$$[T]_{\mathcal{P}_1,\mathcal{B}_U} = \begin{bmatrix} 1 & 0 \\ 0 & 2 \end{bmatrix}.$$

Notice that we were informed that T was invertible to start with and, consequently, did not have to check this. However, we can see it by ourselves now, since $[T]_{\mathcal{P}_1,\mathcal{B}_U}$ is an invertible matrix.

It now follows that

$$[T^{-1}]_{\mathcal{B}_U,\mathcal{P}_1} = \begin{bmatrix} 1 & 0 \\ 0 & \frac{1}{2} \end{bmatrix},$$

yielding

$$[T^{-1}(b_0 + b_1 t)]_{\mathcal{B}_U} = [T^{-1}]_{\mathcal{B}_U,\mathcal{P}_1} \begin{bmatrix} b_0 \\ b_1 \end{bmatrix} = \begin{bmatrix} b_0 \\ \frac{1}{2} b_1 \end{bmatrix}.$$

Hence, $T^{-1}(b_0 + b_1 t) = b_0 t + \frac{1}{2} b_1 t^2$.

We have been integrating!

The next exercise suggests a different way of proving (5.10).

Exercise 5.4 *Let $T: U \to V$ be an isomorphism between the vector spaces U, V and consider the inverse linear transformation*

$$\begin{aligned} T^{-1} \colon V &\to U \\ \boldsymbol{y} &\mapsto \boldsymbol{x}, \end{aligned}$$

where $\boldsymbol{y} = T\boldsymbol{x}$.

a) Denoting by A the matrix $[T]_{B_V,B_U}$ and supposing that $\dim U = n$, show that A is a $n \times n$ invertible matrix. (Hint: use Theorem 3.6 and Theorem 5.3.)

b) Use a) and the equality $\boldsymbol{y} = T\boldsymbol{x}$ to write $\boldsymbol{x}$ as a function of $\boldsymbol{y}$ and conclude that $[T^{-1}]_{B_U,B_V} = A^{-1}$.

5.6   Change of Basis

Sometimes we have to choose the bases carefully in order for the matrix associated with a linear transformation to be easily determined or, at least, as easily as possible (see Example 5.12). A change of basis might then be in order. But how do matrices of a given linear transformation relate with each other?

We begin to tackle this question with linear transformations on $\mathbb{K}^n$.

Let $T : \mathbb{K}^n \to \mathbb{K}^n$ be a linear transformation and let $\mathcal{B} = (\boldsymbol{b}_1, \boldsymbol{b}_2, \ldots, \boldsymbol{b}_n)$ be a basis of $\mathbb{K}^n$.

Given a vector $\boldsymbol{x}$ in $\mathbb{K}^n$, the coordinate vector of the image of $\boldsymbol{x}$ can be determined both using the matrix $A = [T]_{\mathcal{E}_n,\mathcal{E}_n}$ and the matrix $B = [T]_{\mathcal{B},\mathcal{B}}$. We have

$$[T(\boldsymbol{x})]_{\mathcal{E}_n} = A[\boldsymbol{x}]_{\mathcal{E}_n}, \qquad [T(\boldsymbol{x})]_{\mathcal{B}} = B[\boldsymbol{x}]_{\mathcal{B}}.$$

On the other hand, we can see in the scheme below that $[T(\boldsymbol{x})]_{\mathcal{E}_n}$ also can be calculated using B and the change of basis matrices between $\mathcal{E}_n$ and $\mathcal{B}$. In fact,

$$\begin{aligned}
[T(\boldsymbol{x})]_{\mathcal{E}_n} &= M^{-1}_{\mathcal{B} \leftarrow \mathcal{E}_n}[T(\boldsymbol{x})]_{\mathcal{B}} \\
&= M^{-1}_{\mathcal{B} \leftarrow \mathcal{E}_n} B[\boldsymbol{x}]_{\mathcal{B}} \\
&= M^{-1}_{\mathcal{B} \leftarrow \mathcal{E}_n} B M_{\mathcal{B} \leftarrow \mathcal{E}_n}[\boldsymbol{x}]_{\mathcal{E}_n}
\end{aligned}$$

$$\begin{array}{ccc}
[\boldsymbol{x}]_{\mathcal{E}_n} & \xmapsto{A} & [T(\boldsymbol{x})]_{\mathcal{E}_n} \\
\downarrow{\scriptstyle M_{\mathcal{B} \leftarrow \mathcal{E}_n}} & & \uparrow{\scriptstyle M_{\mathcal{E}_n \leftarrow \mathcal{B}}} \\
[\boldsymbol{x}]_{\mathcal{B}} & \xmapsto{B} & [T(\boldsymbol{x})]_{\mathcal{B}}
\end{array}$$

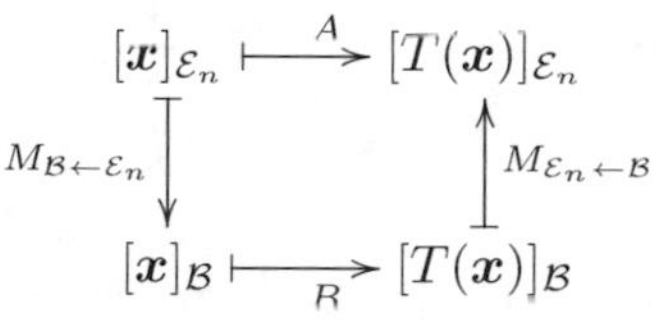

Hence

$$A = M^{-1}_{\mathcal{B} \leftarrow \mathcal{E}_n} B M_{\mathcal{B} \leftarrow \mathcal{E}_n}.$$

Example 5.12 Let $T : \mathbb{R}^2 \to \mathbb{R}^2$ be the reflection relative to the straight line whose equation is $y = 2x$. Find an analytic expression for T.

If you try to determine the matrix $[T]_{\mathcal{E}_2,\mathcal{E}_2}$ of the linear transformation T relative to the standard basis, you will be faced with difficulties. Indeed, you will need to find the images $T(1,0), T(0,1)$ and this is by no means immediate. However, there are vectors whose images are particularly easy to find.

For example, all vectors lying in the line $y = 2x$ are unchanged by the linear transformation. Hence, e.g., $T(1,2) = (1,2)$.

If one looks at the straight line that goes through the origin and is perpendicular to the one given before, then again we have an immediate way of finding the images of the vectors lying in that straight line. That is the case of the vector $(-2,1)$, and we have $T(-2,1) = (2,-1)$.

If we choose the basis $\mathcal{B} = ((1,2),(-2,1))$ of $\mathbb{R}^2$, then

$$[T]_{\mathcal{B},\mathcal{B}} = \begin{bmatrix} 1 & 0 \\ 0 & -1 \end{bmatrix},$$

since

$$T(1,2) = 1(1,2) + 0(-2,1), \qquad T(-2,1) = 0(1,2) - 1(-2,1).$$

It follows that

$$\begin{aligned}
\big[T(x,y)\big]_{\mathcal{E}_2} &= M^{-1}_{\mathcal{B}\leftarrow\mathcal{E}_2}[T]_{\mathcal{B},\mathcal{B}}M_{\mathcal{B}\leftarrow\mathcal{E}_2}\begin{bmatrix} x \\ y \end{bmatrix} \\
&= M_{\mathcal{E}_2\leftarrow\mathcal{B}}\begin{bmatrix} 1 & 0 \\ 0 & -1 \end{bmatrix} M^{-1}_{\mathcal{E}_2\leftarrow\mathcal{B}}\begin{bmatrix} x \\ y \end{bmatrix} \\
&= \begin{bmatrix} 1 & -2 \\ 2 & 1 \end{bmatrix}\begin{bmatrix} 1 & 0 \\ 0 & -1 \end{bmatrix}\begin{bmatrix} 1 & -2 \\ 2 & 1 \end{bmatrix}^{-1}\begin{bmatrix} x \\ y \end{bmatrix} \\
&= \begin{bmatrix} 1 & -2 \\ 2 & 1 \end{bmatrix}\begin{bmatrix} 1 & 0 \\ 0 & -1 \end{bmatrix}\begin{bmatrix} 1/5 & 2/5 \\ -2/5 & 1/5 \end{bmatrix}^{-1}\begin{bmatrix} x \\ y \end{bmatrix} \\
&= 1/5\begin{bmatrix} -3x+4y \\ 4x+3y \end{bmatrix}
\end{aligned}$$

Finally, we have that

$$T(x,y) = \left(-\frac{3}{5}x + \frac{4}{5}y, \frac{4}{5}x + \frac{3}{5}y\right).$$

Consider now the general case of an arbitrary vector space U endowed with two bases $\mathcal{B}_1 = (\boldsymbol{b}_1, \boldsymbol{b}_2, \ldots, \boldsymbol{b}_n)$ and $\mathcal{B}_2 = (\boldsymbol{v}_1, \boldsymbol{v}_2, \ldots, \boldsymbol{v}_n)$ and let $A = [T]_{\mathcal{B}_1,\mathcal{B}_1}$ and $B = [T]_{\mathcal{B}_2,\mathcal{B}_2}$. A reasoning similar to that above yields

$$\begin{aligned}
\big[T\big]_{\mathcal{B}_1,\mathcal{B}_1} &= M_{\mathcal{B}_1\leftarrow\mathcal{B}_2}[T]_{\mathcal{B}_2,\mathcal{B}_2}M_{\mathcal{B}_2\leftarrow\mathcal{B}_1} \\
M_{\mathcal{B}_1\leftarrow\mathcal{B}_2} &= M^{-1}_{\mathcal{B}_2\leftarrow\mathcal{B}_1}
\end{aligned}$$

$$\begin{array}{ccc}
[\boldsymbol{x}]_{\mathcal{B}_1} & \stackrel{A}{\longmapsto} & [T(\boldsymbol{x})]_{\mathcal{B}_1} \\
{\scriptstyle M_{\mathcal{B}_2\leftarrow\mathcal{B}_1}}\downarrow & & \uparrow{\scriptstyle M_{\mathcal{B}_1\leftarrow\mathcal{B}_2}} \\
[\boldsymbol{x}]_{\mathcal{B}_2} & \underset{B}{\longmapsto} & [T(\boldsymbol{x})]_{\mathcal{B}_2}
\end{array}$$

Hence

$$\begin{aligned}[T(\boldsymbol{x})]_{\mathcal{B}_1} &= M^{-1}_{\mathcal{B}_2\leftarrow\mathcal{B}_1}[T(\boldsymbol{x})]_{\mathcal{B}_2}\\ &= M^{-1}_{\mathcal{B}_2\leftarrow\mathcal{B}_1}B[\boldsymbol{x}]_{\mathcal{B}_2}\\ &= M^{-1}_{\mathcal{B}_2\leftarrow\mathcal{B}_1}BM_{\mathcal{B}_2\leftarrow\mathcal{B}_1}[\boldsymbol{x}]_{\mathcal{B}_1}\end{aligned}$$

from which follows that

$$A = M^{-1}_{\mathcal{B}_2\leftarrow\mathcal{B}_1}BM_{\mathcal{B}_2\leftarrow\mathcal{B}_1}. \tag{5.11}$$

Example 5.13 *Let U be the subspace of the 2×2 complex matrices having null trace, and let $T\colon U\to U$ be the transposition. Consider the bases of U*

$$\mathcal{B}_1 = \left(\begin{bmatrix}1 & 0\\ 0 & -1\end{bmatrix}, \begin{bmatrix}0 & 1\\ 0 & 0\end{bmatrix}, \begin{bmatrix}0 & 0\\ 1 & 0\end{bmatrix}\right),$$

$$\mathcal{B}_2 = \left(\begin{bmatrix}i & 0\\ 0 & -i\end{bmatrix}, \begin{bmatrix}0 & -i\\ i & 0\end{bmatrix}, \begin{bmatrix}0 & i\\ i & 0\end{bmatrix}\right).$$

Suppose that $A = [T]_{\mathcal{B}_1,\mathcal{B}_1}$ and $B = [T]_{\mathcal{B}_2,\mathcal{B}_2}$. Then,

$$A = \begin{bmatrix}1 & 0 & 0\\ 0 & 0 & 1\\ 0 & 1 & 0\end{bmatrix}, \qquad B = \begin{bmatrix}1 & 0 & 0\\ 0 & -1 & 0\\ 0 & 0 & 1\end{bmatrix}.$$

Check that

$$A = M^{-1}_{\mathcal{B}_2\leftarrow\mathcal{B}_1}BM_{\mathcal{B}_2\leftarrow\mathcal{B}_1} = \begin{bmatrix}i & 0 & 0\\ 0 & \frac{i}{2} & -\frac{i}{2}\\ 0 & -\frac{i}{2} & -\frac{i}{2}\end{bmatrix}^{-1} B \begin{bmatrix}i & 0 & 0\\ 0 & \frac{i}{2} & -\frac{i}{2}\\ 0 & -\frac{i}{2} & -\frac{i}{2}\end{bmatrix}.$$

In (5.11), we proved the following result.

Proposition 5.9 *Let U be a vector space over $\mathbb{K}$, let $T\colon U\to U$ be a linear transformation and let $\mathcal{B}_1, \mathcal{B}_2$ be bases of U. Then, the matrices $[T]_{\mathcal{B}_1,\mathcal{B}_1}$ and $[T]_{\mathcal{B}_2,\mathcal{B}_2}$ are similar.*

To end this section, we tackle finally the general case where $T\colon U\to V$ is a linear transformation whose domain and codomain do not have to coincide.

Let U, V be vector spaces over $\mathbb{K}$, let $\mathcal{B}_1$ and $\mathcal{B}'_1$ be bases of U and let $\mathcal{B}_2$ and $\mathcal{B}'_2$ be bases of V.

Analogously to what has been done so far in this section, we have

$$[T]_{\mathcal{B}_2,\mathcal{B}_1} = M_{\mathcal{B}_2\leftarrow\mathcal{B}'_2}[T]_{\mathcal{B}'_2,\mathcal{B}'_1}M_{\mathcal{B}'_1\leftarrow\mathcal{B}_1}$$

$$\begin{array}{ccc} [\boldsymbol{x}]_{\mathcal{B}_1} & \xrightarrow{A} & [T(\boldsymbol{x})]_{\mathcal{B}_2}\\ {\scriptstyle M_{\mathcal{B}'_1\leftarrow\mathcal{B}_1}}\downarrow & & \uparrow{\scriptstyle M_{\mathcal{B}_2\leftarrow\mathcal{B}'_2}}\\ [\boldsymbol{x}]_{\mathcal{B}'_1} & \xrightarrow[B]{} & [T(\boldsymbol{x})]_{\mathcal{B}'_2}\end{array}$$

where $A = [T]_{\mathcal{B}_2,\mathcal{B}_1}$ and $B = [T]_{\mathcal{B}'_2,\mathcal{B}'_1}$.

It follows that

$$\begin{aligned}[T(\boldsymbol{x})]_{\mathcal{B}_2} &= M_{\mathcal{B}_2 \leftarrow \mathcal{B}_2'}[T(\boldsymbol{x})]_{\mathcal{B}_2'} \\ &= M_{\mathcal{B}_2 \leftarrow \mathcal{B}_2'} B [\boldsymbol{x}]_{\mathcal{B}_1'} \\ &= M_{\mathcal{B}_2 \leftarrow \mathcal{B}_2'} B M_{\mathcal{B}_1' \leftarrow \mathcal{B}_1} [\boldsymbol{x}]_{\mathcal{B}_1}\end{aligned}$$

Hence

$$A = M_{\mathcal{B}_2 \leftarrow \mathcal{B}_2'} B M_{\mathcal{B}_1' \leftarrow \mathcal{B}_1}.$$

How to relate the matrices of a linear transformation in different bases

Let U, V be vector spaces over $\mathbb{K}$, let $\mathcal{B}_1$ and $\mathcal{B}_1'$ be bases of U and let $\mathcal{B}_2$ and $\mathcal{B}_2'$ be bases of V. Consider a linear transformation $T\colon U \to V$. If $A = [T]_{\mathcal{B}_2, \mathcal{B}_1}$ and $B = [T]_{\mathcal{B}_2', \mathcal{B}_1'}$, then

- Determine the change of basis matrices $M_{\mathcal{B}_1' \leftarrow \mathcal{B}_1}$ and $M_{\mathcal{B}_2 \leftarrow \mathcal{B}_2'}$.
- We have
$$A = M_{\mathcal{B}_2 \leftarrow \mathcal{B}_2'} B M_{\mathcal{B}_1' \leftarrow \mathcal{B}_1}.$$
Given one of the matrices A or B, one can use this equality to obtain the other.
- Determine one (the easiest) of these matrices, should you not have one of them to start with. One can now obtain the other matrix using the above equality.

5.7 Spectrum and Diagonalisation

We can extend to linear transformations what was done for matrices in Chapter 4. Not at all a surprise! Indeed, we saw already that there exists a bijection between linear transformations and matrices. To be precise, if $\mathcal{B}_1, \mathcal{B}_2$ are bases of U and V, respectively, and $\dim U = n, \dim V = k$, then the mapping

$$\mathcal{T}(U, V) \ni T \quad \mapsto \quad [T]_{\mathcal{B}_2, \mathcal{B}_1} \in M_{k,n}(\mathbb{K}),$$

where $\mathcal{T}(U, V) = \{T\colon U \to V\colon T \text{ is a linear transformation}\}$, is a bijection. In fact, this mapping is itself a linear transformation. Notice that the set $\{T\colon U \to V\colon T \text{ is a linear transformation}\}$ is a vector space over $\mathbb{K}$.

Proposition 5.10 *Let U, V be vector spaces over $\mathbb{K}$ and let $\mathcal{B}_1, \mathcal{B}_2$ be bases of U and V, respectively. Let T, S be linear transformations from U to V, let A and B be the matrices representing T and S relative to the bases $\mathcal{B}_1, \mathcal{B}_2$, and let $\alpha \in \mathbb{K}$. Then:*

(i) $T + S$ and αT are linear transformations;

(ii) $[T + S]_{\mathcal{B}_2,\mathcal{B}_1} = A + B$ and $[\alpha T]_{\mathcal{B}_2,\mathcal{B}_1} = \alpha A$.

Moreover, $\{T : U \to V : T$ is a linear transformation$\}$ is a vector space over $\mathbb{K}$.

Proof *Exercise. Here, for $\boldsymbol{x} \in U$ and $\alpha \in \mathbb{K}$, we define as usual $(T + S)(\boldsymbol{x}) = T(\boldsymbol{x}) + S(\boldsymbol{x})$ and $(\alpha T)(\boldsymbol{x}) = \alpha T(\boldsymbol{x})$.*

Definition 47 *Let U be a vector space over $\mathbb{K}$ and let $T : U \to U$ be a linear transformation. A non-zero vector $\boldsymbol{x} \in U$, is called an* ***eigenvector*** *of T if there exists $\lambda \in \mathbb{K}$ such that*

$$T(\boldsymbol{x}) = \lambda \boldsymbol{x}.$$

Under these conditions, λ is called the ***eigenvaluc*** *of A associated with $\boldsymbol{x}$.*

The **spectrum of** T, denoted by $\sigma(T)$, is the set of eigenvalues of the linear transformation T.

For an eigenvalue λ, the **eigenspace** $E(\lambda)$ corresponding to the eigenvalue λ is the null space of the linear transformation $T - \lambda I$, where I is the identity map on U. That is,

$$E(\lambda) = N(T - \lambda I).$$

Given a basis $\mathcal{B}$ of U and the matrix $A = [T]_{\mathcal{B},\mathcal{B}}$ of T relative to the basis $\mathcal{B}$, we have

$$T(\boldsymbol{x}) - \lambda \boldsymbol{x} = \mathbf{0} \qquad \text{if and only if} \qquad (A - \lambda I)[\boldsymbol{x}]_{\mathcal{B}} = \mathbf{0}\ .$$

Hence,

$$\sigma(T) = \sigma(A),$$

and

$$E(\lambda) = \{\boldsymbol{x} \in U : (\boldsymbol{x})_{\mathcal{B}} \in N(A - \lambda I)\}.$$

We have just proved the following.

Proposition 5.11 *Let U be a vector space over $\mathbb{K}$, let $\mathcal{B}$ be a basis of U and let $T : U \to U$ be a linear transformation. If $A = [T]_{\mathcal{B},\mathcal{B}}$ is the matrix of T relative to the basis $\mathcal{B}$, then $\sigma(T) = \sigma(A)$, and*

$$E(\lambda) = \{\boldsymbol{x} \in U : (\boldsymbol{x})_{\mathcal{B}} \in N(A - \lambda I)\}.$$

Example 5.14 *Find the eigenvalues and the eigenvectors of the reflection relative to the straight line in* $\mathbb{R}^2$ *whose cartesian equation is* $y = x$.

Solution: The matrix of this reflection relative to the basis $\mathcal{B} = ((1,1),(1,-1))$ *of* $\mathbb{R}^2$ *is*

$$A = \begin{bmatrix} 1 & 0 \\ 0 & -1 \end{bmatrix}.$$

Hence, $\sigma(T) = \sigma(A) = \{-1, 1\}$. *The eigenspaces are* $E(-1) = \{(x,-x)\colon x \in \mathbb{R}\}$ *and* $E(1) = \{(x,y) \in \mathbb{R}^2 \colon y = x\}$.

To determine the spectrum of T*, it does not matter which basis we use: different bases correspond to similar matrices which, as we know, share a lot of things, namely, their spectrum, the algebraic and geometric multiplicities of the eigenvalues, and the characteristic polynomial, for example (see Proposition 5.9 and Theorem 4.2).*

Given the 'identification' of linear transformations and matrices alluded to at the beginning of this section, concepts existing for matrices can and are transferred to linear transformations.

We define in an obvious way the **characteristic polynomial** of a linear transformation $T \colon U \to U$, the **algebraic and geometric multiplicities** of its eigenvalues and will not repeat ourselves.

We will however speak of the **diagonalisation** of a linear transformation by means of an example. Not all linear transformations are diagonalisable but the one below is.

Example 5.15 *Diagonalise the reflection of Example 5.14. In other words, find a basis of* $\mathbb{R}^2$ *for which the matrix representing the reflection is a diagonal matrix. Find an analytic expression for this linear transformation.*

Solution: We know already from Example 5.14 the eigenvalues and eigenspaces of T*. In fact, we know also that*

$$[T]_{\mathcal{B},\mathcal{B}} = \begin{bmatrix} 1 & 0 \\ 0 & -1 \end{bmatrix}$$

is the matrix of T *relative to the basis* $\mathcal{B} = ((1,1),(1,-1))$ *of eigenvectors. According to what we saw when diagonalising matrices in Chapter 4,*

$$[T]_{\mathcal{E}_2,\mathcal{E}_2} = \begin{bmatrix} 1 & 1 \\ 1 & -1 \end{bmatrix} \begin{bmatrix} 1 & 0 \\ 0 & -1 \end{bmatrix} \begin{bmatrix} 1 & 1 \\ 1 & -1 \end{bmatrix}^{-1} = \begin{bmatrix} 1 & 1 \\ 1 & -1 \end{bmatrix} \begin{bmatrix} 1 & 0 \\ 0 & -1 \end{bmatrix} \begin{bmatrix} \frac{1}{2} & \frac{1}{2} \\ \frac{1}{2} & -\frac{1}{2} \end{bmatrix}.$$

Hence,

$$T(x,y) = [T]_{\mathcal{E}_2,\mathcal{E}_2} \begin{bmatrix} x \\ y \end{bmatrix} = \begin{bmatrix} 0 & 1 \\ 1 & 0 \end{bmatrix} \begin{bmatrix} x \\ y \end{bmatrix},$$

and we have $T(x,y) = (y,x)$.

We end this section with

A word of caution: Not all linear transformations have a non-empty spectrum. For example, a counter-clockwise rotation around the origin of $\mathbb{R}^2$ by an angle θ may or may not have eigenvalues. If one chooses an angle $\theta = \frac{\pi}{2}$, it does not, but if $\theta = \pi$, then the spectrum is $\{-1\}$ (see EX 5.8.13*).*

5.8 Exercises

EX 5.8.1. Are the following functions linear transformations?

a) $T\colon \mathbb{R}^2 \to \mathbb{R}^3$, $T(x,y) = (2x - y, x, y - x)$

b) $T\colon \mathbb{C}^3 \to \mathbb{C}^3$, $T(z_1, z_2, z_3) = (-iz_2, (5-3i)z_3 - z_2, 3z_1)$

c) $T\colon \mathbb{R}^3 \to \mathbb{R}^3$, $T(x,y,z) = (x, x+y+z, 2x-1)$

d) $T\colon \mathbb{C}^2 \to \mathbb{C}^2$, $T(z_1, z_2) = (\overline{z_2}, (2-i)z_1)$

e) $T\colon M_2(\mathbb{K}) \to \mathbb{K}$, $T(A) = \operatorname{tr}(AB^2)$, with $B = E_{21}(-5)$.

f) $T\colon \mathbb{P}_3 \to \mathbb{P}_2$, T is the derivative

g) $T\colon M_{2,3}(\mathbb{R}) \to \mathbb{P}_2$,

$$T(A) = \sum_{i=1,2, j=1,2,3} a_{ij} + \begin{bmatrix} 1 & 1 \end{bmatrix} A \begin{bmatrix} 1 \\ 1 \\ 1 \end{bmatrix} t + a_{11}t^2$$

EX 5.8.2. Find the matrix representing T in Exercises EX 5.8.1 a) and b) relative to the standard bases. Find their null spaces and images. Are they isomorphisms?

EX 5.8.3. Let $S\colon \mathbb{R}^2 \to \mathbb{R}^2$ be the clockwise rotation around the origin by an angle $\theta = \frac{\pi}{2}$, and let $T\colon \mathbb{R}^2 \to \mathbb{R}^2$ be the linear transformation such that

$$[T]_{\mathcal{E}_2,\mathcal{E}_2} = \begin{bmatrix} 1 & 1 \\ 1 & -1 \end{bmatrix}.$$

Find the null space and the image of the composite linear transformation TS. Verify that the Rank-nullity Theorem holds. Is this transformation an isomorphism?

EX 5.8.4. Consider the linear transformation $T\colon \mathbb{R}^2 \to \mathbb{R}^2$ defined by $T(x,y) = (3x + y, -x + 3y)$. Is T invertible? If it is, find the matrix representing T^{-1} relative to the standard basis.

EX 5.8.5. Let $T\colon \mathbb{R}^2 \to \mathbb{R}^2$ be a linear transformation such that $T(x,y) = (\frac{3}{2}x - \frac{1}{2}y, x)$. Let $B = ((1,-1),(1,1))$ be a basis of $\mathbb{R}^2$. Find $[T]_{\mathcal{E}_2,B}$.

EX 5.8.6. Let $T\colon V \to \mathbb{R}^2$ be the linear transformation defined by $T(x, y) = (-9x, -6x)$, where V is the straight line $2x = y$ in $\mathbb{R}^2$. Find a matrix representing T relative to basis of your choice. Find the null space and the image of T. Is T injective or surjective?

EX 5.8.7. Let $T\colon M_2(\mathbb{R}) \to M_2(\mathbb{R})$ be defined by $T(A) = A^T$. Find $[T]_{B_s,B_s}$ and, use the appropriate change of basis matrices to determine $[T]_{B,B}$, where B_s is the standard basis and

$$\mathcal{B} = \left(\begin{bmatrix} 1 & 1 \\ 0 & 0 \end{bmatrix}, \begin{bmatrix} 0 & 1 \\ -1 & 0 \end{bmatrix}, \begin{bmatrix} 0 & 0 \\ 1 & 1 \end{bmatrix}, \begin{bmatrix} 0 & 0 \\ 0 & 1 \end{bmatrix} \right).$$

EX 5.8.8. Let $T\colon \mathbb{P}_2 \to M_2(\mathbb{R})$ be a linear transformation such that

$$[T]_{B,\mathcal{P}_2} = \begin{bmatrix} 1 & -1 \\ 1 & -1 \\ 1 & -1 \\ -1 & 1 \end{bmatrix},$$

where $B = (t - 1, t)$. Find the null space and the image of T.

EX 5.8.9. Let $T\colon \mathbb{P}_2 \to \mathbb{P}_2$ be the derivative, i.e, $T(p) = p'$. Let $B = (1+t, t-t^2, 1)$ be a basis of $\mathbb{P}_2$. Find $[T]_{B,B}$ using $[T]_{\mathcal{P}_2,\mathcal{P}_2}$ and appropriate change of basis matrices.

EX 5.8.10. Let U and V be real vector spaces such that $\dim U = n$ and $\dim V = k$ with $n \leq k$.

True or false?

a) A linear transformation on U is surjective if and only if it is an isomorphism.

b) A linear transformation on V is an isomorphism if and only if it is injective.

c) $k - n$ is the dimension of the null space of any surjective linear transformation from V to U.

d) $k - n$ is the dimension of the null space of any surjective linear transformation from U to V.

e) If $n = k$, then U and V are isomorphic.

f) If $n < k$, then U is isomorphic to a subspace of V.

g) If $n < k$, then U is isomorphic to a unique subspace of V.

EX 5.8.11. Let U be a vector space over $\mathbb{K}$ and let $T : U \to U$ be a linear transformation. A subspace W of U is called an **invariant subspace** of T if $T(W) \subseteq W$. Show that the subspaces $\{\mathbf{0}\}$, U and the eigenspaces of T, should they exist, are examples of invariant subspaces of T.

EX 5.8.12. Let $T\colon \mathbb{R}^3 \to \mathbb{R}^3$ be the orthogonal projection on the plane $z = 0$. Find the spectrum and the eigenspaces of T. Find the invariant subspaces of T.

EX 5.8.13. Let $T, S\colon \mathbb{R}^2 \to \mathbb{R}^2$ be, respectively, the counter-clockwise rotation around the origin by an angle $\theta_T = \frac{\pi}{2}$ and $\theta_S = \pi$.

a) Find the spectra and the eigenspaces of T and S.

b) If A is the matrix that represents T relative to the standard basis, find $\sigma(A) \subseteq \mathbb{C}$ and bases for the eigenspaces.

EX 5.8.14. Consider the linear transformations $T_1, T_2\colon M_2(\mathbb{R}) \to M_2(\mathbb{R})$ defined by

$$T_1(A) = \frac{A + A^T}{2} \qquad\qquad T_2(A) = \frac{A - A^T}{2}.$$

(a) Find the matrices representing T_1 and T_2 relative to the standard basis B_s.

(b) Find the spectra and the eigenspaces of these linear transformations.

EX 5.8.15. Show that the only linear transformation $T\colon M_n(\mathbb{C}) \to \mathbb{C}$ satisfying (i) $T(AB) = T(BA)$, for all $A, B \in M_n(\mathbb{C})$, and (ii) $T(I) = n$ is the trace.

5.9 At a Glance

A linear transformation is an additive and homogeneous function between vector spaces. In other words, it transforms sums of vectors in sums of their images and multiplication of vectors by scalars in multiplication of their images by scalars.

Fixing bases in the domain U and codomain V, a linear transformation T is represented relative to these bases by a $k \times n$ matrix A. Here we assume that the dimension of the domain is n and of the codomain is k. The image of $\boldsymbol{x} \in U$ is then calculated as $[T(\boldsymbol{x})]_{\mathcal{B}_V} = A\mathbf{x}_{\mathcal{B}_U}$.

We have thus an induced linear transformation from $\mathbb{K}^n$ to $\mathbb{K}^k$ given by $\mathbf{z} \mapsto A\mathbf{z}$. If we understand this transformation, then we understand T. Moreover, if $k = n$, then the eigenvalues of A and T coincide, and A and T are only diagonalisable simultaneously.

$N(A)$ consists of the coordinate vectors of the vectors in the null space $N(T)$ and $C(A)$ consists of the coordinate vectors of the vectors in the image $I(T)$. The null space $N(T)$ is isomorphic as a vector space to $N(A)$, and the same applies to $I(T)$ and $C(A)$. As a consequence, T is invertible or, in other words, an isomorphism, if and only if A is invertible.

The fundamental formula linking these spaces is

$$n = \dim N(T) + \dim I(T).$$

If $\dim U = \dim V$, then T is neither surjective nor injective or is both simultaneously.

If $U = V$, then $N(T)$ and $I(T)$ are subspaces of U. If $N(T) \cap I(T) = \{\mathbf{0}\}$, then, by means of the formula above, the transformation splits U into the direct sum $U = N(T) \oplus I(T)$.

Matrices representing T relative to different bases are similar.

Chapter 6

Inner Product Spaces

6.1 Real Inner Product Spaces 205
6.2 Complex Inner Product Spaces 214
6.3 Orthogonal Sets 218
6.3.1 Orthogonal complement 220
6.3.2 Orthogonal projections 228
6.3.3 Gram–Schmidt process 235
6.4 Orthogonal and Unitary Diagonalisation 238
6.5 Singular Value Decomposition 245
6.6 Affine Subspaces of $\mathbb{R}^n$ 249
6.7 Exercises 253
6.8 At a Glance 256

Up to here we have been dealing with purely algebraic structures. For example, we do not have just yet a notion of distance between elements of a vector space, and implicitly we are overlooking the geometric aspects of spaces. We will fix that in this chapter by introducing the inner product spaces. Unlike the previous chapters where we have treated simultaneously the real and complex vector spaces, here we treat the real and complex inner products separately. This is because the definitions of these inner products are different in an essential manner.

6.1 Real Inner Product Spaces

Definition 48 *Let V be a real vector space. A real function*

$$\begin{aligned}\langle\cdot,\cdot\rangle :& V\times V\to\mathbb{R}\\ &(\boldsymbol{x},\boldsymbol{y})\mapsto\langle\boldsymbol{x},\boldsymbol{y}\rangle\end{aligned}$$

is said to be an **inner product** *if, for all $\boldsymbol{x},\boldsymbol{y},\boldsymbol{z}\in V$ and $\alpha\in\mathbb{R}$,*

(i) $\langle\boldsymbol{x},\boldsymbol{y}\rangle=\langle\boldsymbol{y},\boldsymbol{x}\rangle$;

(ii) $\langle\alpha\boldsymbol{x},\boldsymbol{y}\rangle=\alpha\langle\boldsymbol{x},\boldsymbol{y}\rangle$;

DOI: 10.1201/9781351243452-6

(iii) $\langle \boldsymbol{x} + \boldsymbol{y}, \boldsymbol{z} \rangle = \langle \boldsymbol{x}, \boldsymbol{z} \rangle + \langle \boldsymbol{y}, \boldsymbol{z} \rangle$;

(iv) $\langle \boldsymbol{x}, \boldsymbol{x} \rangle \geq 0 \quad \wedge \quad (\langle \boldsymbol{x}, \boldsymbol{x} \rangle = 0 \Rightarrow \boldsymbol{x} = 0)$.

A real vector space V endowed with a inner product is said to be a **real inner product space**.

If one considers the zero vector $\mathbf{0}$, by (ii) above,

$$\langle \mathbf{0}, \mathbf{0} \rangle = \langle 0\mathbf{0}, \mathbf{0} \rangle = 0\langle \mathbf{0}, \mathbf{0} \rangle = 0.$$

Hence, (iv) could be equivalently replaced by

$$\langle \boldsymbol{x}, \boldsymbol{x} \rangle \geq 0 \quad \wedge \quad (\langle \boldsymbol{x}, \boldsymbol{x} \rangle = 0 \Leftrightarrow \boldsymbol{x} = 0). \tag{6.1}$$

A function $f \colon V \times V \to \mathbb{R}$ satisfying, for all $\boldsymbol{x}, \boldsymbol{y}, \boldsymbol{z} \in V$ and $\alpha \in \mathbb{R}$,

(a) $\langle \boldsymbol{x}, \alpha \boldsymbol{y} \rangle = \alpha \langle \boldsymbol{x}, \boldsymbol{y} \rangle$;

(b) $\langle \alpha \boldsymbol{x}, \boldsymbol{y} \rangle = \alpha \langle \boldsymbol{x}, \boldsymbol{y} \rangle$;

(c) $\langle \boldsymbol{x} + \boldsymbol{y}, \boldsymbol{z} \rangle = \langle \boldsymbol{x}, \boldsymbol{z} \rangle + \langle \boldsymbol{y}, \boldsymbol{z} \rangle$.

is said to be a **bilinear function**, meaning that it is linear separately in each variable.

We see that an inner product is a bilinear function which, because it also satisfies (i) in Definition 48, is said to be a **symmetric** bilinear function. Condition (iv) further classifies this symmetric bilinear function as **positive definite**.

Example 6.1 *The usual scalar products in $\mathbb{R}^2$ and $\mathbb{R}^3$ are examples of inner products. Recall that, for vectors $\boldsymbol{x} = (x_1, x_2), \boldsymbol{y} = (y_1, y_2)$ on the plane $\mathbb{R}^2$, the scalar product is defined by*

$$\begin{aligned} \langle \boldsymbol{x}, \boldsymbol{y} \rangle &= \|\boldsymbol{x}\| \|\boldsymbol{y}\| \cos \theta \\ &= x_1 y_1 + x_2 y_2, \end{aligned}$$

where $\theta \in [0, \pi]$ is the angle between the vectors $\boldsymbol{x}$ and $\boldsymbol{y}$ and $\| \cdot \|$ denotes the norm of a vector (see Figure 6.1).

If we consider now vectors $\boldsymbol{x} = (x_1, x_2, x_3), \boldsymbol{y} = (y_1, y_2, y_3)$ in space, then we have

$$\begin{aligned} \langle \boldsymbol{x}, \boldsymbol{y} \rangle &= \|\boldsymbol{x}\| \|\boldsymbol{y}\| \cos \theta \\ &= x_1 y_1 + x_2 y_2 + x_3 y_3. \end{aligned}$$

Here again $\theta \in [0, \pi]$ is the angle between the vectors $\boldsymbol{x}$ and $\boldsymbol{y}$ in $\mathbb{R}^3$ and $\| \cdot \|$ denotes the norm. It is an easy exercise to verify that all conditions of Definition 48 are satisfied by both scalar products.

In both cases, the norm of a vector $\boldsymbol{x}$ satisfies

$$\|\boldsymbol{x}\| = \sqrt{\langle \boldsymbol{x}, \boldsymbol{x} \rangle}.$$

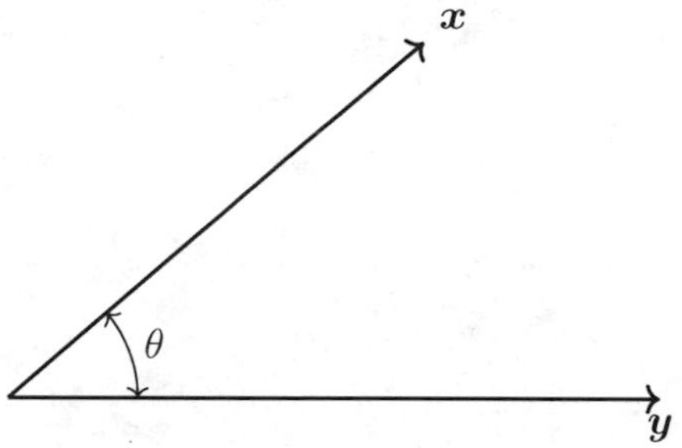

FIGURE 6.1: Vectors in $\mathbb{R}^2$ or $\mathbb{R}^3$ with an angle θ between them.

Generalising in the obvious way the above example to $\mathbb{R}^n$, we define next an inner product on this space.

Definition 49 *Let* $\boldsymbol{x} = (x_1, \dots, x_n), \boldsymbol{y} = (y_1, \dots, y_n) \in \mathbb{R}^n$. *The inner product* $\langle \boldsymbol{x}, \boldsymbol{y} \rangle$ *is defined by*

$$\langle \boldsymbol{x}, \boldsymbol{y} \rangle = x_1 y_1 + x_2 y_2 + \cdots + x_n y_n \tag{6.2}$$

or, alternatively,

$$\langle \boldsymbol{x}, \boldsymbol{y} \rangle = \mathbf{y}^T \mathbf{x} = \mathbf{x}^T \mathbf{y}.$$

We shall refer to $\mathbb{R}^n$ *together with this inner product as the real* **Euclidean space** $\mathbb{R}^n$.

Analogously, to the plane and space cases, we define the **norm** *of a vector* $\boldsymbol{x}$ *by*

$$\|\boldsymbol{x}\| = \sqrt{\langle \boldsymbol{x}, \boldsymbol{x} \rangle} = \sqrt{x_1^2 + x_2^2 + \cdots + x_n^2}.$$

Also generalising what is usual in the plane and space, the **distance** $d(\boldsymbol{x}, \boldsymbol{y})$ *between the points* $\boldsymbol{x}, \boldsymbol{y} \in \mathbb{R}^n$ *is defined by*

$$d(\boldsymbol{x}, \boldsymbol{y}) = \|\boldsymbol{x} - \boldsymbol{y}\|.$$

Exercise 6.1 *Show that* (6.2) *is an inner product in* $\mathbb{R}^n$.

The next example deals with an inner product on a matrix space.

Example 6.2 *For* $A = [a_{ij}], B = [b_{ij}]$ *in the space* $\mathrm{M}_2(\mathbb{R})$ *of the* 2×2 *real matrices, define the inner product*

$$\begin{aligned}\langle A, B \rangle &= \operatorname{tr}(B^T A)\\ &= \sum_{i,j=1}^{2} a_{ij} b_{ij}.\end{aligned}$$

Observe that, for the standard basis $\mathcal{B}_c$ *of* $\mathrm{M}_{2\times 2}(\mathbb{R})$, *we have*

$$\langle A, B \rangle_{\mathrm{M}_2(\mathbb{R})} = \langle A_{\mathcal{B}_c}, B_{\mathcal{B}_c} \rangle_{\mathbb{R}^4},$$

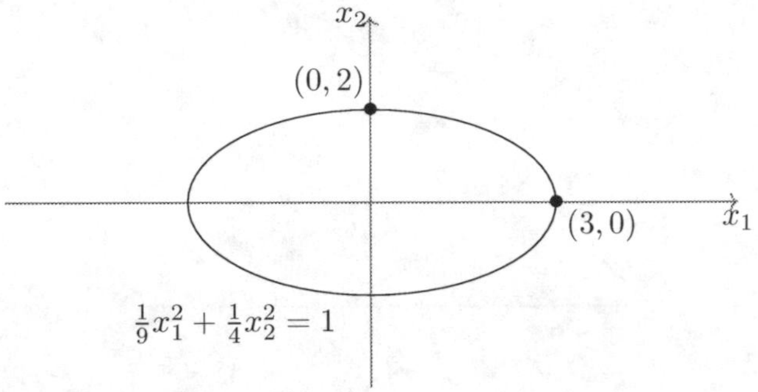

FIGURE 6.2: Ellipse depicting the points at distance 1 from $(0, 0)$ with respect to the inner product of Exercise 6.2.

where $A_{\mathcal{B}_c}, B_{\mathcal{B}_c} \in \mathbb{R}^4$ are, respectively, the coordinate vectors of A and B relative to the standard basis $\mathcal{B}_c$ of $\mathrm{M}_2(\mathbb{R})$. Hence, the inner product defined above respects the isomorphism $A \mapsto A_{\mathcal{B}_c}$ existing between $\mathrm{M}_2(\mathbb{R})$ and $\mathbb{R}^4$.

It is worth noticing that, since

$$\mathrm{tr}(B^T A) = \mathrm{tr}(A^T B),$$

one could alternatively have defined the inner product by

$$\langle A, B\rangle = \mathrm{tr}(A^T B),$$

as some authors do.

Exercise 6.2 *Show that*

$$\langle (x_1, x_2), (y_1, y_2)\rangle = \tfrac{1}{9}x_1y_1 + \tfrac{1}{4}x_2y_2$$

defines an inner product in $\mathbb{R}^2$ and, for this inner product, find the circle C centered at $(0, 0)$ and with radius 1, i.e.,

$$C = \{(x_1, x_2) \in \mathbb{R}^2 \; : \; \|(x_1, x_2)\| = 1\}.$$

Solution. It is easily seen that this function satisfies all the conditions in Definition 48. To find the circle C, we have to determine which points (x_1, x_2) lie at a distance 1 from $(0, 0)$, that is,

$$1 = \|(x_1, x_2)\|^2 = \langle (x_1, x_2), (x_1, x_2)\rangle = \tfrac{1}{9}x_1^2 + \tfrac{1}{4}x_2^2.$$

Hence, the points at distance 1 from $(0, 0)$ lie in an ellipse (see Figure 6.2).

This exercise illustrates how the geometry of a space can be changed by endowing it with different inner products. Here, the 'usual' circle has been changed into an ellipse.

Building on what is known for the Euclidean space, we define the norm of a vector in an (any) inner product space.

Definition 50 *Let V be a real inner product space with inner product $\langle \cdot, \cdot \rangle$ and let $\boldsymbol{x} \in V$. The* **norm** *of $\boldsymbol{x}$ is defined by*

$$\|\boldsymbol{x}\| = \sqrt{\langle \boldsymbol{x}, \boldsymbol{x} \rangle}. \tag{6.3}$$

The **distance** $\mathrm{d}(\boldsymbol{x}, \boldsymbol{y})$ *between $\boldsymbol{x}, \boldsymbol{y} \in V$ is defined by*

$$\mathrm{d}(\boldsymbol{x}, \boldsymbol{y}) = \|\boldsymbol{x} - \boldsymbol{y}\|.$$

Proposition 6.1 *Let V be a real inner product space and let the norm of a vector be defined as in (6.3). Then the function defined by*

$$\begin{aligned} \|\cdot\| : & V \to \mathbb{R} \\ & \boldsymbol{x} \mapsto \|\boldsymbol{x}\| \end{aligned}$$

is such that, for all $\boldsymbol{x}, \boldsymbol{y} \in V, \alpha \in \mathbb{R}$,

(i) $\|\boldsymbol{x}\| \geq 0$, *and* $\|\boldsymbol{x}\| = 0$ *if and only if* $\boldsymbol{x} = 0$;

(ii) $\|\alpha \boldsymbol{x}\| = |\alpha| \|\boldsymbol{x}\|$;

(iii)

$$\|\boldsymbol{x} + \boldsymbol{y}\| \leq \|\boldsymbol{x}\| + \|\boldsymbol{y}\|. \qquad \text{Triangle inequality} \tag{6.4}$$

We prove (i) and (ii) here and leave the proof of (iii) for later, as we shall use in its proof another inequality, the Cauchy–Schwarz inequality, that will be proved in the sequel.

Proof *(i) It is clear that $\|\boldsymbol{x}\| \geq 0$, from the very definition of norm (cf. Definition 50). Moreover, by (6.1), we have that $\langle \boldsymbol{x}, \boldsymbol{x} \rangle = 0$ if and only if $\boldsymbol{x} = \mathbf{0}$.*

(ii) By Definition 48 (ii), for $\boldsymbol{x} \in V, \alpha \in \mathbb{R}$,

$$\|\alpha \boldsymbol{x}\| = \sqrt{\langle \alpha \boldsymbol{x}, \alpha \boldsymbol{x} \rangle} = \sqrt{\alpha^2 \langle \boldsymbol{x}, \boldsymbol{x} \rangle} = |\alpha| \sqrt{\langle \boldsymbol{x}, \boldsymbol{x} \rangle} = |\alpha| \|\boldsymbol{x}\|,$$

as required.

Exercise 6.3 *Let V be an inner product space. Show that for all $\boldsymbol{x}, \boldsymbol{y}, \boldsymbol{z} \in V$,*

(i) $\mathrm{d}(\boldsymbol{x}, \boldsymbol{y}) \geq 0$, *and* $\mathrm{d}(\boldsymbol{x}, \boldsymbol{y}) = 0$ *if and only if* $\boldsymbol{x} = \boldsymbol{y}$;

(ii) $\mathrm{d}(\boldsymbol{x}, \boldsymbol{y}) = \mathrm{d}(\boldsymbol{y}, \boldsymbol{x})$

(iii) $\mathrm{d}(\boldsymbol{x}, \boldsymbol{z}) \leq \mathrm{d}(\boldsymbol{x}, \boldsymbol{y}) + \mathrm{d}(\boldsymbol{y}, \boldsymbol{z})$.

Hint: use Propositin 6.1.

In $\mathbb{R}^2$ and $\mathbb{R}^3$, we have, for any vectors $\boldsymbol{x}, \boldsymbol{y}$ and the usual scalar product that,

$$\langle \boldsymbol{x}, \boldsymbol{y} \rangle = \|\boldsymbol{x}\| \|\boldsymbol{y}\| \cos\theta,$$

yielding

$$|\langle \boldsymbol{x}, \boldsymbol{y} \rangle| \leq \|\boldsymbol{x}\| \|\boldsymbol{y}\|,$$

since $|\cos\theta| \leq 1$. It turns out that this inequality holds in every inner product space.

Theorem 6.1 (Cauchy–Schwarz inequality) *Let V be a real inner product space and let $\boldsymbol{x}, \boldsymbol{y} \in V$. Then*

$$|\langle \boldsymbol{x}, \boldsymbol{y} \rangle| \leq \|\boldsymbol{x}\| \|\boldsymbol{y}\| \tag{6.5}$$

and

$$|\langle \boldsymbol{x}, \boldsymbol{y} \rangle| = \|\boldsymbol{x}\| \|\boldsymbol{y}\| \tag{6.6}$$

if and only if $\{\boldsymbol{x}, \boldsymbol{y}\}$ is a linearly dependent set.

Proof *The theorem holds trivially if $\boldsymbol{y} = \boldsymbol{0}$. Suppose now that $\boldsymbol{y} \neq \boldsymbol{0}$. Then, for $\alpha \in \mathbb{R}$ and $\boldsymbol{x}, \boldsymbol{y} \in V$,*

$$0 \leq \langle \boldsymbol{x} - \alpha\boldsymbol{y}, \boldsymbol{x} - \alpha\boldsymbol{y} \rangle = \langle \boldsymbol{x}, \boldsymbol{x} \rangle - \alpha\langle \boldsymbol{x}, \boldsymbol{y} \rangle - \alpha\langle \boldsymbol{y}, \boldsymbol{x} \rangle + \alpha^2\langle \boldsymbol{y}, \boldsymbol{y} \rangle \tag{6.7}$$

$$= \langle \boldsymbol{x}, \boldsymbol{x} \rangle - \alpha\langle \boldsymbol{x}, \boldsymbol{y} \rangle - \alpha\big(\langle \boldsymbol{x}, \boldsymbol{y} \rangle - \alpha\|\boldsymbol{y}\|^2\big). \tag{6.8}$$

If we set

$$\alpha = \frac{\langle \boldsymbol{x}, \boldsymbol{y} \rangle}{\|\boldsymbol{y}\|^2},$$

then

$$0 \leq \langle \boldsymbol{x}, \boldsymbol{x} \rangle - \frac{\langle \boldsymbol{x}, \boldsymbol{y} \rangle^2}{\|\boldsymbol{y}\|^2} - \frac{\langle \boldsymbol{x}, \boldsymbol{y} \rangle}{\|\boldsymbol{y}\|^2}\left(\langle \boldsymbol{x}, \boldsymbol{y} \rangle - \frac{\langle \boldsymbol{x}, \boldsymbol{y} \rangle}{\|\boldsymbol{y}\|^2}\|\boldsymbol{y}\|^2\right),$$

that is,

$$0 \leq \langle \boldsymbol{x}, \boldsymbol{x} \rangle - \frac{\langle \boldsymbol{x}, \boldsymbol{y} \rangle}{\|\boldsymbol{y}\|^2}\langle \boldsymbol{x}, \boldsymbol{y} \rangle.$$

Hence,

$$0 \leq \|\boldsymbol{x}\|^2\|\boldsymbol{y}\|^2 - \langle \boldsymbol{x}, \boldsymbol{y} \rangle^2$$

from which follows that

$$\langle \boldsymbol{x}, \boldsymbol{y} \rangle^2 \leq \|\boldsymbol{x}\|^2\|\boldsymbol{y}\|^2.$$

Taking square roots,

$$|\langle \boldsymbol{x}, \boldsymbol{y} \rangle| \leq \|\boldsymbol{x}\| \|\boldsymbol{y}\|$$

which proves (6.5).

We show now that

$$|\langle \boldsymbol{x}, \boldsymbol{y} \rangle| = \|\boldsymbol{x}\| \|\boldsymbol{y}\| \tag{6.9}$$

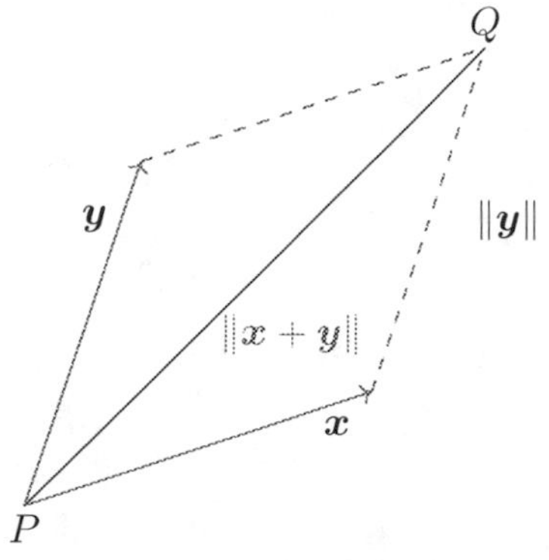

FIGURE 6.3: Triangle inequality: the shortest distance between P and Q is the length of the straight line segment connecting these points.

if and only if the vectors $\boldsymbol{x}, \boldsymbol{y}$ are linearly dependent. If $\boldsymbol{x} = \alpha \boldsymbol{y}$ or $\boldsymbol{y} = \alpha \boldsymbol{x}$, a direct substitution shows that (6.6) *holds. If, on the other hand,* (6.6) *holds and $\boldsymbol{y} \neq \boldsymbol{0}$, then going back retracing our steps from* (6.9) *to* (6.8), (6.7), *we have that*

$$\boldsymbol{x} - \frac{\langle \boldsymbol{x}, \boldsymbol{y} \rangle}{\|\boldsymbol{y}\|^2} \boldsymbol{y} = \boldsymbol{0},$$

which shows that $\boldsymbol{x}, \boldsymbol{y}$ are linearly dependent.

Having proved Theorem 6.1, we are now ready to prove the triangle inequality.

Proof of Proposition 6.1 (iii). Let $\boldsymbol{x}, \boldsymbol{y}$ be vectors in V. Then,

$$\begin{aligned}
\|\boldsymbol{x} + \boldsymbol{y}\|^2 &= \langle \boldsymbol{x} + \boldsymbol{y}, \boldsymbol{x} + \boldsymbol{y} \rangle \\
&= \langle \boldsymbol{x}, \boldsymbol{x} \rangle + 2\langle \boldsymbol{x}, \boldsymbol{y} \rangle + \langle \boldsymbol{y}, \boldsymbol{y} \rangle \\
&= \|\boldsymbol{x}\|^2 + 2\langle \boldsymbol{x}, \boldsymbol{y} \rangle + \|\boldsymbol{y}\|^2 \\
&\leq \|\boldsymbol{x}\|^2 + 2|\langle \boldsymbol{x}, \boldsymbol{y} \rangle| + \|\boldsymbol{y}\|^2 \\
&\leq \|\boldsymbol{x}\|^2 + 2\|\boldsymbol{x}\|\|\boldsymbol{y}\| + \|\boldsymbol{y}\|^2 \qquad \leftarrow \quad \text{using Cauchy–Schwarz inequality} \\
&= (\|\boldsymbol{x}\| + \|\boldsymbol{y}\|)^2.
\end{aligned}$$

Hence,

$$\|\boldsymbol{x} + \boldsymbol{y}\| \leq \|\boldsymbol{x}\| + \|\boldsymbol{y}\|,$$

concluding the proof.

Proposition 6.2 (Parallelogram law) *Let $\boldsymbol{x}, \boldsymbol{y}$ be vectors in a real inner product space V. Then,*

$$\|\boldsymbol{x} + \boldsymbol{y}\|^2 + \|\boldsymbol{x} - \boldsymbol{y}\|^2 = 2(\|\boldsymbol{x}\|^2 + \|\boldsymbol{y}\|^2).$$

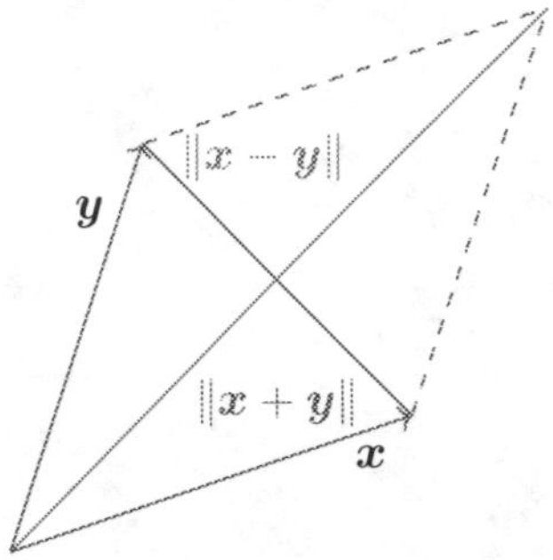

FIGURE 6.4: Parallelogram law.

Proof *For $\boldsymbol{x}, \boldsymbol{y} \in V$, we have*

$$\begin{aligned}\|\boldsymbol{x}+\boldsymbol{y}\|^2 + \|\boldsymbol{x}-\boldsymbol{y}\|^2 &= \langle \boldsymbol{x}+\boldsymbol{y}, \boldsymbol{x}+\boldsymbol{y}\rangle + \langle \boldsymbol{x}-\boldsymbol{y}, \boldsymbol{x}-\boldsymbol{y}\rangle \\ &= 2\langle \boldsymbol{x}, \boldsymbol{x}\rangle + 2\langle \boldsymbol{y}, \boldsymbol{y}\rangle + 2\langle \boldsymbol{x}, \boldsymbol{y}\rangle - 2\langle \boldsymbol{x}, \boldsymbol{y}\rangle \\ &= 2\|\boldsymbol{x}\|^2 + 2\|\boldsymbol{y}\|^2,\end{aligned}$$

as required.

Up to now, our approach to inner product spaces has been coordinate free. It is now the time to make use of the fact that vector spaces do have bases and see how they interact with the inner product.

Let V be a real inner product space and let $\mathcal{B} = (\boldsymbol{b}_1, \boldsymbol{b}_2, \dots, \boldsymbol{b}_n)$ be a basis of V. For $\boldsymbol{x}, \boldsymbol{y} \in V$ such that the coordinate vectors of $\boldsymbol{x}$ and $\boldsymbol{y}$ relative to $\mathcal{B}$ are, respectively, $\boldsymbol{x}_\mathcal{B} = (\alpha_1, \alpha_2, \dots, \alpha_n)$ and $\boldsymbol{y}_\mathcal{B} = (\beta_1, \beta_2, \dots, \beta_n)$, we have

$$\begin{aligned}\langle \boldsymbol{x}, \boldsymbol{y}\rangle &= \langle \alpha_1\boldsymbol{b}_1 + \alpha_2\boldsymbol{b}_2 + \cdots + \alpha_n\boldsymbol{b}_n, \beta_1\boldsymbol{b}_1 + \beta_2\boldsymbol{b}_2 + \cdots + \beta_n\boldsymbol{b}_n\rangle \\ &= \beta_1\langle \boldsymbol{b}_1, \boldsymbol{b}_1\rangle\alpha_1 + \beta_1\langle \boldsymbol{b}_2, \boldsymbol{b}_1\rangle\alpha_2 + \dots \beta_1\langle \boldsymbol{b}_n, \boldsymbol{b}_1\rangle\alpha_n + \\ &\quad + \beta_2\langle \boldsymbol{b}_1, \boldsymbol{b}_2\rangle\alpha_1 + \beta_2\langle \boldsymbol{b}_2, \boldsymbol{b}_2\rangle\alpha_2 + \dots \beta_2\langle \boldsymbol{b}_n, \boldsymbol{b}_2\rangle\alpha_n + \\ &\qquad\qquad \vdots \\ &\quad + \beta_n\langle \boldsymbol{b}_1, \boldsymbol{b}_n\rangle\alpha_1 + \beta_n\langle \boldsymbol{b}_2, \boldsymbol{b}_n\rangle\alpha_2 + \dots \beta_n\langle \boldsymbol{b}_n, \boldsymbol{b}_n\rangle\alpha_n \\ &= \begin{bmatrix}\beta_1 & \beta_2 & \dots & \beta_n\end{bmatrix} \underbrace{\begin{bmatrix}\langle \boldsymbol{b}_1, \boldsymbol{b}_1\rangle & \langle \boldsymbol{b}_2, \boldsymbol{b}_1\rangle & \dots & \langle \boldsymbol{b}_n, \boldsymbol{b}_1\rangle \\ \langle \boldsymbol{b}_1, \boldsymbol{b}_2\rangle & \langle \boldsymbol{b}_2, \boldsymbol{b}_2\rangle & \dots & \langle \boldsymbol{b}_n, \boldsymbol{b}_2\rangle \\ \vdots & \vdots & \ddots & \vdots \\ \langle \boldsymbol{b}_1, \boldsymbol{b}_n\rangle & \langle \boldsymbol{b}_2, \boldsymbol{b}_n\rangle & \dots & \langle \boldsymbol{b}_n, \boldsymbol{b}_n\rangle\end{bmatrix}}_{G} \begin{bmatrix}\alpha_1 \\ \alpha_2 \\ \vdots \\ \alpha_n\end{bmatrix}.\end{aligned}$$

Hence, given an inner product on a real vector space V with $\dim V = n$ and a basis $\mathcal{B}$ of V, it is possible to find an $n \times n$ real matrix

$$G = \begin{bmatrix} \langle \boldsymbol{b}_1, \boldsymbol{b}_1 \rangle & \langle \boldsymbol{b}_2, \boldsymbol{b}_1 \rangle & \dots & \langle \boldsymbol{b}_n, \boldsymbol{b}_1 \rangle \\ \langle \boldsymbol{b}_1, \boldsymbol{b}_2 \rangle & \langle \boldsymbol{b}_2, \boldsymbol{b}_2 \rangle & \dots & \langle \boldsymbol{b}_n, \boldsymbol{b}_2 \rangle \\ \vdots & \vdots & \ddots & \vdots \\ \langle \boldsymbol{b}_1, \boldsymbol{b}_n \rangle & \langle \boldsymbol{b}_2, \boldsymbol{b}_n \rangle & \dots & \langle \boldsymbol{b}_n, \boldsymbol{b}_n \rangle \end{bmatrix} \tag{6.10}$$

such that

$$\langle \boldsymbol{x}, \boldsymbol{y} \rangle = \mathbf{y}_{\mathcal{B}}^T G \mathbf{x}_{\mathcal{B}}.$$

This matrix $G = [g_{ij}]$, where for all $i, j = 1, \dots, n$, we have $g_{ij} = \langle b_j, b_i \rangle$, is said to be the **Gram matrix** of the set of vectors $\{\boldsymbol{b}_1, \boldsymbol{b}_2, \dots, \boldsymbol{b}_n\}$.

Definition 51 *A real symmetric matrix of order n is said to be a* **positive definite matrix** *if, for all non-zero vectors $\boldsymbol{x} \in \mathbb{R}^n$,*

$$\mathbf{x}^T A \mathbf{x} > 0.$$

The proof of the following proposition is left as an exercise.

Proposition 6.3 *Let V be a real inner product space of dimension n and let $\mathcal{B} = (\boldsymbol{b}_1, \boldsymbol{b}_2, \dots, \boldsymbol{b}_n)$ be a basis of V. Let $\boldsymbol{x}, \boldsymbol{y}$ be vectors in V. Then, there exists uniquely an $n \times n$ real matrix G such that*

$$\langle \boldsymbol{x}, \boldsymbol{y} \rangle = \mathbf{y}_{\mathcal{B}}^T G \mathbf{x}_{\mathcal{B}}, \tag{6.11}$$

where $\boldsymbol{x}_{\mathcal{B}}, \boldsymbol{y}_{\mathcal{B}}$ are, respectively, the coordinate vectors of $\boldsymbol{x}, \boldsymbol{y}$ relative to the basis $\mathcal{B}$. Moreover, the matrix G is symmetric and positive definite, that is, for $\boldsymbol{x} \neq 0$,

$$\mathbf{x}_{\mathcal{B}}^T G \mathbf{x}_{\mathcal{B}} > 0. \tag{6.12}$$

Exercise 6.4 *Consider the Euclidean space $\mathbb{R}^2$ and its standard basis $\mathcal{E}_2$. Find the Gram matrix of $\mathcal{E}_2$. What is the Gram matrix if now one considers an inner product $\langle \cdot, \cdot \rangle_1$ which is that of Exercise 6.2? Calculate $\langle (1,2), (3,1) \rangle_1$ using this latter matrix.*

Solution. *An easy application of* (6.10) *leads to the Gram matrix*

$$G = \begin{bmatrix} 1 & 0 \\ 0 & 1 \end{bmatrix},$$

in the first case. For the second inner product, we have

$$G = \begin{bmatrix} \frac{1}{9} & 0 \\ 0 & \frac{1}{4} \end{bmatrix}.$$

It follows that

$$\langle (1,2), ((3,1) \rangle_1 = \begin{bmatrix} 3 & 1 \end{bmatrix} \begin{bmatrix} \frac{1}{9} & 0 \\ 0 & \frac{1}{4} \end{bmatrix} \begin{bmatrix} 1 \\ 2 \end{bmatrix} = \frac{5}{6}$$

Proposition 6.4 *Let A be a real matrix of order n. The following are equivalent.*

(i) The expression

$$\langle \boldsymbol{x}, \boldsymbol{y} \rangle = \mathbf{y}^T A\mathbf{x}$$

defines an inner product on $\mathbb{R}^n$.

(ii) A is a real positive definite matrix.

Proof *We prove only that (i) implies (ii) and leave the other implication as an easy exercise.*

We see now that A is symmetric. For $i, j = 1, \ldots, n$,

$$\langle \mathbf{e}_i, \mathbf{e}_j \rangle = \mathbf{e}_j^T A\mathbf{e}_i = a_{ji}, \quad \langle \mathbf{e}_j, \mathbf{e}_i \rangle = \mathbf{e}_i^T A\mathbf{e}_j = a_{ij}.$$

Hence, $a_{ij} = a_{ji}$ and, therefore, A is symmetric. On the other hand, given $\boldsymbol{x} \in \mathbb{R}^n$,

$$\mathbf{x}^T A\mathbf{x} = \langle \boldsymbol{x}, \boldsymbol{x} \rangle \geq 0$$

and equals 0 only when $\boldsymbol{x} = \mathbf{0}$.

We saw in Chapter 4 that real symmetric matrices have real eigenvalues (see Corollary 4.1). We can say more.

Proposition 6.5 *A real symmetric matrix is positive definite if and only if its eigenvalues are positive numbers.*

For another result on real positive definite matrices, see Corollary 6.3.

Proof *Suppose that A is a positive definite matrix. Let λ be an eigenvalue of A and let $\mathbf{x}$ be and associated eigenvector. Then*

$$0 < \mathbf{x}^T A\mathbf{x} = \lambda \mathbf{x}^T \mathbf{x} = \lambda \|\mathbf{x}\|^2,$$

from which follows that $\lambda > 0$. The proof of the converse is postponed until §6.4.

An immediate consequence of this proposition is that a positive definite matrix is invertible as its spectrum does not contain 0. In other words, the null space of such a matrix is $\{\mathbf{0}\}$. In the same trait, we can conclude that Gram matrices are invertible matrices.

6.2 Complex Inner Product Spaces

As said at the beginning of this chapter, the real and complex inner products differ in an essential way. We shall see that in the definition below.

Definition 52 *Let V be a complex vector space. A complex function*

$$\begin{aligned}\langle\cdot,\cdot\rangle : & V\times V\to\mathbb{C}\\ & (\boldsymbol{x},\boldsymbol{y})\mapsto\langle\boldsymbol{x},\boldsymbol{y}\rangle\end{aligned}$$

is said to be an ***inner product*** *if, for all $\boldsymbol{x},\boldsymbol{y},\boldsymbol{z}\in V$ and $\alpha\in\mathbb{C}$,*

(i) $\langle\boldsymbol{x},\boldsymbol{y}\rangle=\overline{\langle\boldsymbol{y},\boldsymbol{x}\rangle}$;

(ii) $\langle\alpha\boldsymbol{x},\boldsymbol{y}\rangle=\alpha\langle\boldsymbol{x},\boldsymbol{y}\rangle$;

(iii) $\langle\boldsymbol{x}+\boldsymbol{y},\boldsymbol{z}\rangle=\langle\boldsymbol{x},\boldsymbol{z}\rangle+\langle\boldsymbol{y},\boldsymbol{z}\rangle$;

(iv) $\langle\boldsymbol{x},\boldsymbol{x}\rangle\geq 0 \quad\wedge\quad (\langle\boldsymbol{x},\boldsymbol{x}\rangle=0\Rightarrow\boldsymbol{x}=0)$.

A complex vector space V endowed with a inner product it is said to be an **complex inner product space**.

Considerations similar to those leading to (6.1) yield that (iv) can be equivalently replaced by

$$\langle\boldsymbol{x},\boldsymbol{x}\rangle\geq 0 \quad\wedge\quad (\langle\boldsymbol{x},\boldsymbol{x}\rangle=0\Leftrightarrow\boldsymbol{x}=0). \tag{6.13}$$

We see that, apart from the inner product of two vectors being now a complex number, the remaining difference between real and complex inner products is condition (i) above.

By conditions (i) and (ii),

$$\langle\boldsymbol{x},\alpha\boldsymbol{y}\rangle=\overline{\langle\alpha\boldsymbol{y},\boldsymbol{x}\rangle}=\overline{\alpha\langle\boldsymbol{y},\boldsymbol{x}\rangle}=\overline{\alpha}\,\overline{\langle\boldsymbol{y},\boldsymbol{x}\rangle}=\overline{\alpha}\langle\boldsymbol{x},\boldsymbol{y}\rangle,$$

yielding that the inner product is conjugate linear in the second variable.

A function $f\colon V\times V\to\mathbb{C}$ satisfying, for all $\boldsymbol{x},\boldsymbol{y},\boldsymbol{z}\in V$ and $\alpha\in\mathbb{R}$,

(a) $\langle\alpha\boldsymbol{x},\boldsymbol{y}\rangle=\alpha\langle\boldsymbol{x},\boldsymbol{y}\rangle$;

(b) $\langle\boldsymbol{x},\alpha\boldsymbol{y}\rangle=\bar{\alpha}\langle\boldsymbol{x},\boldsymbol{y}\rangle$;

(c) $\langle\boldsymbol{x}+\boldsymbol{y},\boldsymbol{z}\rangle=\langle\boldsymbol{x},\boldsymbol{z}\rangle+\langle\boldsymbol{y},\boldsymbol{z}\rangle$.

is said to be a **sesquilinear function**, meaning that it is linear in the first variable and linear by 'half' in the second variable.

Hence, a complex inner product is a sesquilinear function which, because it also satisfies condition (iv), is a **positive definite** sesquilinear function.

Similarly to §6.1, we define the **norm** of a vector $\boldsymbol{x}\in\mathbb{C}^n$ by

$$\|\boldsymbol{x}\|=\sqrt{\langle\boldsymbol{x},\boldsymbol{x}\rangle}, \tag{6.14}$$

and the **distance** between $\boldsymbol{x},\boldsymbol{y}\in\mathbb{C}^n$ by

$$\mathrm{d}(\boldsymbol{x},\boldsymbol{y})=\|\boldsymbol{x}-\boldsymbol{y}\|. \tag{6.15}$$

Example 6.3 *There is a usual inner product on* $\mathbb{C}^n$. *Namely, for vectors* $\boldsymbol{x} = (x_1, x_2, \ldots, x_n)$ *and* $\boldsymbol{y} = (y_1, y_2, \ldots, y_n)$ *in* $\mathbb{C}^n$, *define*

$$\langle \boldsymbol{x}, \boldsymbol{y} \rangle = x_1\overline{y}_1 + x_2\overline{y}_2 + \cdots + x_n\overline{y}_n.$$

Consequently, we have

$$\langle \boldsymbol{x}, \boldsymbol{y} \rangle = \overline{\mathbf{y}}^T\mathbf{x}.$$

It follows from (6.14),

$$\|\boldsymbol{x}\|^2 = \langle \boldsymbol{x}, \boldsymbol{x} \rangle = x_1\bar{x}_1 + x_2\bar{x}_2 + \cdots + x_n\bar{x}_n,$$

that is,

$$\|\boldsymbol{x}\| = \sqrt{\langle \boldsymbol{x}, \boldsymbol{x} \rangle} = \sqrt{|x_1|^2 + |x_2|^2 + \cdots + |x_n|^2}$$

The space $\mathbb{C}^n$ *together with this inner product is said to be the complex* **Euclidean space**.

The norm defined by (6.14) is a non-negative real function defined on V such that, for all $\boldsymbol{x}, \boldsymbol{y} \in V, \alpha \in \mathbb{C}$,

(i) $\|\boldsymbol{x}\| \geq 0$, and $\|\boldsymbol{x}\| = 0$ if and only if $\boldsymbol{x} = 0$;

(ii) $\|\alpha\boldsymbol{x}\| = |\alpha|\|\boldsymbol{x}\|$;

(iii)

$$\|\boldsymbol{x} + \boldsymbol{y}\| \leq \|\boldsymbol{x}\| + \|\boldsymbol{y}\|. \qquad \textit{Triangle inequality} \qquad (6.16)$$

Showing that (i)–(iii) are properties of the norm can be done similarly to the real case. In fact, the fundamental Cauchy–Schwarz inequality and parallelogram law do hold also in this setting and we make a note of this in the next theorem, whose proof can be easily adapted from the corresponding real case.

Theorem 6.2 *Let* V *be a complex inner product space. The following hold.*

(i) For $\boldsymbol{x}, \boldsymbol{y} \in V$,

$$|\langle \boldsymbol{x}, \boldsymbol{y} \rangle| \leq \|\boldsymbol{x}\|\|\boldsymbol{y}\| \qquad \textit{(Cauchy–Schwarz inequality)}$$

and

$$|\langle \boldsymbol{x}, \boldsymbol{y} \rangle| = \|\boldsymbol{x}\|\|\boldsymbol{y}\|$$

if and only if $\{\boldsymbol{x}, \boldsymbol{y}\}$ *is a linearly dependent set.*

(ii) For $\boldsymbol{x}, \boldsymbol{y} \in V$,

$$\|\boldsymbol{x} + \boldsymbol{y}\|^2 + \|\boldsymbol{x} - \boldsymbol{y}\|^2 = 2(\|\boldsymbol{x}\|^2 + \|\boldsymbol{y}\|^2). \qquad \textit{(Parallelogram law)}$$

In a complex inner product space V one can also make a coordinate approach to the inner product. We shall also get a Gram matrix corresponding to a fixed basis of V.

If $\mathcal{B} = (\boldsymbol{b}_1, \boldsymbol{b}_2, \dots, \boldsymbol{b}_n)$ is a basis of V, then, given $\boldsymbol{x}, \boldsymbol{y} \in V$ such that $\boldsymbol{x}_\mathcal{B} = (\alpha_1, \alpha_2, \dots, \alpha_n)$ and $\boldsymbol{y}_\mathcal{B} = (\beta_1, \beta_2, \dots, \beta_n)$, we have

$$\begin{aligned}
\langle \boldsymbol{x}, \boldsymbol{y} \rangle &= \langle \alpha_1 \boldsymbol{b}_1 + \alpha_2 \boldsymbol{b}_2 + \cdots + \alpha_n \boldsymbol{b}_n, \beta_1 \boldsymbol{b}_1 + \beta_2 \boldsymbol{b}_2 + \cdots + \beta_n \boldsymbol{b}_n \rangle \\
&= \overline{\beta}_1 \langle \boldsymbol{b}_1, \boldsymbol{b}_1 \rangle \alpha_1 + \overline{\beta}_1 \langle \boldsymbol{b}_2, \boldsymbol{b}_1 \rangle \alpha_2 + \dots \overline{\beta}_1 \langle \boldsymbol{b}_n, \boldsymbol{b}_1 \rangle \alpha_n + \\
&\quad + \overline{\beta}_2 \langle \boldsymbol{b}_1, \boldsymbol{b}_2 \rangle \alpha_1 + \overline{\beta}_2 \langle \boldsymbol{b}_2, \boldsymbol{b}_2 \rangle \alpha_2 + \dots \overline{\beta}_2 \langle \boldsymbol{b}_n, \boldsymbol{b}_2 \rangle \alpha_n + \\
&\quad \vdots \\
&\quad + \overline{\beta}_n \langle \boldsymbol{b}_1, \boldsymbol{b}_n \rangle \alpha_1 + \overline{\beta}_n \langle \boldsymbol{b}_2, \boldsymbol{b}_n \rangle \alpha_2 + \dots \overline{\beta}_n \langle \boldsymbol{b}_n, \boldsymbol{b}_n \rangle \alpha_n \\
&= \begin{bmatrix} \overline{\beta}_1 & \overline{\beta}_2 & \dots & \overline{\beta}_n \end{bmatrix} \underbrace{\begin{bmatrix} \langle \boldsymbol{b}_1, \boldsymbol{b}_1 \rangle & \langle \boldsymbol{b}_2, \boldsymbol{b}_1 \rangle & \dots & \langle \boldsymbol{b}_n, \boldsymbol{b}_1 \rangle \\ \langle \boldsymbol{b}_1, \boldsymbol{b}_2 \rangle & \langle \boldsymbol{b}_2, \boldsymbol{b}_2 \rangle & \dots & \langle \boldsymbol{b}_n, \boldsymbol{b}_2 \rangle \\ \vdots & \vdots & \ddots & \vdots \\ \langle \boldsymbol{b}_1, \boldsymbol{b}_n \rangle & \langle \boldsymbol{b}_2, \boldsymbol{b}_n \rangle & \dots & \langle \boldsymbol{b}_n, \boldsymbol{b}_n \rangle \end{bmatrix}}_{G} \begin{bmatrix} \alpha_1 \\ \alpha_2 \\ \vdots \\ \alpha_n \end{bmatrix}.
\end{aligned}$$

Hence, we can find a unique $n \times n$ complex matrix

$$G = \begin{bmatrix} \langle \boldsymbol{b}_1, \boldsymbol{b}_1 \rangle & \langle \boldsymbol{b}_2, \boldsymbol{b}_1 \rangle & \dots & \langle \boldsymbol{b}_n, \boldsymbol{b}_1 \rangle \\ \langle \boldsymbol{b}_1, \boldsymbol{b}_2 \rangle & \langle \boldsymbol{b}_2, \boldsymbol{b}_2 \rangle & \dots & \langle \boldsymbol{b}_n, \boldsymbol{b}_2 \rangle \\ \vdots & & & \\ \langle \boldsymbol{b}_1, \boldsymbol{b}_n \rangle & \langle \boldsymbol{b}_2, \boldsymbol{b}_n \rangle & \dots & \langle \boldsymbol{b}_n, \boldsymbol{b}_n \rangle \end{bmatrix}$$

such that

$$\langle \boldsymbol{x}, \boldsymbol{y} \rangle = \overline{\mathbf{y}}_\mathcal{B}^T G \mathbf{x}_\mathcal{B}.$$

The matrix $G = [g_{ij}]$, where, for all $i, j = 1, \dots, n$, we have $g_{ij} = \langle b_j, b_i \rangle$, is said to be the **Gram matrix** of the set of vectors $\{\boldsymbol{b}_1, \boldsymbol{b}_2, \dots, \boldsymbol{b}_n\}$.

Definition 53 *A hermitian matrix A of order n is said to be* **positive definite** *if, for all non-zero vectors $\boldsymbol{x} \in \mathbb{C}^n$,*

$$\overline{\mathbf{x}}^T A \mathbf{x} > 0.$$

Similarly to the real case, the next proposition, whose proof is left as an exercise, collects some properties of the Gram matrix.

Proposition 6.6 *Let V be a complex inner product space of dimension n and let $\mathcal{B} = (\boldsymbol{b}_1, \boldsymbol{b}_2, \dots, \boldsymbol{b}_n)$ be a basis of V. Let $\boldsymbol{x}, \boldsymbol{y}$ in be vectors in V. Then, there exists uniquely an $n \times n$ complex matrix G such that*

$$\langle \boldsymbol{x}, \boldsymbol{y} \rangle = \overline{\mathbf{y}}_\mathcal{B}^T G \mathbf{x}_\mathcal{B}, \tag{6.17}$$

where $\boldsymbol{x}_{\mathcal{B}}, \boldsymbol{y}_{\mathcal{B}}$ are, respectively, the coordinate vectors of $\boldsymbol{x}, \boldsymbol{y}$ relative to the basis $\mathcal{B}$. Moreover, G is a hermitian positive definite matrix, i.e., for $\boldsymbol{x} \neq 0$,

$$\overline{\mathbf{x}}_{\mathcal{B}}^T G \mathbf{x}_{\mathcal{B}} > 0. \tag{6.18}$$

The next two results are a counterpart of Proposition 6.4 and Proposition 6.5 for the complex inner product. Their proofs are an easy adaptation to the complex setting of those propositions and, for this reason, are left as an exercise.

Proposition 6.7 *Let A be a complex $n \times n$ matrix. The following are equivalent.*

(i) The expression

$$\langle \boldsymbol{x}, \boldsymbol{y} \rangle = \overline{\mathbf{y}}^T A \mathbf{x}$$

defines an inner product in $\mathbb{C}^n$.

(ii) A is a positive definite matrix.

The next proposition can be proved similarly to Proposition 6.5.

Proposition 6.8 *A hermitian matrix is positive definite if and only if its eigenvalues are positive numbers.*

This proposition shows that, when one considers a complex inner product space V, the Gram matrix corresponding to some basis of V is an invertible matrix.

6.3 Orthogonal Sets

This section is concerned with an intrinsically geometric notion: that of the angle between vectors. More to the point, a particular emphasis is given to vectors which are, in some sense, 'perpendicular' to each other.

Here, real and complex inner product spaces will be treated simultaneously. For this reason, we shall mostly refer to either as inner product spaces.

Definition 54 *Let $\boldsymbol{x}, \boldsymbol{y}$ be non-zero vectors in a real inner product space V. The* **angle** *between $\boldsymbol{x}$ and $\boldsymbol{y}$ is $0 \leq \theta \leq \pi$ such that*

$$\cos\theta = \frac{\langle \boldsymbol{x}, \boldsymbol{y} \rangle}{\|\boldsymbol{x}\|\|\boldsymbol{y}\|}.$$

Notice that the Cauchy–Schwarz inequality guarantees that $|\cos\theta| \leq 1$ and, therefore, the angle is well-defined.

Exercise 6.5 *Find the angle θ between the vectors $(1, 1, -1, 0), (0, 0, 0, 1)$ in the Euclidean space $\mathbb{R}^4$.*

Solution. Using Definition 54,

$$\cos\theta = \frac{\langle(1,1,-1,0),(0,0,0,1)\rangle}{\|(1,1,-1,0)\|\|(0,0,0,1)\|} = \frac{0}{\sqrt{3}} = 0.$$

Hence $\theta = \frac{\pi}{2}$.

Definition 55 *Let $\boldsymbol{x}, \boldsymbol{y}$ be vectors in a real or complex inner product space V. The vector $\boldsymbol{x}$ is said to be* **orthogonal** *to the vector $\boldsymbol{y}$, denoted by $\boldsymbol{x} \perp \boldsymbol{y}$, if $\langle \boldsymbol{x}, \boldsymbol{y}\rangle = 0$.*

Since clearly $\boldsymbol{x}$ is orthogonal to $\boldsymbol{y}$ if and only if $\boldsymbol{y}$ is orthogonal to $\boldsymbol{x}$, we say simply that the vectors $\boldsymbol{x}, \boldsymbol{y}$ are orthogonal.

In Exercise 6.5, we found that the vectors $(1, 1, -1, 0), (0, 0, 0, 1)$ in $\mathbb{R}^4$ are orthogonal, i.e., $(1, 1, -1, 0) \perp (0, 0, 0, 1)$.

Although we defined the angle between vectors only in real inner product spaces, the notion of orthogonality is valid in both real and complex inner product spaces. In the real case, two vectors are orthogonal if the angle between them is $\frac{\pi}{2}$.

Exercise 6.6 *Which vectors are orthogonal to $(1, 1, 0)$, in the Euclidean space $\mathbb{R}^3$?*

Solution. We want to find the vectors $(x, y, z) \in \mathbb{R}^3$ such that

$$0 = \langle(x,y,z),(1,1,0)\rangle = \begin{bmatrix}1 & 1 & 0\end{bmatrix}\begin{bmatrix}x\\y\\z\end{bmatrix} = x + y.$$

Hence the vectors orthogonal to $(1, 1, 0)$ are those in the plane whose cartesian equation is $x + y = 0$.

Theorem 6.3 (Pythagorean theorem) *Let $\boldsymbol{x}$ and $\boldsymbol{y}$ be orthogonal vectors in an inner product space V over $\mathbb{K}$. Then,*

$$\|\boldsymbol{x} + \boldsymbol{y}\|^2 = \|\boldsymbol{x}\|^2 + \|\boldsymbol{y}\|^2.$$

Proof *Let $\boldsymbol{x}, \boldsymbol{y}$ be vectors in V such that $\boldsymbol{x} \perp \boldsymbol{y}$. We have*

$$\begin{aligned}\|\boldsymbol{x} + \boldsymbol{y}\|^2 &= \langle \boldsymbol{x} + \boldsymbol{y}, \boldsymbol{x} + \boldsymbol{y}\rangle \\ &= \langle \boldsymbol{x}, \boldsymbol{x}\rangle + \langle \boldsymbol{y}, \boldsymbol{y}\rangle + \langle \boldsymbol{x}, \boldsymbol{y}\rangle + \langle \boldsymbol{y}, \boldsymbol{x}\rangle \\ &= \langle \boldsymbol{x}, \boldsymbol{x}\rangle + \langle \boldsymbol{y}, \boldsymbol{y}\rangle = \|\boldsymbol{x}\|^2 + \|\boldsymbol{y}\|^2.\end{aligned}$$

6.3.1 Orthogonal complement

We define next the notion of a vector orthogonal to a set.

Definition 56 *Let V be an inner product space and let X be a subset of V. A vector $\boldsymbol{x}$ is said to be* **orthogonal to** X*, denoted by $\boldsymbol{x} \perp X$, if $\boldsymbol{x}$ is orthogonal to all elements in X.*

For example, in Exercise 6.6 we saw that $(1,1,0)$ is orthogonal to the plane having the equation $x+y=0$.

Definition 57 *Let W be a subspace of a real or complex inner product space V. The* ***orthogonal complement*** $W^\perp$ *of the subspace W is defined by*

$$W^\perp = \{\boldsymbol{x} \in V : \boldsymbol{x} \perp W\}.$$

The orthogonal complement $W^\perp$ contains always the zero vector.

Exercise 6.7 *Find the orthogonal complement of the straight line U spanned by $(1,1,0)$.*

Solution. We want to find the vectors $(x,y,z) \in \mathbb{R}^3$ such that, for all $\alpha \in \mathbb{R}$, $(x,y,z) \perp \alpha(1,1,0)$. Hence,

$$0 = \langle \alpha(1,1,0), (x,y,z)\rangle = \alpha\langle (1,1,0), (x,y,z)\rangle = \alpha(x+y)$$

from which follows that the orthogonal complement $U^\perp$ of U is the plane with equation $x+y=0$.

Proposition 6.9 *Let V be an inner product space and let $W \subseteq V$ be a subspace. The orthogonal complement $W^\perp$ of W is a subspace of V.*

Proof *We need to show that $W^\perp$ is closed under vector addition and scalar multiplication. Let $\boldsymbol{x}, \boldsymbol{y}$ be vectors in $W^\perp$ and let $\boldsymbol{w} \in W$. Then*

$$\langle \boldsymbol{x}+\boldsymbol{y}, \boldsymbol{w}\rangle = \langle \boldsymbol{x}, \boldsymbol{w}\rangle + \langle \boldsymbol{y}, \boldsymbol{w}\rangle = 0+0 = 0,$$

which shows that $\boldsymbol{x}+\boldsymbol{y} \in W^\perp$.

Now let $\alpha \in \mathbb{K}$. Then

$$\langle \alpha\boldsymbol{x}, \boldsymbol{w}\rangle = \alpha\langle \boldsymbol{x}, \boldsymbol{w}\rangle = \alpha 0 = 0.$$

We have shown that $W^\perp \neq \emptyset$ is closed under vector addition and scalar multiplication thus concluding the proof that $W^\perp$ is a subspace of V.

Proposition 6.10 *Let V be an inner product space, let $W \subseteq V$ be a subspace and let $\{\boldsymbol{x}_1, \dots, \boldsymbol{x}_k\}$ be a spanning set for W. Then $\boldsymbol{y} \in V$ is orthogonal to W if and only if $\boldsymbol{y}$ is orthogonal to the set $\{\boldsymbol{x}_1, \dots, \boldsymbol{x}_k\}$.*

Proof *It is clear that, if $\boldsymbol{y} \in V$ is orthogonal to W, then it is also orthogonal to $\{\boldsymbol{x}_1, \dots, \boldsymbol{x}_k\}$. We show now that the converse also holds.*

Suppose that $\boldsymbol{y}$ is orthogonal to $\{\boldsymbol{x}_1, \boldsymbol{x}_2, \dots, \boldsymbol{x}_k\}$ and let $\boldsymbol{w} \in W$. Since $\{\boldsymbol{x}_1, \boldsymbol{x}_2, \dots, \boldsymbol{x}_k\}$ is a spanning set for W, it follows that there exist $\alpha_1, \dots, \alpha_k \in \mathbb{K}$ with

$$\boldsymbol{w} = \alpha_1 \boldsymbol{x}_1 + \cdots + \alpha_k \boldsymbol{x}_k.$$

Hence

$$\langle \boldsymbol{w}, \boldsymbol{y} \rangle = \alpha_1 \langle \boldsymbol{w}, \boldsymbol{x}_1 \rangle + \cdots + \alpha_k \langle \boldsymbol{w}, \boldsymbol{x}_k \rangle = 0,$$

which shows that $\boldsymbol{y}$ is orthogonal to all vectors in W.

The next result is an immediate corollary of this proposition.

Corollary 6.1 *Let V be an inner product space and let $W \subseteq V$ be a subspace. Then, $\boldsymbol{x} \in V$ is orthogonal to W if and only if it is orthogonal to a basis of W.*

Example 6.4 *Find the orthogonal complement of the plane W in $\mathbb{R}^3$ with cartesian equation $x = y$.*

Using Corollary 6.1, it suffices to find a basis of W and all the vectors in $\mathbb{R}^3$ orthogonal to that basis. The set $\{(1,1,0),(0,0,1)\}$ is a basis of W and (x,y,z) is orthogonal to this basis if and only if

$$\begin{cases} \langle (x,y,z),(1,1,0) \rangle = \begin{bmatrix} 1 & 1 & 0 \end{bmatrix} \begin{bmatrix} x \\ y \\ z \end{bmatrix} = 0 \\ \langle (x,y,z),(0,0,1) \rangle = \begin{bmatrix} 0 & 0 & 1 \end{bmatrix} \begin{bmatrix} x \\ y \\ z \end{bmatrix} = 0. \end{cases}$$

Hence, we have to solve the homogeneous system of linear equations

$$\begin{bmatrix} 1 & 1 & 0 \\ 0 & 0 & 1 \end{bmatrix} \begin{bmatrix} x \\ y \\ z \end{bmatrix} = \begin{bmatrix} 0 \\ 0 \end{bmatrix}.$$

It follows that $W^\perp$ is the straight line of which we give three possible sets of equations

$$\begin{cases} x = -y \\ z = 0 \end{cases} \qquad \textit{cartesian equations}$$

or

$$(x,y,z) = t(-1,1,0) \qquad (t \in \mathbb{R}) \qquad \textit{vector equation}$$

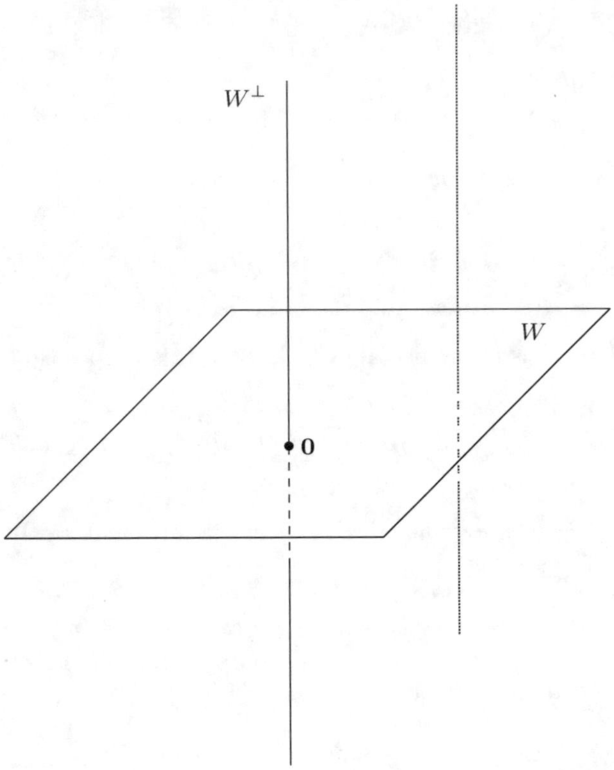

FIGURE 6.5: A plane W in $\mathbb{R}^3$, its orthogonal complement $W^\perp$ and a straight line perpendicular to the plane W.

or

$$\begin{cases} x = -t \\ y = t \\ z = 0 \end{cases} \qquad (t \in \mathbb{R}) \qquad \textit{parametric equations}$$

A common mistake. It is frequent that the orthogonal complement of a plane in $\mathbb{R}^3$ *(containing* $(0,0,0)$*), is confused with a (any) straight line perpendicular to that plane. Observe that, as seen in Proposition 6.9, the orthogonal complement is itself a subspace. Hence the orthogonal complement must contain* $(0,0,0)$.

The orthogonal complement of a plane in $\mathbb{R}^3$ ***is the straight line perpendicular to the plane which goes through*** $(0,0,0)$.

How to find the orthogonal complement of a subspace of the Euclidean space $\mathbb{K}^n$

Let W be a subspace of the Euclidean space $\mathbb{K}^n$ with $\dim W = k$. To determine the orthogonal complement of $W^\perp$ follow the next steps.

1. Find a basis $\{\boldsymbol{b}_1, \dots, \boldsymbol{b}_k\}$ of W.
2. Solve the homogeneous system of linear equations

$$\underbrace{\begin{bmatrix} \overline{\mathbf{b}_1}^T \\ \vdots \\ \overline{\mathbf{b}_k}^T \end{bmatrix}}_{k \times n} \underbrace{\begin{bmatrix} x_1 \\ \vdots \\ x_n \end{bmatrix}}_{n \times 1} = \underbrace{\begin{bmatrix} 0 \\ \vdots \\ 0 \end{bmatrix}}_{k \times 1}.$$

3. $W^\perp$ is the solution set of this system:

$$W^\perp = N\left(\begin{bmatrix} \overline{\mathbf{b}_1}^T \\ \vdots \\ \overline{\mathbf{b}_k}^T \end{bmatrix}\right). \tag{6.19}$$

Notice that, when dealing with $\mathbb{R}^n$, the bar over the vectors is irrelevant.

Observe that, since $W^\perp$ is the null space of

$$A = \begin{bmatrix} \overline{\mathbf{b}_1}^T \\ \vdots \\ \overline{\mathbf{b}_k}^T \end{bmatrix}$$

whose rank is $\operatorname{rank}(A) = k$, by Proposition 3.11 (i),

$$\dim W^\perp = n - \operatorname{rank}(A) = n - k.$$

Hence,

$$\dim W + \dim W^\perp = n. \tag{6.20}$$

We shall come back to this equality in Lemma 6.1.

Proposition 6.11 *Let W be a subspace of an inner product space V. Then, $W \cap W^\perp = \{\mathbf{0}\}$.*

Proof *Suppose that $\boldsymbol{x}$ is a vector lying in $W \cap W^\perp$. Then, $\boldsymbol{x}$ must be orthogonal to itself, i.e., $\boldsymbol{x} \perp \boldsymbol{x}$. Hence*

$$0 = \langle \boldsymbol{x}, \boldsymbol{x} \rangle = \|\boldsymbol{x}\|^2,$$

and we have that $\boldsymbol{x} = \mathbf{0}$.

Lemma 6.1 *Let V be an n-dimensional inner product space, let W be a subspace of V and let $W^\perp$ be its orthogonal complement. Then, if $\mathcal{B}_W$ and $\mathcal{B}_{W^\perp}$ are basis of W and $W^\perp$, respectively, then $\mathcal{B}_W \cup \mathcal{B}_{W^\perp}$ is a basis of V and*

$$\dim W + \dim W^\perp = n. \tag{6.21}$$

Proof *We show firstly that $\mathcal{B} = \mathcal{B}_W \cup \mathcal{B}_{W^\perp}$ is a linearly independent set. Let $\mathcal{B}_W = \{\boldsymbol{b}_1, \dots, \boldsymbol{b}_k\}$ and $\mathcal{B}_{W^\perp} = \{\boldsymbol{u}_1, \dots, \boldsymbol{u}_r\}$. Consider the linear combination*

$$\sum_{i=1}^{k} \alpha_i \boldsymbol{b}_i + \sum_{j=1}^{r} \beta_j \boldsymbol{u}_j = \mathbf{0}.$$

We must show that, for $i = 1, \dots, k$ and $j = 1, \dots, r$, all the scalars α_i, β_j are zero. Observe that the above equality is equivalent to

$$\sum_{i=1}^{k} \alpha_i \boldsymbol{b}_i = \sum_{j=1}^{r} -\beta_j \boldsymbol{u}_j.$$

It follows from Proposition 6.11 that

$$\sum_{i=1}^{k} \alpha_i \boldsymbol{b}_i = \mathbf{0} = \sum_{j=1}^{r} -\beta_j \boldsymbol{u}_j.$$

Since both sets $\mathcal{B}_W, \mathcal{B}_{W^\perp}$ are linearly independent, we have finally that, for $i = 1, \dots, k$ and $j = 1, \dots, r$, all the scalars $\alpha_i = 0$ and $\beta_j = 0$.

To finish the proof, it suffices to show that $r = n - k$. Since $\mathcal{B}$ is a linearly independent set, by Theorem 3.5 (ii), we already know that $r + k \leq n$.

Suppose that $r < n - k$. We show next that, in this case, it is possible to find a non-zero vector $\boldsymbol{v} \in W \cap W^\perp$. Observe that, by Proposition 6.11, this is a contradiction.

We want to find all the vectors $\boldsymbol{v} \in V$ such that $\boldsymbol{v} \perp W$ and $\boldsymbol{v} \perp W^\perp$. Hence, by Corollary 6.1, we want to find the solutions of the problem

$$\begin{cases} \langle \boldsymbol{v}, \boldsymbol{b}_i \rangle = 0, & i = 1, \dots, k \\ \langle \boldsymbol{v}, \boldsymbol{u}_j \rangle = 0, & j = 1, \dots, r. \end{cases}$$

If $\mathcal{B}'$ is a basis of V, this can be written in terms of the coordinate vectors relative to $\mathcal{B}'$ as the homogeneous system of linear equations

$$\underbrace{\overline{\begin{bmatrix} [\mathbf{b}_1]_{B'}^T \\ \vdots \\ [\mathbf{b}_k]_{B'}^T \\ [\mathbf{u}_1]_{B'}^T \\ \vdots \\ [\mathbf{u}_r]_{B'}^T \end{bmatrix}}}_{A} G\mathbf{v}_{\mathcal{B}'} = \mathbf{0},$$

where G is the Gram matrix relative to the basis $\mathcal{B}'$. Observe that, when V is a real inner product space, the bar over the first matrix is not relevant: it does not change anything, since the conjugate of a real number is that same number.

One can re-write the equation above as $A\mathbf{x} = \mathbf{0}$, where $\mathbf{x} := G\mathbf{v}_{\mathcal{B}'}$ lies in $\mathbb{K}^n$.

Since $\mathcal{B}$ is linearly independent, that is, the rows of the $(k+r)\times n$ matrix A are linearly independent, $\operatorname{rank}(A) = k + r$ *(see Exercise 6.8 below). In other words, the null space $N(A)$ has dimension $n - k - r \geq 1$. Consequently, $N(A) \neq \{\mathbf{0}\}$. Hence, we have a non-zero solution $\boldsymbol{v}$ of our initial problem. That is,*

$$\boldsymbol{v} \in W \cap W^{\perp} \neq \{\mathbf{0}\},$$

contradicting Proposition 6.11. It follows that $r = n - k$, as required.

Exercise 6.8 *Let B be a complex matrix. Show that* $\operatorname{rank}(B) = \operatorname{rank}(\overline{B})$.

Proposition 6.12 *Let W be a subspace of an inner product V and let $W^{\perp\perp} := (W^{\perp})^{\perp}$. Then, $W^{\perp\perp} = W$.*

Proof *It is clear from the definition that $W \subseteq W^{\perp\perp}$. Suppose now that* $\dim V = n$ *and* $\dim W = k$. *Hence, by* (6.21),

$$k = \dim W \leq \dim W^{\perp\perp} = n - (n - k) = k.$$

It follows that W is a subspace of $W^{\perp\perp}$ having the same dimension. Consequently, $W = W^{\perp\perp}$.

Theorem 6.4 *Let V be an inner product space and let W be a subspace of V. Then, for $\boldsymbol{x} \in V$, there exist uniquely $\boldsymbol{x}_W \in W$ and $\boldsymbol{x}_{W^{\perp}} \in W^{\perp}$ such that*

$$\boldsymbol{x} = \boldsymbol{x}_W + \boldsymbol{x}_{W^{\perp}}. \tag{6.22}$$

Notice that this theorem implies that every subspace W induces a splitting of V into a sum of two subspaces, that is, $V = W + W^{\perp}$.

Proof *By Lemma 6.1, we know that any $\boldsymbol{x} \in V$ is a linear combination of the vectors of the basis $\mathcal{B} = \mathcal{B}_W \cup \mathcal{B}_{W^{\perp}}$. It follows that*

$$\boldsymbol{x} = \sum_{i=1}^{k} \alpha_i \boldsymbol{b}_i + \sum_{j=1}^{n-k} \beta_j \boldsymbol{u}_j = \boldsymbol{x}_W + \boldsymbol{x}_{W^{\perp}}, \tag{6.23}$$

where we use the same notation as in the proof of Lemma 6.1. It only remains to prove that this decomposition of $\boldsymbol{x}$ relative to W and $W^{\perp}$ is unique. Suppose that $\boldsymbol{x} = \boldsymbol{x}_W + \boldsymbol{x}_{W^{\perp}}$ and $\boldsymbol{x} = \boldsymbol{x}'_W + \boldsymbol{x}'_{W^{\perp}}$, for some $\boldsymbol{x}'_W \in W$ and $\boldsymbol{x}'_{W^{\perp}} \in W^{\perp}$. Then,

$$\mathbf{0} = (\boldsymbol{x}_W - \boldsymbol{x}'_W) + (\boldsymbol{x}_{W^{\perp}} - \boldsymbol{x}'_{W^{\perp}}).$$

By the Pythagorean Theorem 6.3,

$$0 = \|\boldsymbol{x}_W - \boldsymbol{x}'_W\|^2 + \|\boldsymbol{x}_{W^\perp} - \boldsymbol{x}'_{W^\perp}\|^2$$

from which follows that

$$\boldsymbol{x}_W - \boldsymbol{x}'_W = \boldsymbol{0} = \boldsymbol{x}_{W^\perp} - \boldsymbol{x}'_{W^\perp}.$$

Hence, $\boldsymbol{x}_W = \boldsymbol{x}'_W$ *and* $\boldsymbol{x}_{W^\perp} = \boldsymbol{x}'_{W^\perp}$, *proving the uniqueness part of the theorem.*

Theorem 6.4 tells us that V is the direct sum of a(ny) subspace W of V and its orthogonal complement $W^\perp$, that is,

$$V = W \oplus W^\perp. \tag{6.24}$$

We aim now at finding a formula to calculate the parts $\boldsymbol{x}_W$ and $\boldsymbol{x}_{W^\perp}$ of a given vector $\boldsymbol{x}$ in an inner product space V. To that effect, it is convenient to choose bases of W and $W^\perp$ 'easy' to deal with. Having this in mind, we begin with a crucial notion, that of an orthogonal set.

Definition 58 *A subset X of an inner product space V is said to be an* **orthogonal set** *if, given any $\boldsymbol{x}, \boldsymbol{y} \in X$ with $\boldsymbol{x} \neq \boldsymbol{y}$, then $\boldsymbol{x} \perp \boldsymbol{y}$.*

Examples of orthogonal sets in $\mathbb{R}^2$ are the standard basis $\mathcal{E}_2$, $\{(1,2),(-2,1)\}$, $\{(1,2),(-2,1),(0,0)\}$, any set having a single element and the empty set.

Observe also, as hinted in the previous paragraph, that the zero vector $\boldsymbol{0}$ in an inner product space V is orthogonal to any vector and, hence, whenever included in an orthogonal set, the new set is still an orthogonal set.

Exercise 6.9 *Can you propose an answer (even if not a formal one) to the next questions?*

a) Let $X \subseteq \mathbb{R}^2$ be an orthogonal set not containing $(0,0)$. How many vectors, at most, lie in X?

b) Let $X \subseteq \mathbb{R}^3$ be an orthogonal set not containing $(0,0,0)$. How many vectors, at most, lie in X?

Proposition 6.13 *Let V be an inner product space and let $X = \{\boldsymbol{v}_1, \dots, \boldsymbol{v}_k\}$ be an orthogonal subset of V not containing $\boldsymbol{0}$. Then, X is a linearly independent set.*

Proof *We must show that, given $\alpha_1, \dots, \alpha_k \in \mathbb{K}$,*

$$\alpha_1\boldsymbol{v}_1 + \cdots + \alpha_k\boldsymbol{v}_k = \boldsymbol{0} \Rightarrow \alpha_1, \dots, \alpha_k = 0.$$

Suppose then that $\alpha_1 \boldsymbol{v}_1 + \cdots + \alpha_k \boldsymbol{v}_k = \mathbf{0}$. *Since* X *is an orthogonal set, for any* $j \in \{1, \ldots, k\}$, *we have*

$$\langle \alpha_1 \boldsymbol{v}_1 + \cdots + \alpha_k \boldsymbol{v}_k, \boldsymbol{v}_j \rangle = \alpha_j^2 \|\boldsymbol{v}_j\|^2 = 0.$$

Hence, since $\boldsymbol{v}_j \neq \mathbf{0}$, *it follows that* $\alpha_j = 0$, *concluding the proof.*

An obvious consequence of this proposition is that in an inner product space V, with $\dim V = n$, any orthogonal subset of V consisting of n non-zero vectors is a basis of V.

Now we are ready to answer the questions in Exercise 6.9. How many are they?
We can say confidently that: a) 2; b) 3.

The proof of the following corollary is left as an exercise.

Corollary 6.2 *Let* V *be an inner product space of dimension* n *and let* $X = \{\boldsymbol{v}_1, \ldots, \boldsymbol{v}_k\}$ *be an orthogonal subset of* V *not containing* $\mathbf{0}$. *Then* $k \leq n$. *Moreover, if* $k = n$, *then* X *is a basis of* V.

The form of the orthogonal complements of the subspaces associated with a real matrix are given in the next result.

Proposition 6.14 *Let* A *be a* $n \times k$ *real matrix and let* $\mathbb{R}^n$ *and* $\mathbb{R}^k$ *be endowed with the usual inner products* (6.2). *Then the following hold.*

(i) $L(A)^\perp = N(A)$.

(ii) $N(A)^\perp = L(A)$.

(iii) $C(A)^\perp = N(A^T)$.

(iv) $N(A^T)^\perp = C(A)$.

Proof *(i) Since the rows of* A *are a spanning set for* $L(A)$, *by Proposition 6.10, the orthogonal complement of* $L(A)$ *consists of the solution set of the system* $A\mathbf{x} = \mathbf{0}$. *In other words,* $L(A)^\perp = N(A)$.

(ii) By Proposition 6.12 and (i) of this proposition, we have

$$N(A)^\perp = L(A)^{\perp\perp} = L(A).$$

(iii) Observing that $C(A) = L(A^T)$, *by (i) of this proposition,*

$$C(A)^\perp = L(A^T)^\perp = N(A^T).$$

(iv) Using (iii) of this proposition and Proposition 6.12,

$$N(A^T)^\perp = C(A)^{\perp\perp} = C(A).$$

6.3.2 Orthogonal projections

We saw in the previous section that an orthogonal set of non-zero vectors in an inner product space V is linearly independent and, therefore, if it has enough vectors then it is, in fact, a basis of V. At this point, we are interested in analysing precisely those bases which are also orthogonal sets.

Definition 59 *Let V be an inner product space. A basis $\mathcal{B}$ of V is said to be an* **orthogonal basis** *if $\mathcal{B}$ is an orthogonal set and is said to be an* **orthonormal basis** *if $\mathcal{B}$ is an orthogonal basis whose vectors are all norm one vectors.*

For example, the standard basis $\mathcal{E}_n = (\boldsymbol{e}_1, \dots, \boldsymbol{e}_n)$ of $\mathbb{K}^n$ is an orthonormal basis, since its vectors are pairwise orthogonal and, for all $j = 1, \dots, n$, $\|\boldsymbol{e}_j\| = 1$.

A key feature of an orthogonal basis is how simple it is to calculate the coordinates of any vector relative to said basis. Let $\boldsymbol{x}$ be a vector in the inner product space V and let $\mathcal{B} = (\boldsymbol{b}_1, \dots, \boldsymbol{b}_n)$ be an ordered orthogonal basis of V. Suppose that the coordinate vector of $\boldsymbol{x}$ relative to $\mathcal{B}$ is

$$\boldsymbol{x}_{\mathcal{B}} = (\alpha_1, \dots, \alpha_n),$$

that is,

$$\boldsymbol{x} = \alpha_1 \boldsymbol{b}_1 + \cdots + \alpha_n \boldsymbol{b}_n.$$

Since $\mathcal{B}$ is an orthogonal set, we have, for $i = 1, \dots, n$,

$$\langle \boldsymbol{x}, \boldsymbol{b}_i \rangle = \sum_{j=1}^{n} \langle \alpha_j \boldsymbol{b}_j, \boldsymbol{b}_i \rangle = \alpha_i \|\boldsymbol{b}_i\|^2.$$

Hence, for $i = 1, \dots, n$,

$$\alpha_i = \frac{\langle \boldsymbol{x}, \boldsymbol{b}_i \rangle}{\|\boldsymbol{b}_i\|^2}. \tag{6.25}$$

Moreover, if $\mathcal{B}$ is an orthonormal basis, then we have even a simpler formula to calculate the coordinate vector of $\boldsymbol{x}$. For all $i = 1, \dots, n$,

$$\alpha_i = \langle \boldsymbol{x}, \boldsymbol{b}_i \rangle. \tag{6.26}$$

Example 6.5 *Find the coordinate vector of $(1, 2, 3)$ relative to the orthogonal basis $\mathcal{B} = ((1, 1, 0), (0, 0, 1), (1, -1, 0))$ of $\mathbb{R}^3$.*

Let $(1, 2, 3)_{\mathcal{B}} = (\alpha_1, \alpha_2, \alpha_3)$. Then, by (6.25), we have

$$\alpha_1 = \frac{\langle (1, 2, 3), ((1, 1, 0) \rangle}{2} = \frac{3}{2};$$

$$\alpha_2 = \frac{\langle (1, 2, 3), ((0, 0, 1) \rangle}{1} = 3;$$

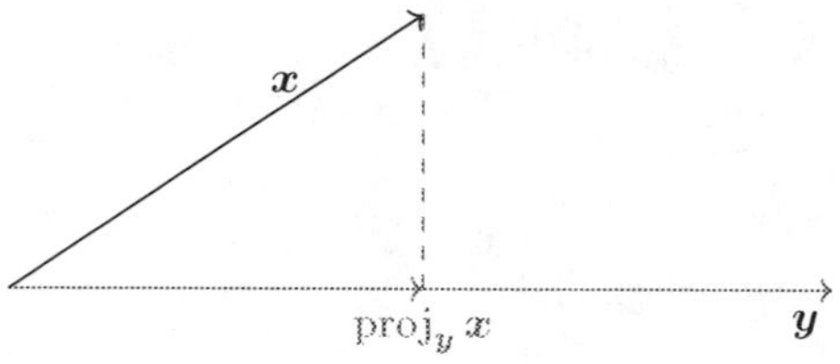

FIGURE 6.6: Orthogonal projection of $\boldsymbol{x}$ on $\boldsymbol{y}$.

$$\alpha_3 = \frac{\langle (1,2,3), ((1,-1,0)\rangle}{2} = -\frac{1}{2}.$$

Hence, $(1,2,3)_{\mathcal{B}} = (\frac{3}{2}, 3, -\frac{1}{2})$.

When dealing with orthogonal basis, as we have seen, it is very easy to obtain the coordinates of any given vector. However, one might ask 'Does an inner product space always possess an orthogonal basis?' The answer is 'Yes'. We shall see how to construct such a basis in §6.3.3. For the moment, we shall keep on developing the properties of an inner product space assuming that it has an orthogonal basis.

At this point, an observation is in order. We can get an orthonormal basis out of an orthogonal one. Indeed, let $\mathcal{B} = \{\boldsymbol{b}_1, \dots, \boldsymbol{b}_n\}$ be an orthogonal basis of V. Then, for $i = 1, \dots, n$, by a property of the norm

$$\left\| \frac{1}{\|\boldsymbol{b}_i\|} \boldsymbol{b}_i \right\| = \frac{1}{\|\boldsymbol{b}_i\|} \|\boldsymbol{b}_i\| = 1.$$

Hence, the set

$$\mathcal{B}' = \left\{ \frac{1}{\|\boldsymbol{b}_1\|} \boldsymbol{b}_1, \dots, \frac{1}{\|\boldsymbol{b}_n\|} \boldsymbol{b}_n \right\}$$

is an orthonormal basis of V.

If we apply this to the orthogonal basis of Example 6.5, we have that the basis

$$\mathcal{B}' = ((\tfrac{1}{\sqrt{2}}, \tfrac{1}{\sqrt{2}}, 0), (0,0,1), (\tfrac{1}{\sqrt{2}}, -\tfrac{1}{\sqrt{2}}, 0))$$

is an orthonormal basis of $\mathbb{R}^3$.

Summing up:

An orthogonal basis can be transformed into an orthonormal basis.

Now we know that orthonormal bases are easy to come by, provided we have orthogonal bases to start with.

Definition 60 *Let V be an inner product space and let $\boldsymbol{x}$ and $\boldsymbol{y}$ be vectors in V, with $\boldsymbol{y} \neq 0$. The* **orthogonal projection of $\boldsymbol{x}$ on $\boldsymbol{y}$** *is the vector defined by*

$$\operatorname{proj}_{\boldsymbol{y}} \boldsymbol{x} = \frac{\langle \boldsymbol{x}, \boldsymbol{y} \rangle}{\|\boldsymbol{y}\|^2} \boldsymbol{y}.$$

(Compare with (6.25), and check that this is exactly what we have in $\mathbb{R}^2$ and $\mathbb{R}^3$ for the usual inner product.)

Let W be a k-dimensional subspace of V and let

$$\mathcal{B}_W = \{\boldsymbol{b}_1, \ldots, \boldsymbol{b}_k\}, \qquad B_{W^\perp} = \{\boldsymbol{u}_1, \ldots, \boldsymbol{u}_{n-k}\}$$

be orthogonal bases for W and $W^\perp$, respectively. Then, by (6.23),

$$\boldsymbol{x} = \underbrace{\sum_{i=1}^{k} \alpha_i \boldsymbol{b}_i}_{\boldsymbol{x}_W} + \underbrace{\sum_{j=1}^{n-k} \beta_j \boldsymbol{u}_j}_{\boldsymbol{x}_{W^\perp}}.$$

Hence, by (6.25), we have that

$$\boldsymbol{x} = \underbrace{\sum_{i=1}^{k} \frac{\langle \boldsymbol{x}, \boldsymbol{b}_i \rangle}{\|\boldsymbol{b}_i\|^2} \boldsymbol{b}_i}_{\boldsymbol{x}_W} + \underbrace{\sum_{j=1}^{n-k} \frac{\langle \boldsymbol{x}, \boldsymbol{u}_j \rangle}{\|\boldsymbol{u}_j\|^2} \boldsymbol{u}_j}_{\boldsymbol{x}_{W^\perp}}. \tag{6.27}$$

In other words,

$$\boldsymbol{x} = \underbrace{\operatorname{proj}_{\boldsymbol{b}_1} \boldsymbol{x} + \cdots + \operatorname{proj}_{\boldsymbol{b}_k} \boldsymbol{x}}_{\boldsymbol{x}_W} + \underbrace{\operatorname{proj}_{\boldsymbol{u}_1} \boldsymbol{x} + \cdots + \operatorname{proj}_{\boldsymbol{u}_{n-k}} \boldsymbol{x}}_{\boldsymbol{x}_{W^\perp}}.$$

Definition 61 *Let V be an inner product space, let W be a subspace of V and let $\boldsymbol{x} \in V$. We define the* ***orthogonal projection*** $\operatorname{proj}_W \boldsymbol{x}$ *of the vector $\boldsymbol{x}$ on W by* $\operatorname{proj}_W \boldsymbol{x} = \boldsymbol{x}_W$.

Hence, we have

$$\boldsymbol{x} = \underbrace{\operatorname{proj}_{\boldsymbol{b}_1} \boldsymbol{x} + \cdots + \operatorname{proj}_{\boldsymbol{b}_k} \boldsymbol{x}}_{\operatorname{proj}_W} + \underbrace{\operatorname{proj}_{\boldsymbol{u}_1} \boldsymbol{x} + \cdots + \operatorname{proj}_{\boldsymbol{u}_{n-k}} \boldsymbol{x}}_{\operatorname{proj}_{W^\perp}}. \tag{6.28}$$

It is worth to point out three facts which are consequences of the definitions of orthogonal projections:
(i) In the extreme case $W = \{\boldsymbol{0}\}$, the orthogonal projection $proj_W \boldsymbol{x}$ of any given vector $\boldsymbol{x}$ is $\boldsymbol{0}$;
(ii) If W is spanned by a single non-zero vector $\boldsymbol{y}$, then

$$proj_W \boldsymbol{x} = \operatorname{proj}_{\boldsymbol{y}} \boldsymbol{x},$$

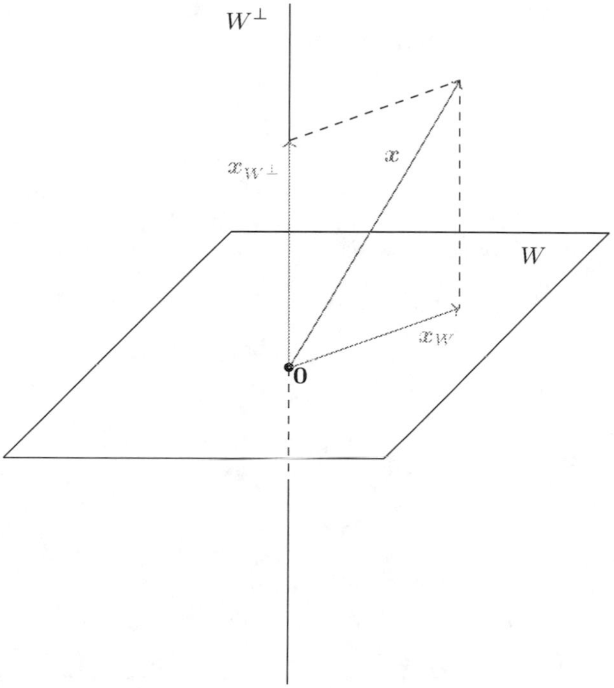

FIGURE 6.7: Decomposition of a vector $\boldsymbol{x} \in \mathbb{R}^3$ as a sum of orthogonal projections.

according to (6.28). Hence, for a 1-dimensional subspace W, the orthogonal projection $\boldsymbol{x}_W$ of $\boldsymbol{x}$ on W is, in fact, the orthogonal projection of $\boldsymbol{x}$ on the basis vector $\boldsymbol{y}$ (see Definition 60);
(iii) If W is a subspace of V

$$\boldsymbol{x} = \operatorname{proj}_W \boldsymbol{x} + \operatorname{proj}_{W^{\perp}} \boldsymbol{x} \tag{6.29}$$

Example 6.6 *Let W be the plane in the Euclidean space $\mathbb{R}^3$ whose cartesian equation is $x - y$. Find* $\operatorname{proj}_W(1,2,3)$ *and* $\operatorname{proj}_{W^{\perp}}(1,2,3)$.

We begin by determining $\operatorname{proj}_{W^{\perp}}(1,2,3)$. *Notice that* $\dim W = 2$ *and, by Lemma 6.1,*

$$\dim W^{\perp} = 3 - 2 = 1.$$

To use (6.27) and (6.29), we need to find an orthogonal basis of $W^{\perp}$ which is easy since any set having a single element is an orthogonal set.

Observe that, for $(x, y, z) \in W$,

$$\begin{bmatrix} x \\ y \\ z \end{bmatrix} = x \begin{bmatrix} 1 \\ 1 \\ 0 \end{bmatrix} + z \begin{bmatrix} 0 \\ 0 \\ 1 \end{bmatrix}.$$

Hence, a basis of W is

$$\mathcal{B}_W = \{(1,1,0),(0,0,1)\}.$$

Since

$$W^\perp = N\left(\begin{bmatrix}1 & 1 & 0\\ 0 & 0 & 1\end{bmatrix}\right)$$

(cf. (6.19)), it follows that one basis of $W^\perp$ is $\mathcal{B}_{W^\perp} = \{(1,-1,0)\}$. This is of course an orthogonal basis and, therefore,

$$\mathrm{proj}_{W^\perp}(1,2,3) = \frac{\langle (1,2,3),(1,-1,0)\rangle}{2}(1,-1,0) = \left(-\frac{1}{2},\frac{1}{2},0\right).$$

Finally, we have

$$\mathrm{proj}_W(1,2,3) = (1,2,3) - \left(-\frac{1}{2},\frac{1}{2},0\right) = \left(\frac{3}{2},\frac{3}{2},3\right).$$

In this particularly simple example, the basis of W is also an orthogonal basis.

How to find the orthogonal decomposition of a vector relative to a subspace

Let V be an inner product space, let W be a subspace of V and let $\mathcal{B}_W = \{\boldsymbol{b}_1,\dots,\boldsymbol{b}_k\}$ be an orthogonal basis of W. Given a vector $\boldsymbol{x}$ in V, to obtain the orthogonal projections $\mathrm{proj}_W \boldsymbol{x}$ and $\mathrm{proj}_{W^\perp} \boldsymbol{x}$:

1. Calculate $\mathrm{proj}_W \boldsymbol{x}$ using (6.27), i.e.,

$$\mathrm{proj}_W \boldsymbol{x} = \sum_{i=1}^{k} \frac{\langle \boldsymbol{x},\boldsymbol{b}_i\rangle}{\|\boldsymbol{b}_i\|^2}\boldsymbol{b}_i;$$

2. $\mathrm{proj}_{W^\perp} \boldsymbol{x} = \boldsymbol{x} - \mathrm{proj}_W \boldsymbol{x}$;

3. Finally $\boldsymbol{x} = \mathrm{proj}_W \boldsymbol{x} + \mathrm{proj}_{W^\perp} \boldsymbol{x}$.

Consider the Euclidean space $\mathbb{R}^n$. Suppose that the subspace $W \subseteq \mathbb{R}^n$ is endowed with an orthonormal basis $(\mathbf{b}_1,\mathbf{b}_2,\dots,\mathbf{b}_k)$. Then, given $\mathbf{x} \in \mathbb{R}^n$,

$$\mathrm{proj}_W \mathbf{x} = \langle \mathbf{x},\mathbf{b}_1\rangle\mathbf{b}_1 + \langle \mathbf{x},\mathbf{b}_2\rangle\mathbf{b}_2 + \cdots + \langle \mathbf{x},\mathbf{b}_k\rangle\mathbf{b}_k.$$

Let B be the $n\times k$ matrix whose columns are the vectors $\mathbf{b}_1,\mathbf{b}_2,\dots,\mathbf{b}_k$. Then the equality above can be seen as a linear combination of the columns of B

whose coefficients are, respectively, $\mathbf{b}_1^T\mathbf{x}, \ldots, \mathbf{b}_k^T\mathbf{x}$. It follows that we can write the equality as

$$\text{proj}_W\, \mathbf{x} = BB^T\mathbf{x}.$$

In other words, one can construct a linear transformation, the projection P_W onto the subspace W, $P_W : \mathbb{R}^n \to \mathbb{R}^n$ defined by

$$P_W(\mathbf{x}) = BB^T\mathbf{x} = \mathbf{b}_1\mathbf{b}_1^T\mathbf{x} + \mathbf{b}_2\mathbf{b}_2^T\mathbf{x} + \cdots + \mathbf{b}_k\mathbf{b}_k^T\mathbf{x}, \tag{6.30}$$

where, for each $i \in \{1, 2, \ldots, k\}$, the $n \times n$ matrix $\mathbf{b}_i\mathbf{b}_i^T$ is a matrix corresponding to the projection onto the subspace spanned by $\mathbf{b}_i$. Each of these matrices has rank 1, since its columns are multiples of the first column.

Notice that the second equality in (6.30) reiterates that the orthogonal projection onto W is the sum of the orthogonal projections on the basis vectors (see (6.28)).

Exercise 6.10 *Show that the linear transformation $P_W : \mathbb{R}^n \to \mathbb{R}^n$ is such that $P_W^2 = P_W$ and $P_W^T = P_W$. That is, the matrix*

$$A := \sum_{i=1}^{k} \mathbf{b}_i\mathbf{b}_i^T$$

corresponding to the linear transformation P_W is symmetric ($A^T = A$) and idempotent ($A^2 = A$).

A real or complex matrix A such that $A^2 = A$ is said to be ***idempotent***. Matrix A above corresponds to the orthogonal projection onto W.

Exercise 6.11 *Let W be a subspace of the Euclidean space $\mathbb{C}^n$ with a orthonormal basis $(\mathbf{b}_1, \mathbf{b}_2, \ldots, \mathbf{b}_k)$. Show that the projection $P_W : \mathbb{C}^n \to \mathbb{C}^n$ onto the subspace W is given by the hermitian idempotent matrix*

$$A := B\overline{B}^T = \sum_{i=1}^{k} \mathbf{b}_i\overline{\mathbf{b}}_i^T, \tag{6.31}$$

where B is the $n \times k$ matrix whose columns are, respectively, $\mathbf{b}_1, \mathbf{b}_2, \ldots, \mathbf{b}_k$. This matrix is hermitian and idempotent: it corresponds to the orthogonal projection onto W.

In (6.31), for each $i \in \{1, 2, \ldots, k\}$, the $n \times n$ matrix $\mathbf{b}_i\overline{\mathbf{b}}_i^T$ is a matrix corresponding to the projection onto the subspace spanned by $\mathbf{b}_i$. Each of these matrices have rank 1, since we have again that each column is obtained by multiplying the first column by a complex number.

In the *How to determine the matrix of the orthogonal projection onto a subspace* box of Section 7.1, we give a formula for the projection matrix

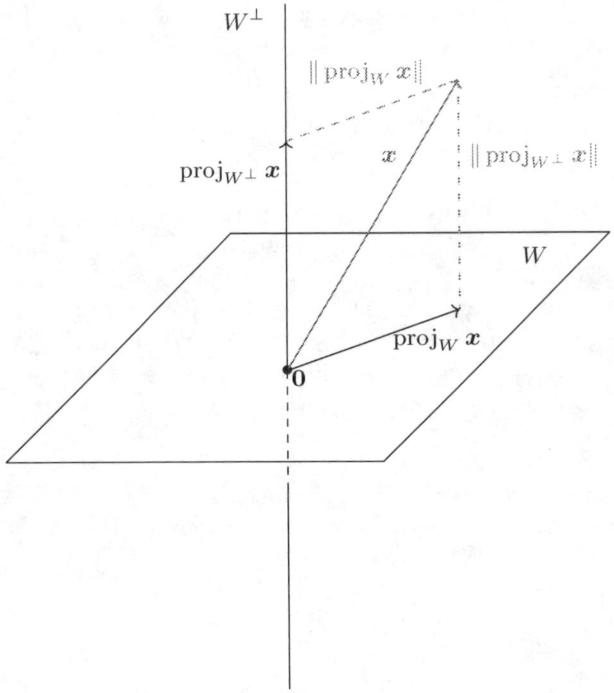

FIGURE 6.8: Best approximation of a vector $\boldsymbol{x}$ of $\mathbb{R}^3$ in a plane W and in the straight line $W^\perp$.

corresponding to the orthogonal projection onto a subspace, given a (not necessarily orthonormal) basis of this subspace.

In applications, it is sometimes necessary to find which is the element of a subspace W of an inner product space V that is the closest to a given vector $\boldsymbol{x} \in V$ (see Section 7.1). We show next that such a best approximation of $\boldsymbol{x}$ exists in W and shall devise a way to calculate it.

Let W be a subspace of an inner product space V and let $\boldsymbol{x} \in V$. Given $\boldsymbol{y} \in W$, the distance between $\boldsymbol{x}$ and $\boldsymbol{y}$ satisfies, by Theorem 6.3,

$$\begin{aligned}
\mathrm{d}(\boldsymbol{x}, \boldsymbol{y})^2 = \|\boldsymbol{x} - \boldsymbol{y}\|^2 &= \|\operatorname{proj}_W \boldsymbol{x} + \operatorname{proj}_{W^\perp} \boldsymbol{x} - \boldsymbol{y}\|^2 \\
&= \|\underbrace{\operatorname{proj}_{W^\perp} \boldsymbol{x}}_{\in W^\perp} + \underbrace{(\operatorname{proj}_W \boldsymbol{x} - \boldsymbol{y})}_{\in W}\|^2 \\
&= \|\operatorname{proj}_{W^\perp} \boldsymbol{x}\|^2 + \|(\operatorname{proj}_W \boldsymbol{x} - \boldsymbol{y})\|^2.
\end{aligned}$$

It follows that the element $\boldsymbol{y} \in W$ closest to $\boldsymbol{x}$ is that which annihilates $\|(\operatorname{proj}_W \boldsymbol{x} - \boldsymbol{y})\|$, that is, $\boldsymbol{y} = \operatorname{proj}_W \boldsymbol{x}$.

Definition 62 *Let V be an inner product space and let $W \subseteq V$ be a subspace. For $\boldsymbol{x}$ in V, the* **best approximation** *of $\boldsymbol{x}$ in W is $\operatorname{proj}_W \boldsymbol{x}$.*

The **distance of the point $\boldsymbol{x}$ to the subspace W** *is*

$$\mathrm{d}(\boldsymbol{x}, W) = \| \operatorname{proj}_{W^\perp} \boldsymbol{x} \|.$$

Example 6.7 *Using Example 6.6, the best approximation of $(1,2,3)$ in W is*

$$\operatorname{proj}_W(1,2,3) = \left(\frac{3}{2}, \frac{3}{2}, 3\right)$$

and

$$\mathrm{d}((1,2,3), W) = \left\| \left(-\frac{1}{2}, \frac{1}{2}, 0\right) \right\| = \sqrt{\frac{1}{4} + \frac{1}{4}} = \frac{\sqrt{2}}{2}.$$

Similarly, the best approximation of $(1,2,3)$ in $W^\perp$ is

$$\operatorname{proj}_{W^\perp}(1,2,3) = \left(-\frac{1}{2}, \frac{1}{2}, 0\right)$$

and

$$\mathrm{d}((1,2,3), W^\perp) = \left\| \left(\frac{3}{2}, \frac{3}{2}, 3\right) \right\| = \sqrt{\frac{9}{4} + \frac{9}{4} + 9} = \frac{\sqrt{54}}{2}.$$

Alternatively, using Theorem 6.3,

$$\mathrm{d}((1,2,3), W^\perp)^2 = \|\boldsymbol{x}\|^2 - \left(\frac{\sqrt{2}}{2}\right)^2 = 14 - \frac{2}{4},$$

and we have again

$$\mathrm{d}((1,2,3), W^\perp) = \frac{\sqrt{54}}{2}.$$

6.3.3 Gram–Schmidt process

The formula to calculate a projection of a vector on a subspace (cf. Definition 61) needs that we have an orthogonal basis of that subspace. In Example 6.6, we luckily avoided this because $W^\perp$ was a 1-dimensional subspace where, therefore, all bases are orthogonal.

We shall derive a method to transform a basis into an orthogonal basis. This is will be done using the so-called Gram–Schmidt Process. This method will be motivated with a concrete example.

Let $X = \{(-1,0,1,0), (1,-2,0,-1), (0,0,1,1)\}$ be a spanning set of a subspace S of the Euclidean space $\mathbb{R}^4$, and consider the problem of finding a orthonormal basis $\mathcal{B}$ for S. The tougher question is to find an orthogonal basis, since we saw how to obtain an orthonormal basis from any given orthogonal basis (see Section 6.3.2).

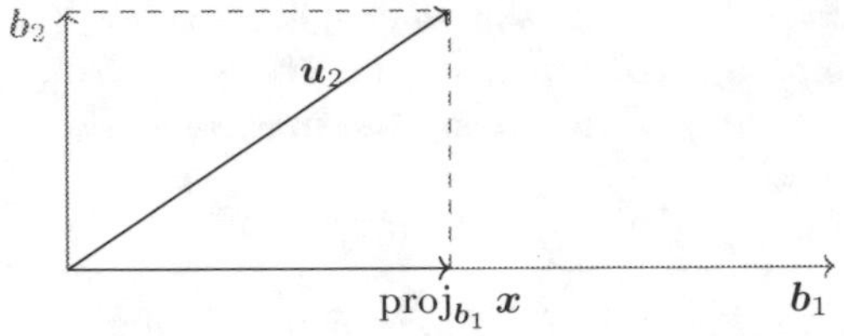

FIGURE 6.9: Step 1 – orthogonal projection of $\boldsymbol{u}_2$ on $\boldsymbol{b}_1$.

Hence, we will first solve the problem of finding an orthogonal basis $\mathcal{B}' = \{\boldsymbol{b}_1, \boldsymbol{b}_2, \boldsymbol{b}_3\}$ of S. Observe that X is linearly independent (although not orthogonal) and, therefore, $\mathcal{B}'$ must have also three vectors.

Set the notation:

$$\boldsymbol{u}_1 = (-1, 0, 1, 0),\ \boldsymbol{u}_2 = (1, -2, 0, -1),\ \boldsymbol{u}_3 = (0, 0, 1, 1).$$

Step 1. Choose a (any) vector from X, say, $\boldsymbol{u}_1 = (-1, 0, 1, 0)$, and set it as the first vector of the new basis: $\boldsymbol{b}_1 = \boldsymbol{u}_1 = (-1, 0, 1, 0)$.

Step 2. Let $\boldsymbol{b}_2 = \boldsymbol{u}_2 - \text{proj}_{\boldsymbol{u}_1} \boldsymbol{u}_2$. Notice that

$$\text{span}\{\boldsymbol{b}_1, \boldsymbol{b}_2\} = \text{span}\{\boldsymbol{b}_1, \boldsymbol{u}_2\} = \text{span}\{\boldsymbol{u}_1, \boldsymbol{u}_2\}.$$

It is easily seen that the first equality holds (hint: $\text{proj}_{\boldsymbol{u}_1} \boldsymbol{u}_2$ is spanned by $\boldsymbol{u}_1$). The second equality is obvious.

Observe that, by Theorem 6.4, $\boldsymbol{b}_2$ lies in $(\text{span}\{\boldsymbol{b}_1\})^\perp$. Hence, $\{\boldsymbol{b}_1, \boldsymbol{b}_2\}$ is an orthogonal basis of $\text{span}\{\boldsymbol{u}_1, \boldsymbol{u}_2\}$.

Since

$$\text{proj}_{\boldsymbol{u}_1} \boldsymbol{u}_2 = \frac{\langle \boldsymbol{u}_2, \boldsymbol{u}_1 \rangle}{\|\boldsymbol{u}_1\|^2} \boldsymbol{u}_1 = -\tfrac{1}{2}\boldsymbol{u}_1 = -\tfrac{1}{2}(-1, 0, 1, 0),$$

we have that

$$\boldsymbol{b}_2 = (1, -2, 0, -1) - (\tfrac{1}{2}, 0, -\tfrac{1}{2}, 0) = (\tfrac{1}{2}, -2, \tfrac{1}{2}, -1)$$

(check that $\boldsymbol{b}_1 \perp \boldsymbol{b}_2$).

By now, probably it is already clear from Figure 6.10 the way things are going.

Step 3. Let

$$\boldsymbol{b}_3 = \boldsymbol{u}_3 - \text{proj}_{\text{span}\{\boldsymbol{b}_1, \boldsymbol{b}_2\}} \boldsymbol{u}_3 = \boldsymbol{u}_3 - (\text{proj}_{\boldsymbol{b}_1} \boldsymbol{u}_3 + \text{proj}_{\boldsymbol{b}_2} \boldsymbol{u}_3).$$

Notice that

$$\text{span}\{\boldsymbol{b}_1, \boldsymbol{b}_2, \boldsymbol{b}_3\} = \text{span}\{\boldsymbol{b}_1, \boldsymbol{b}_2, \boldsymbol{u}_3\} = S$$

and that, by Theorem 6.4, $\boldsymbol{b}_3$ lies in $(\text{span}\{\boldsymbol{b}_1, \boldsymbol{b}_2\})^\perp$. We have

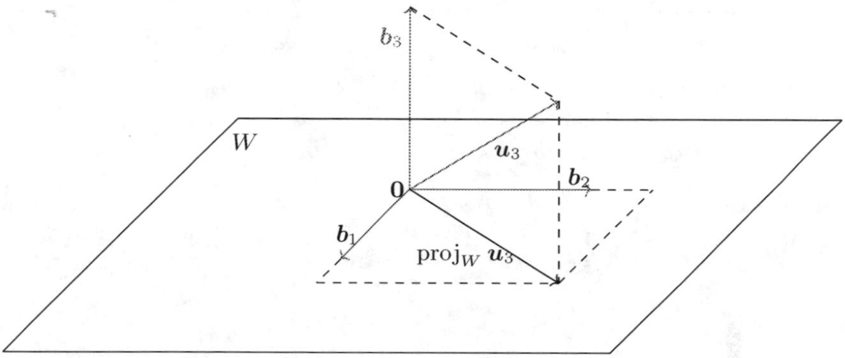

FIGURE 6.10: Step 2 – orthogonal projection of $\boldsymbol{u}_3$ on $W = \text{span}\{\boldsymbol{b}_1, \boldsymbol{b}_2\}$.

$$\begin{aligned}\boldsymbol{b}_3 &= \boldsymbol{u}_3 - (\text{proj}_{\boldsymbol{b}_1} \boldsymbol{u}_3 + \text{proj}_{\boldsymbol{b}_2} \boldsymbol{u}_3) \\ &= \boldsymbol{u}_3 - \left(\frac{\langle \boldsymbol{u}_3, \boldsymbol{b}_1 \rangle}{\|\boldsymbol{b}_1\|^2} \boldsymbol{b}_1 + \frac{\langle \boldsymbol{u}_3, \boldsymbol{b}_2 \rangle}{\|\boldsymbol{b}_2\|^2} \boldsymbol{b}_2 \right) \\ &= \left(\tfrac{6}{11}, -\tfrac{2}{11}, \tfrac{6}{11}, \tfrac{10}{11}\right).\end{aligned}$$

Hence, the set $\mathcal{B}' = \{\boldsymbol{b}_1, \boldsymbol{b}_2, \boldsymbol{b}_3\}$ is an orthogonal basis of S. Consequently,

$$\mathcal{B} = \left\{ \tfrac{1}{\sqrt{2}}(-1, 0, 1, 0), \tfrac{2}{11}\left(\tfrac{1}{2}, -2, \tfrac{1}{2}, -1\right), \tfrac{1}{176}(6, -2, 6, 10) \right\}$$

is an orthonormal basis for S.

Gram–Schmidt process

Let V be an inner product space and let S be a subspace of V spanned by a linearly independent subset $X = \{\boldsymbol{u}_1, \boldsymbol{u}_2, \boldsymbol{u}_3, \dots, \boldsymbol{u}_{k-1}, \boldsymbol{u}_k\}$ of V. To obtain a orthogonal basis $\{\boldsymbol{b}_1, \boldsymbol{b}_2, \boldsymbol{b}_3, \dots, \boldsymbol{b}_{k-1}, \boldsymbol{b}_k\}$ of S, proceed as follows.

1. Fix a(ny) vector, say, $\boldsymbol{b}_1 = \boldsymbol{u}_1$.
2. The second vector of the new basis is
$$\boldsymbol{b}_2 = \boldsymbol{u}_2 - \operatorname{proj}_{\boldsymbol{b}_1} \boldsymbol{u}_2,$$
that is, remove from $\boldsymbol{u}_2$ its component along $\boldsymbol{b}_1$ (already spanned by $\boldsymbol{b}_1$).
3. The next vector $\boldsymbol{b}_3$ coincides with $\boldsymbol{u}_3$ minus its projection on the subspace spanned by $\boldsymbol{b}_1, \boldsymbol{b}_2$, that is
$$\boldsymbol{b}_3 = \boldsymbol{u}_3 - (\operatorname{proj}_{\boldsymbol{b}_1} \boldsymbol{u}_3 + \operatorname{proj}_{\boldsymbol{b}_2} \boldsymbol{u}_3)$$
4. Continue this process up to the last vector
$$\boldsymbol{b}_k = \boldsymbol{u}_k - (\operatorname{proj}_{\boldsymbol{b}_1} \boldsymbol{u}_k + \operatorname{proj}_{\boldsymbol{b}_2} \boldsymbol{u}_k + \cdots + \operatorname{proj}_{\boldsymbol{b}_{k-1}} \boldsymbol{u}_k).$$

An orthonormal basis of S is
$$\left\{\frac{1}{\|\boldsymbol{b}_1\|}\boldsymbol{b}_1, \frac{1}{\|\boldsymbol{b}_2\|}\boldsymbol{b}_2, \frac{1}{\|\boldsymbol{b}_3\|}\boldsymbol{b}_3, \dots, \frac{1}{\|\boldsymbol{b}_{k-1}\|}\boldsymbol{b}_{k-1}, \frac{1}{\|\boldsymbol{b}_k\|}\boldsymbol{b}_k\right\}.$$

6.4 Orthogonal and Unitary Diagonalisation

Definition 63 *A matrix S in $\mathrm{M}_n(\mathbb{R})$ is said to be an* **orthogonal matrix** *if $SS^T = I$.*

If S is an orthogonal matrix, then S and S^T are inverses of each other. By Proposition 1.17, we have that a matrix S is orthogonal if and only if
$$SS^T = I = S^T S.$$

Proposition 6.15 *Let S be a matrix in $\mathrm{M}_n(\mathbb{R})$. Then the following are equivalent.*

(i) S is an orthogonal matrix.

(ii) $S^T S = I$.

(iii) $SS^T = I$.

(iv) The columns of S *are an orthonormal basis of* $\mathbb{R}^n$.

(v) The rows of S *are an orthonormal basis of* $\mathbb{R}^n$.

Proof *The equivalence between (i), (ii), and (iii) has been proved above. Notice that*

$$(S^T S)_{ij} = \mathbf{c}_i^T \mathbf{c}_j = \langle \mathbf{c}_i, \mathbf{c}_j \rangle,$$

where $\mathbf{c}_i, \mathbf{c}_j$ *are, respectively, the columns* i *and* j *of* S. *Hence, (ii) yields that* $\langle \mathbf{c}_i, \mathbf{c}_j \rangle = 0$, *for* $i \neq j$, *and*

$$\langle \mathbf{c}_i, \mathbf{c}_i \rangle = \|\mathbf{c}_i\|^2 = 1.$$

It follows that the columns of S *form an orthonormal set having* n *vectors which, therefore, are an orthonormal basis of* $\mathbb{R}^n$. *Similarly (iv) implies (ii).*

We prove likewise that (v) holds if and only if (iii) holds, one has only to consider equality $SS^T = I$ *and use a similar reasoning.*

Definition 64 *A real square matrix is said to be* **orthogonally diagonalisable,** *if there exist a diagonal matrix* D *and an orthogonal matrix* S *such that* $D = S^T AS$.

If a real matrix A is orthogonally diagonalisable, then

$$A^T = (SDS^T)^T = SDS^T = A,$$

that is, A is a real symmetric matrix. In fact, the converse is also true. Before proving this, we need an auxiliary result.

We already know that the spectrum of a real symmetric matrix is a non-empty set of real numbers (see Corollary 4.1). Now we add something about its eigenspaces.

Lemma 6.2 *Let* A *be a real symmetric matrix and let* $\mathbf{x}_1, \mathbf{x}_2$ *be eigenvectors of* A *associated with distinct eigenvalues* λ_1, λ_2, *respectively. Then,* $\mathbf{x}_1 \perp \mathbf{x}_2$.

Proof *Let* λ_1, λ_2 *and* $\mathbf{x}_1, \mathbf{x}_2$ *be as above. Then,*

$$\langle \mathbf{x}_1, A\mathbf{x}_2 \rangle = (A\mathbf{x}_2)^T \mathbf{x}_1 = \mathbf{x}_2^T A^T \mathbf{x}_1 = \mathbf{x}_2^T A\mathbf{x}_1 = \langle A\mathbf{x}_1, \mathbf{x}_2 \rangle = \lambda_1 \langle \mathbf{x}_1, \mathbf{x}_2 \rangle.$$

However, since $\langle \mathbf{x}_1, A\mathbf{x}_2 \rangle = \lambda_2 \langle \mathbf{x}_1, \mathbf{x}_2 \rangle$, *we have*

$$(\lambda_1 - \lambda_2)\langle \mathbf{x}_1, \mathbf{x}_2 \rangle = 0.$$

Hence, $\langle \mathbf{x}_1, \mathbf{x}_2 \rangle = 0$.

Theorem 6.5 *A real square matrix is orthogonally diagonalisable if and only if it is symmetric.*

Proof *It only remains to prove that, if A is symmetric, then A is orthogonally diagonalisable. We prove this by induction on the size of the matrix.*

If A is 1×1, the result holds trivially. Suppose now that A is an $n \times n$ matrix, with $n \geq 2$, and that the assertion holds for all square matrices of order $n-1$.

Let λ be an eigenvalue of A and let $\mathbf{x}$ be a norm one eigenvector corresponding to λ.

Let $\mathcal{B} = \{\mathbf{x}\} \cup \mathcal{B}_\perp$ be an orthonormal basis of $\mathbb{R}^n$, where $\mathcal{B}_\perp$ is a basis of $\operatorname{span}\{\mathbf{x}\}^\perp$. *Then, given a vector $\mathbf{y}$ in $\mathcal{B}_\perp$,*

$$\langle A\mathbf{y}, \mathbf{x}\rangle = \mathbf{x}^T A\mathbf{y} = \mathbf{x}^T A^T \mathbf{y} = (A\mathbf{x})^T \mathbf{y} = \lambda(\mathbf{x})^T \mathbf{y} = \lambda\langle \mathbf{x}, \mathbf{y}\rangle = 0.$$

Hence, the matrix S_1 whose columns consist of the vectors of $\mathcal{B}$ is orthogonal and such that

$$S_1^T A S_1 = \begin{bmatrix} \lambda & 0 \\ 0 & M \end{bmatrix}.$$

Since A is symmetric, M is an $n-1 \times n-1$ symmetric matrix. Hence, by the induction hypothesis, there exists an $n-1 \times n-1$ orthogonal matrix N such that $M = ND_1N^T$. In other words,

$$S_1^T A S_1 = \begin{bmatrix} 1 & 0 \\ 0 & N \end{bmatrix} \begin{bmatrix} \lambda & 0 \\ 0 & D_1 \end{bmatrix} \begin{bmatrix} 1 & 0 \\ 0 & N \end{bmatrix}^T.$$

Notice that

$$\begin{bmatrix} 1 & 0 \\ 0 & N \end{bmatrix}$$

is an orthogonal matrix. It now follows that

$$A = S_1 \begin{bmatrix} 1 & 0 \\ 0 & N \end{bmatrix} \begin{bmatrix} \lambda & 0 \\ 0 & D_1 \end{bmatrix} \begin{bmatrix} 1 & 0 \\ 0 & N \end{bmatrix}^T S_1^T,$$

where

$$S = S_1 \begin{bmatrix} 1 & 0 \\ 0 & N \end{bmatrix}$$

is an orthogonal matrix.

Corollary 6.3 *Let A be a real positive definite matrix. Then, there exists a non-singular matrix B such that $A = BB^T$.*

Proof *By the above theorem, we know that A is orthogonally diagonalisable: $A = SDS^T$, for some orthogonal matrix S and a diagonal matrix with positive diagonal entries. Setting $D^{\frac{1}{2}}$ as the matrix whose entries are the square roots of the corresponding entries of D, we have*

$$A = SD^{\frac{1}{2}} D^{\frac{1}{2}} S^T = (SD^{\frac{1}{2}})(SD^{\frac{1}{2}})^T.$$

Hence, $A = BB^T$, where $B = SD^{\frac{1}{2}}$.

We prove now the remaining implication in Proposition 6.5.

Proof *of Proposition 6.5 continued.* *Conversely, let A be a real symmetric matrix with positive eigenvalues only. By Theorem 6.5, a real symmetric matrix is diagonalisable and the diagonalising matrix S is such that $S^{-1} = S^T$. Hence, given a non-zero $\mathbf{x} \in \mathbb{R}^n$,*

$$\mathbf{x}^T A\mathbf{x} = \mathbf{x}^T SDS^T\mathbf{x}.$$

Let D' be the diagonal matrix whose entries are the square root of the corresponding entries of D. Then,

$$\mathbf{x}^T A\mathbf{x} = (S^T\mathbf{x})^T D'D'S^T\mathbf{x} = (D'S^T\mathbf{x})^T(D'S^T\mathbf{x}) = \langle D'S^T\mathbf{x}, D'S^T\mathbf{x}\rangle,$$

that is,

$$\mathbf{x}^T A\mathbf{x} = \|D'S^T\mathbf{x}\|^2 > 0,$$

as required.

If A is orthogonally diagonalisable, then

$$D = \begin{bmatrix} \lambda_1 & & & \\ & \lambda_2 & & \\ & & \ddots & \\ & & & \lambda_n \end{bmatrix}, \tag{6.32}$$

where the diagonal entries of D are the n eigenvalues of A, and if $S = \begin{bmatrix} \mathbf{u}_1 & \mathbf{u}_2 & \dots & \mathbf{u}_n \end{bmatrix}$ is the diagonalising orthogonal matrix, then

$$A = SDS^T = \lambda_1\mathbf{u}_1\mathbf{u}_1^T + \lambda_2\mathbf{u}_2\mathbf{u}_2^T + \cdots + \lambda_n\mathbf{u}_n\mathbf{u}_n^T.$$

Moreover, $\{\mathbf{u}_1, \mathbf{u}_2, \dots, \mathbf{u}_n\}$ is an orthonormal basis of $\mathbb{R}^n$ consisting of the eigenvectors $\mathbf{u}_1, \mathbf{u}_2, \dots, \mathbf{u}_n$ corresponding to the eigenvalues $\lambda_1, \lambda_2, \dots, \lambda_n$, respectively. We have just proved an important theorem:

Theorem 6.6 (Spectral decomposition – real symmetric matrices) *Let A be a real symmetric matrix. Then, A has a* ***spectral decomposition***

$$A = \lambda_1\mathbf{u}_1\mathbf{u}_1^T + \lambda_2\mathbf{u}_2\mathbf{u}_2^T + \cdots + \lambda_n\mathbf{u}_n\mathbf{u}_n^T,$$

where $\lambda_1, \lambda_2, \dots, \lambda_n$ are the eigenvalues of A corresponding, respectively, to the orthonormal set of eigenvectors $\mathbf{u}_1, \mathbf{u}_2, \dots, \mathbf{u}_n$.

Notice that the matrices $\mathbf{u}_1\mathbf{u}_1^T, \mathbf{u}_2\mathbf{u}_2^T, \dots, \mathbf{u}_n\mathbf{u}_n^T$ are $n \times n$ projection matrices onto the subspaces spanned, respectively, by $\mathbf{u}_1, \mathbf{u}_2, \dots, \mathbf{u}_n$ (cf. (6.30)).

Exercise 6.12 *Find the spectral decomposition of*

$$A = \begin{bmatrix} 5 & 1 & 1 \\ 1 & 5 & 1 \\ 1 & 1 & 5 \end{bmatrix} = SDS^T$$

Solution. We need to find first the eigenvalues of A. Since the sum of the entries of each row is 7, we have

$$A \begin{bmatrix} 1 \\ 1 \\ 1 \end{bmatrix} = 7 \begin{bmatrix} 1 \\ 1 \\ 1 \end{bmatrix}.$$

Hence, $\lambda_1 = 7$ is an eigenvalue and has an associated eigenvector $\begin{bmatrix} 1 \\ 1 \\ 1 \end{bmatrix}$. On the other hand, if λ_2, λ_3 are the other eigenvalues, then

$$\lambda_2 + \lambda_3 = \operatorname{tr} A - 7 = 15 - 7 = 8, \qquad \lambda_2\lambda_3 = \tfrac{1}{7}|A| = 16.$$

Solving these equations, gives $\lambda_2 = \lambda_3 = 4$.

To determine the remaining eigenvectors, we need to solve the system $(A-4I)\mathbf{x} = \mathbf{0}$. We know already that the two eigenvectors we need to construct S are orthogonal to $(1, 1, 1)$, which means that they lie in the plane having the cartesian equation $x + y + z = 0$. Then, one possible eigenvector is $(-1, 1, 0)$.

Now we need to find a vector satisfying $x + y + z = 0$, which is also orthogonal to $(-1, 1, 0)$. One possibility is $(-1, -1, 2)$.

Hence,

$$A = \begin{bmatrix} \frac{1}{\sqrt{3}} & -\frac{1}{\sqrt{2}} & -\frac{1}{\sqrt{6}} \\ \frac{1}{\sqrt{3}} & \frac{1}{\sqrt{2}} & -\frac{1}{\sqrt{6}} \\ \frac{1}{\sqrt{3}} & 0 & \frac{2}{\sqrt{6}} \end{bmatrix} \begin{bmatrix} 7 & 0 & 0 \\ 0 & 4 & 0 \\ 0 & 0 & 4 \end{bmatrix} \begin{bmatrix} \frac{1}{\sqrt{3}} & \frac{1}{\sqrt{3}} & \frac{1}{\sqrt{3}} \\ -\frac{1}{\sqrt{2}} & \frac{1}{\sqrt{2}} & 0 \\ -\frac{1}{\sqrt{6}} & -\frac{1}{\sqrt{6}} & \frac{2}{\sqrt{6}} \end{bmatrix}.$$

Finally, we arrive to the spectral decomposition of A:

$$A = 7 \begin{bmatrix} \frac{1}{\sqrt{3}} \\ \frac{1}{\sqrt{3}} \\ \frac{1}{\sqrt{3}} \end{bmatrix} \begin{bmatrix} \frac{1}{\sqrt{3}} & \frac{1}{\sqrt{3}} & \frac{1}{\sqrt{3}} \end{bmatrix} + 4 \begin{bmatrix} -\frac{1}{\sqrt{2}} \\ \frac{1}{\sqrt{2}} \\ 0 \end{bmatrix} \begin{bmatrix} -\frac{1}{\sqrt{2}} & \frac{1}{\sqrt{2}} & 0 \end{bmatrix}$$

$$+ 4 \begin{bmatrix} -\frac{1}{\sqrt{6}} \\ -\frac{1}{\sqrt{6}} \\ \frac{2}{\sqrt{6}} \end{bmatrix} \begin{bmatrix} -\frac{1}{\sqrt{6}} & -\frac{1}{\sqrt{6}} & \frac{2}{\sqrt{6}} \end{bmatrix},$$

that is

$$A = 7 \begin{bmatrix} \frac{1}{3} & \frac{1}{3} & \frac{1}{3} \\ \frac{1}{3} & \frac{1}{3} & \frac{1}{3} \\ \frac{1}{3} & \frac{1}{3} & \frac{1}{3} \end{bmatrix} + 4 \begin{bmatrix} \frac{1}{2} & -\frac{1}{2} & 0 \\ -\frac{1}{2} & \frac{1}{2} & 0 \\ 0 & 0 & 0 \end{bmatrix} + 4 \begin{bmatrix} \frac{1}{6} & \frac{1}{6} & -\frac{2}{6} \\ \frac{1}{6} & \frac{1}{6} & -\frac{2}{6} \\ -\frac{2}{6} & -\frac{2}{6} & \frac{4}{6} \end{bmatrix}.$$

Hence,

$$A = 7\begin{bmatrix} \frac{1}{3} & \frac{1}{3} & \frac{1}{3} \\ \frac{1}{3} & \frac{1}{3} & \frac{1}{3} \\ \frac{1}{3} & \frac{1}{3} & \frac{1}{3} \end{bmatrix} + 4\begin{bmatrix} \frac{2}{3} & -\frac{1}{3} & -\frac{1}{3} \\ -\frac{1}{3} & \frac{2}{3} & -\frac{1}{3} \\ -\frac{1}{3} & -\frac{1}{3} & \frac{2}{3} \end{bmatrix}.$$

Observe that these matrices are symmetric and idempotent as they should be, since they correspond to orthogonal projections (onto the eigenspaces).

This exercise served two purposes: one was to show how to obtain an orthogonal diagonalisation/spectral decomposition of a real symmetric matrix, the other was to use an alternative way of calculating the eigenvalues bypassing the (often not so easy) problem of determining the roots of the characteristic polynomial.

Real symmetric matrices have been characterised as exactly those that are orthogonally diagonalisable. A counterpart for complex matrices refers to those that can be diagonalised by unitary matrices.

Definition 65 *A matrix S in $\mathrm{M}_n(\mathbb{C})$ is said to be a* **unitary matrix** *if $S\overline{S}^T = I$.*

Proposition 6.16 *Let S be a matrix in $\mathrm{M}_n(\mathbb{C})$. Then the following are equivalent.*

(i) S is a unitary matrix.

(ii) $\overline{S}^T S = I$.

(iii) $S\overline{S}^T = I$.

(iv) The columns of S are an orthonormal basis of $\mathbb{C}^n$.

(v) The rows of S are an orthonormal basis of $\mathbb{C}^n$.

The proof of this proposition is left as an easy exercise (see the proof of Proposition 6.15).

Definition 66 *A complex square matrix is said to be* **unitarily diagonalisable,** *if there exist a diagonal matrix D and an unitary matrix S such that $D = \overline{S}^T AS$.*

It is easy to see that, if D is real, then A is hermitian.

The next result refers to the orthogonality of the eigenspaces of a hermitian matrix. The proof is an easy adaptation of that of Lemma 6.2.

Lemma 6.3 *Let A be a hermitian matrix and let $\mathbf{x}_1, \mathbf{x}_2$ be eigenvectors of A associated with distinct eigenvalues λ_1, λ_2, respectively. Then, $\mathbf{x}_1 \perp \mathbf{x}_2$.*

Theorem 6.7 *A hermitian matrix is unitarily diagonalisable.*

Proof *This proof is similar to that of Theorem 6.5.*

Corollary 6.4 *Let A be a complex positive definite matrix. Then, there exists a non-singular matrix such that $A = B\overline{B}^T$.*

Proof *Exercise.*

Finally, we can present the spectral decomposition for hermitian matrices:

Theorem 6.8 (Spectral decomposition – hermitian matrices) *Let A be a hermitian matrix. Then, A has a* ***spectral decomposition***

$$A = \lambda_1 \mathbf{u}_1 \overline{\mathbf{u}}_1^T + \lambda_2 \mathbf{u}_2 \overline{\mathbf{u}}_2^T + \cdots + \lambda_n \mathbf{u}_n \overline{\mathbf{u}}_n^T,$$

where $\lambda_1, \lambda_2, \ldots, \lambda_n$ are the eigenvalues of A corresponding, respectively, to the orthonormal set of eigenvectors $\mathbf{u}_1, \mathbf{u}_2, \ldots, \mathbf{u}_n$.

Notice once again that the matrices $\mathbf{u}_1 \overline{\mathbf{u}}_1^T, \mathbf{u}_2 \overline{\mathbf{u}}_2^T, \ldots, \mathbf{u}_n \overline{\mathbf{u}}_n^T$ are $n \times n$ projection matrices onto the subspaces spanned, respectively, by $\mathbf{u}_1, \mathbf{u}_2, \ldots, \mathbf{u}_n$ (cf. (6.31)).

How to do a spectral decomposition

Let A be a $n \times n$ real symmetric (respectively, hermitian) matrix.

1. Find the eigenvalues $\lambda_1, \lambda_2, \ldots, \lambda_n$ of A, possibly repeated.
2. Find a corresponding orthonormal set of eigenvectors $\mathbf{u}_1, \mathbf{u}_2, \ldots, \mathbf{u}_n$.
3. A spectral decomposition of A is

$$A = \lambda_1 \mathbf{u}_1 \mathbf{u}_1^T + \lambda_2 \mathbf{u}_2 \mathbf{u}_2^T + \cdots + \lambda_n \mathbf{u}_n \mathbf{u}_n^T,$$

if A is real, or

$$A = \lambda_1 \mathbf{u}_1 \overline{\mathbf{u}}_1^T + \lambda_2 \mathbf{u}_2 \overline{\mathbf{u}}_2^T + \cdots + \lambda_n \mathbf{u}_n \overline{\mathbf{u}}_n^T,$$

if A is complex.

6.5 Singular Value Decomposition

The orthogonal diagonalisation of real symmetric matrices can be extended to rectangular matrices, in some sense. The departing point being a very simple fact:

If A is a $k \times n$ real matrix, then $A^T A$ is an $n \times n$ real symmetric matrix.

Then, the spectrum of $A^T A$, besides being a non-empty subset of real numbers (see Corollary 4.1), consists only of non-negative numbers. In fact, if λ is an eigenvalue of $A^T A$ and $\mathbf{v} \in \mathbb{R}^n$ is an associated eigenvector, then

$$0 \leq \langle A\mathbf{v}, A\mathbf{v} \rangle = (A\mathbf{v})^T A\mathbf{v} = \mathbf{v}^T A^T A\mathbf{v} = \lambda \mathbf{v}^T \mathbf{v} = \lambda \|\mathbf{v}\|^2,$$

from which follows that $\lambda \geq 0$.

Suppose that $\{\mathbf{v}_1, \mathbf{v}_2, \ldots, \mathbf{v}_n\}$ is an orthonormal basis of $\mathbb{R}^n$ consisting of eigenvectors of $A^T A$. Then, given $i \in \{1, 2, \ldots, n\}$,

$$\|A\mathbf{v}_i\| = \sqrt{\langle A\mathbf{v}_i, A\mathbf{v}_i \rangle} = \sqrt{\lambda_i \|\mathbf{v}_i\|^2} = \sqrt{\lambda_i}, \tag{6.33}$$

where $\lambda_i \in \sigma(A^T A)$ is the eigenvalue associated with $\mathbf{v}_i$.

We have also that, given $i, j \in \{1, 2, \ldots, n\}$ with $i \neq j$,

$$\langle A\mathbf{v}_i, A\mathbf{v}_j \rangle = \mathbf{v}_j^T A^T A\mathbf{v}_i = \lambda_i \langle \mathbf{v}_i, \mathbf{v}_j \rangle = 0.$$

Define, for all $i \in \{1, 2, \ldots, n\}$, the non-negative real number $\sigma_i = \sqrt{\lambda_i}$, called a ***singular value*** of the matrix A.

Let $\sigma_1 \geq \sigma_2 \geq \cdots \geq \sigma_r > 0$, be all the non-zero singular values, repeated as many times as the corresponding algebraic multiplicities.

Setting, for all $i \in \{1, 2, \ldots, r\}$,

$$\mathbf{u}_i = \frac{1}{\|A\mathbf{v}_i\|} A\mathbf{v}_i = \frac{1}{\sigma_i} A\mathbf{v}_i,$$

we see that $\{\mathbf{u}_1, \mathbf{u}_2, \ldots, \mathbf{u}_r\}$ is an orthogonal subset of the column space $C(A)$ of matrix A. Indeed, $\{\mathbf{u}_1, \mathbf{u}_2, \ldots, \mathbf{u}_r\}$ is an orthonormal basis of $C(A)$, since

$$\begin{aligned} C(A) &= \operatorname{span}\big(\{A\mathbf{v}_i \colon i = 1, 2, \ldots, r\} \cup \{A\mathbf{v}_i \colon i = r+1, \ldots, n\}\big) \\ &= \operatorname{span}\big(\{A\mathbf{v}_i \colon i = 1, 2, \ldots, r\} \cup \{0\}\big) \\ &= \operatorname{span}\{A\mathbf{v}_i \colon i = 1, 2, \ldots, r\}. \end{aligned}$$

Let $\{\mathbf{u}_1, \mathbf{u}_2, \ldots, \mathbf{u}_r, \mathbf{u}_{r+1}, \ldots, \mathbf{u}_k\}$ be an orthonormal basis of $\mathbb{R}^k$ containing the orthonormal basis of $C(A)$ defined above. Then,

$$A \underbrace{\begin{bmatrix} \mathbf{v}_1 & \mathbf{v}_2 & \ldots & \mathbf{v}_n \end{bmatrix}}_{V} = \underbrace{\begin{bmatrix} \mathbf{u}_1 & \mathbf{u}_2 & \ldots & \mathbf{u}_k \end{bmatrix}}_{U} \underbrace{\begin{bmatrix} D & 0 \\ 0 & 0 \end{bmatrix}}_{\Sigma}, \tag{6.34}$$

where D is the diagonal matrix

$$D = \begin{bmatrix} \sigma_1 & & & \\ & \sigma_2 & & \\ & & \ddots & \\ & & & \sigma_r \end{bmatrix}.$$

Hence, we can now make a note of this ***singular value decomposition*** of A.

Theorem 6.9 (Singular value decomposition – real matrix) *Let A be a $k \times n$ real matrix. Then there exist a $k \times k$ orthogonal matrix U and an $n \times n$ orthogonal matrix V such that*

$$A = U\Sigma V^T,$$

where U, V, and the $k \times n$ matrix Σ are as in (6.34).

Example 6.8 *We are going to obtain a singular value decomposition of*

$$A = \begin{bmatrix} 1 & -1 \\ 0 & 1 \\ 1 & 1 \end{bmatrix}.$$

The eigenvalues of

$$A^T A = \begin{bmatrix} 2 & 0 \\ 0 & 3 \end{bmatrix}$$

are $\lambda_1 = 3$ and $\lambda_2 = 2$. Hence, the singular values of A are $\sqrt{3} = \sigma_1 > \sigma_2 = \sqrt{2}$. Norm one eigenvectors of $A^T A$ are, for example, $\boldsymbol{v}_1 = (0,1)$, $\boldsymbol{v}_2 = (1,0)$.

We have that

$$\boldsymbol{u}_1 = \frac{1}{\|A\mathbf{v}_1\|} A\mathbf{v}_1 = \frac{1}{\sqrt{3}}(-1,1,1), \qquad \boldsymbol{u}_2 = \frac{1}{\|A\mathbf{v}_2\|} A\mathbf{v}_2 = \frac{1}{\sqrt{2}}(1,0,1).$$

A norm one vector orthogonal to $\boldsymbol{u}_1, \boldsymbol{u}_2$ is, for example, $\boldsymbol{u}_3 = \frac{1}{\sqrt{6}}(-1,-2,1)$. Finally, the singular value decomposition sought is

$$A = \begin{bmatrix} -\frac{1}{\sqrt{3}} & \frac{1}{\sqrt{2}} & -\frac{1}{\sqrt{6}} \\ \frac{1}{\sqrt{3}} & 0 & -\frac{2}{\sqrt{6}} \\ \frac{1}{\sqrt{3}} & \frac{1}{\sqrt{2}} & \frac{1}{\sqrt{6}} \end{bmatrix} \begin{bmatrix} \sqrt{3} & 0 \\ 0 & \sqrt{2} \\ 0 & 0 \end{bmatrix} \begin{bmatrix} 0 & 1 \\ 1 & 0 \end{bmatrix}.$$

As always, there is a counterpart for complex matrices. Let A be a $k \times n$ complex matrix and let

$$\lambda_1 \geq \lambda_2 \geq \cdots \geq \lambda_n \geq 0$$

be all the eigenvalues of the hermitian matrix $\overline{A}^T A$ (possibly repeated). Notice that, similarly to what we did for real matrices above, it is easily seen that these eigenvalues are all non-negative.

Let $\{\mathbf{v}_1, \mathbf{v}_2, \ldots, \mathbf{v}_n\}$ be an orthonormal basis of $\mathbb{C}^n$ consisting of eigenvectors such that, for all $i = 1, 2, \ldots, n$,

$$\overline{A}^T A\mathbf{v}_i = \lambda_i \mathbf{v}_i.$$

The ***singular values*** of A are defined by

$$\sigma_i = \sqrt{\lambda_i}, \qquad i = 1, 2, \ldots, n.$$

Let $\sigma_1 \geq \sigma_2 \geq \ldots \sigma_r > 0$ be all the non-zero singular values (possibly repeated), and define, for all $i = 1, 2, \ldots, r$,

$$\mathbf{u}_i = \frac{1}{\|\mathbf{v}_i\|} A\mathbf{v}_i = \frac{1}{\sigma_i} A\mathbf{v}_i.$$

Define the unitary matrices

$$U = \begin{bmatrix} \mathbf{u}_1 & \mathbf{u}_2 & \ldots & \mathbf{u}_k \end{bmatrix}, \qquad V = \begin{bmatrix} \mathbf{v}_1 & \mathbf{v}_2 & \ldots & \mathbf{v}_n \end{bmatrix}, \tag{6.35}$$

where $\{\mathbf{u}_1, \mathbf{u}_2, \ldots, \mathbf{u}_r, \mathbf{u}_{r+1}, \ldots, \mathbf{u}_k\}$ is an orthonormal basis of $\mathbb{C}^k$.

Theorem 6.10 (Singular value decomposition – complex matrix) *Let A be a $k \times n$ complex matrix. Then there exist a $k \times k$ unitary matrix U and an $n \times n$ unitary matrix V such that*

$$A = U\Sigma\overline{V}^T,$$

where U, V are as in (6.35) and the $k \times n$ matrix Σ is as in (6.34).

Proof *Exercise.*

Example 6.9 *We are going to obtain a singular value decomposition of*

$$A = \begin{bmatrix} 0 & 1 & i \\ -i & 0 & 0 \end{bmatrix}.$$

The eigenvalues of

$$\overline{A}^T A = \begin{bmatrix} 1 & 0 & 0 \\ 0 & 1 & i \\ 0 & -i & 1 \end{bmatrix}$$

are $\lambda_1 = 2$, $\lambda_2 = 1$, and $\lambda_3 = 0$. An eigenvector corresponding to $\lambda_1 = 2$ is $\boldsymbol{v}_1 = (0, i, 1)$, an eigenvector corresponding to $\lambda_2 = 1$ is $\boldsymbol{v}_1 = (1, 0, 0)$, and eigenvector corresponding to $\lambda_3 = 0$ is $\boldsymbol{v}_3 = (0, -i, 1)$.

We obtain $\boldsymbol{u}_1 = (i, 0), \boldsymbol{u}_2 = (0, -i)$. The singular value decomposition of A is then

$$A = \begin{bmatrix} i & 0 \\ 0 & -i \end{bmatrix} \begin{bmatrix} \sqrt{2} & 0 & 0 \\ 0 & 1 & 0 \end{bmatrix} \begin{bmatrix} 0 & -\frac{i}{\sqrt{2}} & \frac{1}{\sqrt{2}} \\ 1 & 0 & 0 \\ 0 & \frac{i}{\sqrt{2}} & \frac{1}{\sqrt{2}} \end{bmatrix}.$$

How to do a singular value decomposition

Let A be a $k \times n$ real (respectively, complex matrix).

1. Determine the eigenvalues
$$\lambda_1 \geq \lambda_2 \geq \cdots \geq \lambda_n \geq 0$$
of the real symmetric matrix $A^T A$ (respectively, hermitian matrix $\overline{A}^T A$). In the list above, each eigenvalue is repeated as many times as its algebraic multiplicity.

2. The singular values $\sigma_1 \geq \sigma_2 \geq \cdots \geq \sigma_n \geq 0$ of A are the square roots of the eigenvalues:
$$\sigma_1 = \sqrt{\lambda_1} \geq \sigma_2 = \sqrt{\lambda_2} \geq \cdots \geq \sigma_n = \sqrt{\lambda_n} \geq 0 \tag{6.36}$$

3. Find an orthonormal basis $(\mathbf{v}_1, \mathbf{v}_2, \ldots, \mathbf{v}_n)$ of $\mathbb{R}^n$ (respectively, $\mathbb{C}^n$) consisting entirely of eigenvectors of $A^T A$ (respectively, $\overline{A}^T A$) and such that, for all $i = 1, 2, \ldots, n$, the vector $\mathbf{v}_i$ is an eigenvector corresponding to λ_i. Construct the $n \times n$ matrix $V = \begin{bmatrix} \mathbf{v}_1 & \mathbf{v}_2 & \ldots & \mathbf{v}_n \end{bmatrix}$.

4. Let $\sigma_1 \geq \sigma_2 \geq \cdots \geq \sigma_r > 0$ be all the non-zero singular values in (6.36). For all $i = 1, 2, \ldots, r$, define the vectors in $\mathbb{R}^k$ (respectively, $\mathbb{C}^k$) by
$$\mathbf{u}_i = \frac{1}{\sigma_i} A\mathbf{v}_i = \frac{1}{\|A\mathbf{v}_i\|} A\mathbf{v}_i.$$

5. Obtain an orthonormal basis of $\mathbb{R}^k$ (respectively, $\mathbb{C}^k$) containing the vectors in 4. That is, construct an orthonormal basis $(\mathbf{u}_1, \mathbf{u}_2, \ldots, \mathbf{u}_r, \mathbf{u}_{r+1}, \ldots, \mathbf{u}_k)$. (Of course, if $r = k$ to start with, one does not have to do anything and can ignore this part.) Construct the $k \times k$ matrix $U = \begin{bmatrix} \mathbf{u}_1 & \mathbf{u}_2 & \ldots & \mathbf{u}_k \end{bmatrix}$.

6. Construct a $k \times n$ matrix Σ, whose first r diagonal entries are $\sigma_1, \sigma_2, \ldots, \sigma_r$, in this order. Fill the remaining entries in with zeros.

7. The singular value decomposition of A is
$$A = U\Sigma V^T, \qquad \text{if } A \text{ is a real matrix,}$$
and
$$A = U\Sigma \overline{V}^T, \qquad \text{if } A \text{ is a complex matrix.}$$

6.6 Affine Subspaces of $\mathbb{R}^n$

The theory developed up to here is mostly concerned with subspaces of a vector space which, as we know, always go through the zero vector. For example, we know how to obtain the cartesian equations of straight lines and planes in space which contain $(0,0,0)$. What if we consider straight lines and planes in $\mathbb{R}^3$ which do not contain $(0,0,0)$?

Here we present briefly a way to obtain equations for this type of sets and of their counterparts in $\mathbb{R}^n$ which relies on the Euclidean inner product structure.

Definition 67 *Let S be a subset of the Euclidean space $\mathbb{R}^n$. S is said to be an* **affine subspace of dimension** *k if there exists $\mathbf{p}$ in $\mathbb{R}^n$ and a subspace W, with $\dim W = k$, such that*

$$S = \mathbf{p} + W. \tag{6.37}$$

If $k = 0, 1, 2, n-1$, the affine subspace S is said to be a **point**, *a* **straight line**, *a* **plane**, *and a* **hyperplane**, *respectively.*

Affine subspaces can be thought of intuitively as subspaces which were 'moved away' from the origin $\mathbf{0}$. Notice however that subspaces of $\mathbb{R}^n$ are particular cases of affine subspaces.

Given $\boldsymbol{x} = (x_1, x_2, \dots, x_n)$ in S, there exists $\boldsymbol{y}$ em W such that

$$\boldsymbol{x} = \boldsymbol{y} + \boldsymbol{p}$$

or, equivalently,

$$\boldsymbol{y} = \boldsymbol{x} - \boldsymbol{p}. \tag{6.38}$$

The equality (6.38) shows that we can find a vector equation, cartesian equations or parametric equations for S using the corresponding equations of W. Indeed, it suffices to replace $\boldsymbol{y}$ by $\boldsymbol{x} - \boldsymbol{p}$ in those equations.

Example 6.10 *Find a vector equation, cartesian equations, and parametric equations for the straight line S going through $(0,0,1)$ and having the direction of $(1,1,1)$.*

We have that $S = (0,0,1) + W$, where W is the 1-dimensional subspace with basis $\{(1,1,1)\}$. Hence, any $(x_1, x_2, x_3) \in S$ satisfies

$$\begin{bmatrix} x_1 \\ x_2 \\ x_3 \end{bmatrix} = \begin{bmatrix} 0 \\ 0 \\ 1 \end{bmatrix} + t \begin{bmatrix} 1 \\ 1 \\ 1 \end{bmatrix},$$

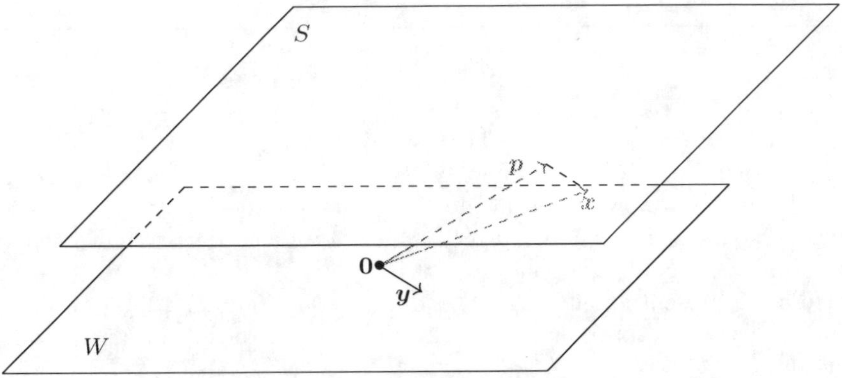

FIGURE 6.11: The affine subspace $S = \boldsymbol{p} + W$.

for some $t \in \mathbb{R}$. This is a vector equation for S which yields the parametric equations for S

$$\begin{cases} x_1 &= 1 \\ x_2 &= 1 \\ x_3 &= 1 + t \qquad (t \in \mathbb{R}). \end{cases}$$

Finally, since W has the cartesian equations $y_1 = y_2 = y_3$, it follows that

$$x_1 = x_2 = x_3 - 1$$

are cartesian equations for S.

Analogously, one can use $W^\perp$ to obtain equations for S which in some circumstances might be more convenient. If $\mathcal{B}_{W^\perp} = (\boldsymbol{b}_1, \boldsymbol{b}_2, \dots, \boldsymbol{b}_{n-k})$ is a basis of the orthogonal complement of W, we have $\boldsymbol{x} - \boldsymbol{p} \in W$ or, equivalently,

$$\underbrace{\begin{bmatrix} \mathbf{b}_1^T \\ \mathbf{b}_2^T \\ \vdots \\ \mathbf{b}_{n-k}^T \end{bmatrix}}_{(n-k)\times n} \underbrace{\begin{bmatrix} x_1 - p_1 \\ x_2 - p_2 \\ \vdots \\ x_n - p_n \end{bmatrix}}_{n\times 1} = \underbrace{\begin{bmatrix} 0 \\ 0 \\ \vdots \\ 0 \end{bmatrix}}_{(n-k)\times 1},$$

where $\boldsymbol{p} = (p_1, p_2, \dots, p_n)$ and $\boldsymbol{x} = (x_1, x_2, \dots, x_n)$. Observe that in this case $\dim W = k$. Letting A be the matrix

$$A = \begin{bmatrix} \mathbf{b}_1^T \\ \mathbf{b}_2^T \\ \vdots \\ \mathbf{b}_{n-k}^T \end{bmatrix},$$

we have the homogeneous system of linear equations

$$A(\mathbf{x} - \mathbf{p}) = \mathbf{0}.$$

or, equivalently, the system of linear equations

$$A\mathbf{x} = A\mathbf{p}.$$

Hence, from a vector equation, cartesian equations or parametric equations of $N(A)$, one can obtain the corresponding equations for S.

Example 6.11 *Find a vector equation, cartesian equations, and parametric equations of the plane containing the point $\boldsymbol{p} = (1, 2, 0)$ and which is perpendicular to the straight line containing $\mathbf{p}$ and whose direction is $\boldsymbol{n} = (5, 1, -2)$.*

Let W be the subspace (of dimension 2) such that $\boldsymbol{n} \perp W$. A cartesian equation for W is

$$5y_1 + y_2 - 2y_3 = 0.$$

Since $S = \mathbf{p} + W$, we have, for $(x_1, x_2, x_3) \in S$,

$$(x_1, x_2, x_3) = (1, 2, 0) + (y_1, y_2, y_3).$$

Hence,

$$5(x_1 - 1) + (x_2 - 2) - 2x_3 = 0$$

from which follows that

$$5x_1 + x_2 - 2x_3 = 7$$

is a cartesian equation for S.

From this equation we have that $x_2 = -5x_1 + 2x_3 + 7$. Hence,

$$\begin{bmatrix} x_1 \\ x_2 \\ x_3 \end{bmatrix} = t \begin{bmatrix} 1 \\ -5 \\ 0 \end{bmatrix} + s \begin{bmatrix} 0 \\ 2 \\ 1 \end{bmatrix} + \begin{bmatrix} 0 \\ 7 \\ 0 \end{bmatrix} \qquad (s, t \in \mathbb{R})$$

is a vector equation for S. Hence, the parametric equations are

$$\begin{cases} x_1 &= t \\ x_2 &= -5t + 2s + 7 \\ x_3 &= s \end{cases} \qquad (s, t \in \mathbb{R}).$$

Continuing with this idea of establishing a parallel between affine subspaces and subspaces, we tackle now the distance from a point to an affine subspace.

Let $S = \boldsymbol{p} + W$ and let $\boldsymbol{q}$ be a point in $\mathbb{R}^n$. Given $\boldsymbol{x}$ in S,

$$\begin{aligned} \mathrm{d}(\boldsymbol{q}, \boldsymbol{x}) &= \|\boldsymbol{q} - \boldsymbol{x}\| \\ &= \|(\boldsymbol{q} - \boldsymbol{p}) + \underbrace{(\boldsymbol{p} - \boldsymbol{x})}_{-\boldsymbol{y} \in W}\| \\ &= \|(\boldsymbol{q} - \boldsymbol{p}) - \boldsymbol{y}\| \\ &= \mathrm{d}(\boldsymbol{q} - \boldsymbol{p}, \boldsymbol{y}). \end{aligned}$$

The minimum value of this distance $\mathrm{d}(\boldsymbol{q} - \boldsymbol{p}, \boldsymbol{y})$ is attained when $\boldsymbol{y} = \mathrm{proj}_W(\boldsymbol{q} - \boldsymbol{p})$, as we have seen in §6.3.3; $\boldsymbol{y}$ is the best approximation of $\boldsymbol{q}$ in S. Naturally, this suggests the next definition.

Definition 68 *Let $S = \mathbf{p} + W$ be an affine subspace of an inner product space V, where $\mathbf{p} \in V$ and $W \subseteq V$ is a subspace. The* **distance of the point $\boldsymbol{q}$ to the affine subspace** S *is*

$$\mathrm{d}(\boldsymbol{q}, S) = \mathrm{d}(\boldsymbol{q} - \mathbf{p}, W) = \|\,\mathrm{proj}_{W^\perp}(\boldsymbol{q} - \mathbf{p})\|. \tag{6.39}$$

Example 6.12 *Calculate the distance of $(3, 2, -1)$ to the plane S of Example 6.11.*

In this case $S = \mathbf{p} + W$, where $\mathbf{p} = (1, 2, 0)$ and W has the cartesian equation

$$5x + y - 2z = 0.$$

Recall that this equation was obtained using the fact that a basis of $W^\perp$ is $\{(5, 1, 2)\}$. Notice that just by looking at the previous equation we would arrive to the same conclusion:

$$\langle (x, y, z), (5, 1, -2)\rangle = 5x + y - 2z = 0,$$

showing again that a basis of $W^\perp$ is $\{(5, 1, 2)\}$.

Using (6.39),

$$\mathrm{d}((3, 2, -1), S) = \|\,\mathrm{proj}_{W^\perp}((3, 2, -1) - (1, 2, 0))\| = \|\,\mathrm{proj}_{W^\perp}((2, 0, -1)\|.$$

We have

$$\mathrm{proj}_{W^\perp}((2, 0, -1) = \frac{\langle (2, 0, -1), (5, 1, 2)\rangle}{\|(2, 0, -1)\|^2}(2, 0, -1) = \frac{8}{\|(2, 0, -1)\|^2}(2, 0, -1).$$

Hence

$$\mathrm{d}((3, 2, -1), S) = \frac{8}{\|(2, 0, -1)\|^2}\|(2, 0, -1)\| = \frac{8}{\|(2, 0, -1)\|} = \frac{8}{\sqrt{5}}.$$

6.7 Exercises

EX 6.7.1. For the vectors $\boldsymbol{u} = (4, 1, 0, -2)$ and $\boldsymbol{v} = (2, -1, 0, 3)$ in $\mathbb{R}^4$, calculate:

(a) $\|\boldsymbol{u} + \boldsymbol{v}\|$

(b) $\|\boldsymbol{u}\| + \|\boldsymbol{v}\|$

(c) $\| - 3\boldsymbol{u}\|$

(d) $\frac{1}{\|\boldsymbol{v}\|}\boldsymbol{v}$

(e) $\left\|\frac{1}{\|\boldsymbol{v}\|}\boldsymbol{v}\right\|$

(f) $\measuredangle(\boldsymbol{u}, \boldsymbol{v})$

(g) $\mathrm{d}(\boldsymbol{u}, \boldsymbol{v})$

EX 6.7.2. Find two norm one vectors orthogonal to $(1, 1, -2)$ and $(-2, 3, -1)$.

EX 6.7.3. Verify that the Cauchy–Schwarz inequality holds for $(1, 1, 2)$ and $(2, 1, 3)$. Verify also that the parallelogram law holds for the same vectors.

EX 6.7.4. Consider the usual inner product in $\mathbb{C}^2$. Find $\mathrm{proj}_{(1,2)}(-i, -i)$.

EX 6.7.5. Use the Gram–Schmidt process to find an orthonormal basis for the subspace spanned by $\{(-2, 2, -2, 2), (1, 1, 3, -1), (0, 0, 0, 1)\}$.

EX 6.7.6. Find a basis and cartesian equations for the orthogonal complement of the subspace $W = \{(x, y, z) \colon y + 2z = 0,\, x - y = z\}$ of $\mathbb{R}^3$.

EX 6.7.7. Let S be the subspace of $\mathbb{R}^4$ defined by $S = \mathcal{L}\{(2, 1, 1, 0)\}$. Find the distance of $\boldsymbol{u} = (1, 1, 1, 1)$ to $S^\perp$. Determine the matrix corresponding to the orthogonal projection onto S, and $\boldsymbol{u}_1 \in S, \boldsymbol{u}_2 \in S^\perp$ such that $\boldsymbol{u} = \boldsymbol{u}_1 + \boldsymbol{u}_2$.

EX 6.7.8. Consider the subspace

$S = \mathcal{L}\left\{\left(\frac{4}{3}, 1, -1\right), (0, -1, 1)\right\}$ of $\mathbb{R}^3$. Suppose that $\mathbb{R}^3$ is endowed with the inner product $\langle \cdot, \cdot \rangle$ whose Gram matrix of the vectors of the standard basis $\mathcal{E}_3$ is

$$G = \begin{bmatrix} 3 & -2 & 1 \\ -2 & 2 & 0 \\ 1 & 0 & 2 \end{bmatrix}.$$

(a) Show that $\left\{\left(\frac{4}{3}, 1, -1\right), (0, -1, 1)\right\}$ is an orthogonal basis of S.

(b) Find the norms $\| \left(\frac{4}{3}, 1, -1\right) \|$, $\|(0, -1, 1)\|$.

(c) For the vector $\boldsymbol{u} = (1, -1, -1)$, determine $\boldsymbol{u}_1 \in S$ and $\boldsymbol{u}_2 \in S^\perp$ such that

$$\boldsymbol{u} = \boldsymbol{u}_1 + \boldsymbol{u}_2 \ .$$

EX 6.7.9. Let A be a 3×4 real matrix such that its column space is the subspace of $\mathbb{R}^3$ consisting of the vectors (x, y, z) satisfying the equation $x + z = 0$. Find a basis of $C(A)^\perp$ and a vector lying in $N(A^T)$.

EX 6.7.10. Let W be the subspace of $\mathbb{C}^4$ defined by $W = \{(x, y, z, w) \in \mathbb{C}^4 : ix + iy + iz = 0\}$, and let $\boldsymbol{p} = (1, 0, -1, 0)$ be a vector in W. Considering the usual inner product on $\mathbb{C}^4$, find:

(a) $\dim(W^\perp)$

(b) $\mathrm{d}(\boldsymbol{p}, W^\perp)$

(c) an orthonormal basis of $W^\perp$

EX 6.7.11. Let A a real square matrix such that its column space is

$$C(A) = \{(x, y, z) \in \mathbb{R}^3 \colon x + y + z = 0 \wedge x - y = z\}.$$

(a) Is the system

$$A \begin{bmatrix} x \\ y \\ z \end{bmatrix} = \begin{bmatrix} 2 \\ 1 \\ -1 \end{bmatrix}$$

consistent?

(b) If $(1, 2, 3)$ is a particular solution of

$$A^T \begin{bmatrix} x \\ y \\ z \end{bmatrix} = \begin{bmatrix} 0 \\ -1 \\ 1 \end{bmatrix},$$

what is the solution set of this system?

EX 6.7.12. Consider the points $P_0 = (1, 0, -1, 0)$ $\quad P_1 = (0, 1, 0, 0)$ $\quad P_2 = (1, 1, 1, 0)$ in $\mathbb{R}^4$.

(a) Find cartesian and parametric equations of the straight line containing P_0 and parallel to $\boldsymbol{u} = (0, -1, -3, 0)$.

(b) Find a cartesian equation for the plane containing P_0 and perpendicular to the straight line which contains P_0 and is parallel to $\boldsymbol{n} = (1, 0, 1, 0)$.

(c) Find a cartesian equation and parametric equations of the plane defined by P_0, P_1, and P_2. Find a normal vector to that plane.

(d) Calculate the distance of $(1, 1, 0, 0)$ to the same plane.

EX 6.7.13. Let U be the subspace of $M_2(\mathbb{R})$ consisting of the anti-symmetric matrices. Consider the real function on $M_2(\mathbb{R})$ defined, for $A, B \in M_2(\mathbb{R})$ by

$$\langle A, B \rangle = \operatorname{tr}(B^T A), \tag{6.40}$$

where tr is the trace.

(a) Show that (6.40) defines an inner product.

(b) Find orthonormal bases for U and $U^\perp$ and their dimension.

(c) Find the orthogonal projections of $\begin{bmatrix} 3 & 3 \\ -3 & 6 \end{bmatrix}$ on U and $U^\perp$.

(d) Which is the anti-symmetric matrix closest to $\begin{bmatrix} 3 & 3 \\ -3 & 6 \end{bmatrix}$? Find the distance of $\begin{bmatrix} 3 & 3 \\ -3 & 6 \end{bmatrix}$ to U.

EX 6.7.14. Find the orthogonal complement of the subspace $\mathcal{L}\{-3(t+1)^2\}$ of $\mathbb{P}_2$ for the inner product

$$\langle a_0 + a_1 t + a_2 t^2, b_0 + b_1 t + b_2 t^2 \rangle = a_0 b_0 + a_1 b_1 + a_2 b_2.$$

EX 6.7.15. Find an orthogonal diagonalisation for

$$\begin{bmatrix} 1 & 1 & 1 \\ 1 & 1 & 1 \\ 1 & 1 & 1 \end{bmatrix}.$$

EX 6.7.16. Let A be an $n \times n$ unitary matrix. For $\mathbf{x}, \mathbf{y} \in \mathbb{C}^n$, show that:

(a) $\langle A\mathbf{x}, A\mathbf{y} \rangle = \langle \mathbf{x}, \mathbf{y} \rangle$

(b) $\|A\mathbf{x}\| = \|\mathbf{x}\|$

(c) $\sigma(A) \subseteq \{\lambda \in \mathbb{C} \colon |\lambda| = 1\}$

EX 6.7.17. Find an unitary diagonalisation for

$$\begin{bmatrix} 7 & 0 & 0 \\ 0 & -\frac{1}{3} & \frac{2}{3} + \frac{2}{3}i \\ 0 & \frac{2}{3} - \frac{2}{3}i & \frac{1}{3} \end{bmatrix}.$$

and its spectral decomposition.

EX 6.7.18. Let A be an $n \times n$ complex matrix such that there exists a orthonormal basis of $\mathbb{C}^n$ consisting exclusively of eigenvectors of A. Prove that, if $\sigma(A) \subseteq \mathbb{R}$, then A is a hermitian matrix.

EX 6.7.19. Find a singular value decomposition of

$$\begin{bmatrix} 3 & 2 & 2 \\ 2 & 3 & -2 \end{bmatrix}.$$

6.8 At a Glance

A real or complex vector space V can be endowed with an inner product, with respect to which each vector has a norm.

The inner product is given by a positive definite matrix, the Gram matrix with respect to the basis of V. The Gram matrix depends on the basis.

By means of the inner product, one defines the distance between two points in the space, the angle, and the notion of orthogonal vectors.

Any subspace W of V leads to a direct sum $V = W \oplus W^{\perp}$, yielding a unique splitting of each vector in V into two summands, one from W and the other from the orthogonal complement $W^{\perp}$ of W.

The orthogonal projection of $\mathbf{x} \in V$ on W is given, with respect to an orthonormal basis of V, by a projection matrix A: $\text{proj}_W \mathbf{x} = A\mathbf{x}$. The matrix A is idempotent and symmetric (respectively, hermitian) if V is a real (respectively, complex) vector space.

It is always possible to have an orthonormal basis, since any basis can be transformed into one using the Gram–Schmidt process. Nevertheless, there is a formula for the projection matrix considering any given basis (see the *How to determine the matrix of the orthogonal projection onto a subspace* in §7.1).

Real symmetric matrices and hermitian matrices are diagonalisable: in the former case, the diagonalising matrix can be chosen to be orthogonal and unitary, in the latter.

Real symmetric matrices and hermitian matrices have a spectral decomposition.

A real (respectively, complex) positive definite matrix A is the product of two non-singular matrices $A = BB^T$ (respectively, $A = B\overline{B}^T$).

The singular values of a real (respectively, complex) $k \times n$ matrix A are the square roots of the eigenvalues of $A^T A$ (respectively $\overline{A}^T A$).

The singular values allow for a factorisation of $A = U\Sigma V^T$ (respectively, $A = U\Sigma\overline{V}^T$), where U, V are orthogonal (respectively, unitary) and Σ is a $k \times n$ matrix having the non-zero singular values of A in its diagonal.

Chapter 7

Special Matrices by Example

7.1 Least Squares Solutions 257
7.2 Markov Chains 260
7.2.1 Google matrix and PageRank 265
7.3 Population Dynamics 266
7.4 Graphs 271
7.5 Differential Equations 275
7.6 Exercises 279
7.7 At a Glance 282

In this chapter, we introduce some types of matrices via examples, the latter chosen from applications of Linear Algebra. Each section is a brief introduction to a special type of matrix always motivated by the analysis of a concrete problem. The main purpose is to highlight some particular matrices arising in applications rather than explore profoundly each application itself. To do so would be beyond the scope of this book.

7.1 Least Squares Solutions

It is often the case that in applications we are faced with inconsistent systems of linear equations. Under these circumstances, it is not so much the case of finding a solution but of finding a 'solution' as close to the ideal situation (which would be having an exact solution) as possible. In this brief introduction to the problem, projection matrices are prominent, and we will obtain a formula for a projection matrix on a subspace.

Suppose that we want to solve the system $A\mathbf{x} = \mathbf{b}$, where A is some $k \times n$ real matrix but $\mathbf{b}$ does not lie in the column space of A. The best we can do is to find a point $\hat{\mathbf{b}}$ in $C(A)$ closest to $\mathbf{b}$ and then solve $A\mathbf{x} = \hat{\mathbf{b}}$.

We know how to determine $\hat{\mathbf{b}}$: it is the orthogonal projection of $\mathbf{b}$ on $C(A)$ (cf. the best approximation in Chapter 6). This orthogonal projection $\hat{\mathbf{b}}$ is, of course, a linear combination of the columns of A. That is,

$$\hat{\mathbf{b}} = A\hat{\mathbf{x}}$$

DOI: 10.1201/9781351243452-7

for some $\hat{\mathbf{x}} \in \mathbb{R}^n$. The vector $\hat{\mathbf{x}}$ is a ***least squares solution*** of the problem $A\mathbf{x} = \mathbf{b}$.

By our discussion of the best approximation in Chapter 6,

$$\mathbf{b} - \hat{\mathbf{b}} = \mathbf{b} - A\hat{\mathbf{x}} \perp C(A) = N(A^T)^{\perp}.$$

It follows that

$$A^T(\mathbf{b} - A\hat{\mathbf{x}}) = \mathbf{0},$$

i.e.,

$$A^T A\hat{\mathbf{x}} = A^T\mathbf{b}. \tag{7.1}$$

Example 7.1 *Find the least squares solution of the system*

$$\begin{bmatrix} 1 & 1 \\ -1 & 0 \\ 0 & -1 \end{bmatrix} \begin{bmatrix} x \\ y \end{bmatrix} = \begin{bmatrix} 1 \\ 2 \\ 3 \end{bmatrix}.$$

Notice that the columns of A span the plane $x+y+z=0$ and that $(1,2,3)$ does not lie in this plane.

By (7.1), we have to solve the system

$$\begin{bmatrix} 1 & -1 & 0 \\ 1 & 0 & -1 \end{bmatrix} \begin{bmatrix} 1 & 1 \\ -1 & 0 \\ 0 & -1 \end{bmatrix} \begin{bmatrix} \hat{x} \\ \hat{y} \end{bmatrix} = \begin{bmatrix} 1 & -1 & 0 \\ 1 & 0 & -1 \end{bmatrix} \begin{bmatrix} 1 \\ 2 \\ 3 \end{bmatrix}.$$

Hence,

$$\begin{bmatrix} 2 & 1 \\ 1 & 2 \end{bmatrix} \begin{bmatrix} \hat{x} \\ \hat{y} \end{bmatrix} = \begin{bmatrix} -1 \\ -2 \end{bmatrix},$$

yielding the least squares solution $(0,-1)$. The **error vector** *is*

$$\mathbf{e} = \mathbf{b} - \hat{\mathbf{b}} = \begin{bmatrix} 1 \\ 2 \\ 3 \end{bmatrix} - 0\begin{bmatrix} 1 \\ -1 \\ 0 \end{bmatrix} - (-1)\begin{bmatrix} 1 \\ 0 \\ -1 \end{bmatrix} = \begin{bmatrix} 2 \\ 2 \\ 2 \end{bmatrix}.$$

The **error** *is $\|\mathbf{e}\| = \sqrt{2^2+2^2+2^2} = 2\sqrt{3}$.*

If the $n \times n$ matrix $A^T A$ in (7.1) is invertible, then

$$\hat{\mathbf{x}} = (A^T A)^{-1} A^T\mathbf{b}$$

and

$$\hat{\mathbf{b}} = A\hat{\mathbf{x}} = \underbrace{A(A^T A)^{-1} A^T}_{P}\,\mathbf{b}.$$

The $k \times k$ matrix P is the projection matrix onto $C(A)$. In fact, P is a projection since P is symmetric and $P^2 = P$:

$$\begin{aligned} P^T = (A(A^TA)^{-1}A^T)^T &= A((A^TA)^{-1})^TA^T \\ &= A((A^TA)^T)^{-1}TA^T \\ &= A(A^TA)^{-1}A^T = P; \end{aligned}$$

$$\begin{aligned} P^2 = (A(A^TA)^{-1}A^T)(A(A^TA)^{-1}A^T) &= A((A^TA)^{-1}A^TA)(A^TA)^{-1}A^T \\ &= A((A^TA)^T)^{-1}A^T = P. \end{aligned}$$

Proposition 7.1 *Let A be a $k \times n$ matrix. A^TA is invertible if and only if the columns of A are linearly independent.*

Proof *We show firstly that $N(A^TA) = N(A)$. Since it is clear that $N(A) \subseteq N(A^TA)$, it remains to show that $N(A) \supseteq N(A^TA)$.*

Suppose the $\mathbf{x} \in N(A^TA)$. Then,

$$\mathbf{x}^T \underbrace{A^TA\mathbf{x}}_{\mathbf{0}} = \langle A\mathbf{x}, A\mathbf{x}\rangle = \|A\mathbf{x}\|^2 = 0$$

which shows that $\mathbf{x} \in N(A)$.

On the other hand, A^TA is invertible only when $N(A^TA) = \{\mathbf{0}\}$. However, since $N(A) = N(A^TA)$, it follows that A^TA is invertible if and only if $N(A) = \{\mathbf{0}\}$. But $N(A) = \{\mathbf{0}\}$ if and only if rank (A) *coincides with the number of its columns. Hence, A^TA is invertible if and only if the columns of A are linearly independent.*

Summing it up: if the columns of A are a basis of $C(A)$, then we obtain a matrix P corresponding to the orthogonal projection onto $C(A)$.

How to determine the matrix of the orthogonal projection onto a subspace

Let W be a subspace of $\mathbb{R}^n$.

1. Construct a matrix A whose columns are a (not necessarily orthogonal) basis of W.
2. The projection matrix is $P = A(A^TA)^{-1}A^T$.
3. Given $\mathbf{y}$ in $\mathbb{R}^n$, the orthogonal projection $\text{proj}_W\, \mathbf{y}$ of $\mathbf{y}$ on W is

$$\text{proj}_W\, \mathbf{y} = P\mathbf{y} = A(A^TA)^{-1}A^T\mathbf{y}.$$

Example 7.2 *We want to find the straight line that, in some sense, best fits the points* $(1,2),(2,5),(3,3),(4,8)$*. More precisely, we want to determine a straight line* $y = \hat{c} + \hat{m}x$ *which gives a least squares solution of the problem*

$$\begin{bmatrix} 1 & 1 \\ 1 & 2 \\ 1 & 3 \\ 1 & 4 \end{bmatrix} \begin{bmatrix} c \\ m \end{bmatrix} = \begin{bmatrix} 2 \\ 5 \\ 3 \\ 8 \end{bmatrix}.$$

We have

$$\begin{bmatrix} \hat{c} \\ \hat{m} \end{bmatrix} = \left(\begin{bmatrix} 1 & 1 & 1 & 1 \\ 1 & 2 & 3 & 4 \end{bmatrix} \begin{bmatrix} 1 & 1 \\ 1 & 2 \\ 1 & 3 \\ 1 & 4 \end{bmatrix} \right)^{-1} \begin{bmatrix} 1 & 1 & 1 & 1 \\ 1 & 2 & 3 & 4 \end{bmatrix} \begin{bmatrix} 2 \\ 5 \\ 3 \\ 8 \end{bmatrix}.$$

The solution is $(\hat{c}, \hat{m}) = (\frac{1}{2}, \frac{8}{5})$ *from which follows that the best fitting straight line is* $y = \frac{1}{2} + \frac{8}{5}x$.

7.2 Markov Chains

We begin with a very simple example. Consider a physical system consisting of a single particle which may be at any given time in one of two different states 1 and 2 with a certain probability. Denote by p_{ij}, for $i, j = 1, 2$, the probability of the particle making a transition from state j to state i. If we let $p_{11} = 0.6$, then

$$p_{21} = 0.4 = 1 - 0.6,$$

since the particle either stays in state 1 or moves to state 2. Suppose further that the particle when in state 2 stays there with probability $p_{22} = 0.7$. Hence we must have

$$p_{12} = 0.3 = 1 - 0.7.$$

Consider a matrix P built with these data

$$P = \begin{bmatrix} 0.6 & 0.3 \\ 0.4 & 0.7 \end{bmatrix},$$

where column 1 corresponds to the probabilities of the particle moving or staying when in state 1 and, similarly, column 2 corresponds to state 2.

We are describing here a process where the probability of the system being in some state at a given observation time t_m only depends on which state it was at the previous observation time t_{m-1}. This is an example of what is called a **Markov chain**.

In the present case, since the Markov chain has only two states, we have a 2×2 matrix P. In general, if one has a Markov chain with n possible states then we have an $n \times n$ **transition matrix** $P = [p_{ij}]$ whose entries are nonnegative and such that the sum of the entries in any given column is equal to 1. In other words, the $n \times n$ matrix P is such that, for all $i, j = 1, \ldots, n$, $p_{ij} \geq 0$ and

$$\sum_{i=1}^{n} p_{ij} = 1.$$

This type of matrix is called a **Markov matrix**, a **stochastic matrix**, or a **probability matrix**. The transition matrix of a Markov chain is, therefore, a Markov matrix.

Suppose that the initial state of our system is 1 and consider the corresponding initial **state vector**

$$\mathbf{x}_0 = \begin{bmatrix} 1 \\ 0 \end{bmatrix},$$

i.e., the particle is in state 1 with probability 1 and has zero probability of being in state 2. It follows from probability theory that the state vector of the next observation is

$$\mathbf{x}_1 = P\mathbf{x}_0 = \begin{bmatrix} 0.6 & 0.3 \\ 0.4 & 0.7 \end{bmatrix} \begin{bmatrix} 1 \\ 0 \end{bmatrix} = \begin{bmatrix} 0.6 \\ 0.4 \end{bmatrix}.$$

Hence, the particle will be in state 1 with 0.6 probability and in state 2 with 0.4 probability.

Similarly, the following observation $\mathbf{x}_2$ satisfies

$$\mathbf{x}_2 = P\mathbf{x}_1 = P^2\mathbf{x}_0 = \begin{bmatrix} 0.6 & 0.3 \\ 0.4 & 0.7 \end{bmatrix}^2 \begin{bmatrix} 1 \\ 0 \end{bmatrix} = \begin{bmatrix} 0.48 \\ 0.52 \end{bmatrix}$$

and, in general,

$$\mathbf{x}_k = P^k\mathbf{x}_0.$$

In a general Markov chain associated with an $n \times n$ Markov matrix P, each **state vector** $\mathbf{x}_0, \mathbf{x}_1, \ldots, \mathbf{x}_k, \ldots$ is calculated according to

$$\mathbf{x}_1 = P\mathbf{x}_0, \quad \mathbf{x}_2 = P^2\mathbf{x}_0, \quad \ldots \quad \mathbf{x}_k = P^k\mathbf{x}_0, \quad \ldots$$

where $\mathbf{x}_0$ is the initial state vector. Notice that, in a generic state vector

$$\mathbf{x}_j = \begin{bmatrix} x_{1j} \\ x_{2j} \\ \vdots \\ x_{nj} \end{bmatrix},$$

the sum of all its entries equals 1, since x_{1j} is the probability of the system being in state 1 at observation j, x_{2j} is the probability of the system being in state 2 at observation j, etc. An $n \times 1$ vector whose entries are non-negative and add up to 1 is said to be a **probability vector**.

In the example that we have been analysing, one can verify that, for $k \geq 6$,

$$\mathbf{x}_k \approx \begin{bmatrix} 0.428 \\ 0.571 \end{bmatrix}, \tag{7.2}$$

if only three decimal places are considered. Hence this system seems to be approaching a steady state. In other words, what seems to be happening is that

$$\lim \mathbf{x}_n \approx \begin{bmatrix} 0.428 \\ 0.571 \end{bmatrix}.$$

The behaviour displayed in this example does not always occur, for a Markov chain may not approach a steady state.

Example 7.3 *If we have the* 2×2 *transition matrix*

$$P = \begin{bmatrix} 0 & 1 \\ 1 & 0 \end{bmatrix}$$

and an initial state vector $\mathbf{x_0} = \begin{bmatrix} 0.2 \\ 0.8 \end{bmatrix}$, *then the system oscillates between the state vectors*

$$\mathbf{x_1} = \begin{bmatrix} 0.8 \\ 0.2 \end{bmatrix}, \quad \mathbf{x_2} = \begin{bmatrix} 0.2 \\ 0.8 \end{bmatrix}.$$

A Markov chain might not approach a steady state.

A **steady-state vector** of a Markov matrix P is a probability vector $\mathbf{x}$ such that $P\mathbf{x} = \mathbf{x}$.

In other words, this steady-state vector $\mathbf{x}$, should it exist, is an eigenvector of P associated with the eigenvalue 1, besides being a probability vector.

If P is an $n \times n$ Markov matrix, then the sum of the entries of each row in P^T equals 1. Hence,

$$P^T \begin{bmatrix} 1 \\ 1 \\ \vdots \\ 1 \end{bmatrix} = \begin{bmatrix} \sum_{j=1}^n p_{j1} \\ \sum_{j=1}^n p_{j2} \\ \vdots \\ \sum_{j=1}^n p_{jn} \end{bmatrix} = \begin{bmatrix} 1 \\ 1 \\ \vdots \\ 1 \end{bmatrix}.$$

It follows that $1 \in \sigma(P^T)$. Since $\sigma(P^T) = \sigma(P)$ (see Proposition 4.3 (i)), the Markov matrix itself has an eigenvalue equal to 1.

Proposition 7.2 *A Markov matrix P has always the eigenvalue $\lambda = 1$ and a steady-state vector.*

Proof *See* EX 7.6.5.

- Is there always a steady-state vector? Yes, as this proposition shows.
- Can there be more than one steady-state vector? Yes.

Example 7.4 *Suppose that a drunkard is walking along a wooden pier with a sea endpoint (1), a cotton candy stall (2), and a photo booth (3). Last Winter's weather damaged the pier. The endpoint (1) railing is broken and at a short distance from the photo booth the wood is damaged and having a hole (4) directly above the beach. These positions (1)–(4) are located in this order. If the drunkard reaches (2) or (3), he will leave to the neighbouring points with equal probability. On the other hand, if the drunkard reaches (1) or (4), he will fall off the pier or be trapped in the hole. His random walk might be described by the transition matrix*

$$P = \begin{bmatrix} 1 & 0.5 & 0 & 0 \\ 0 & 0 & 0.5 & 0 \\ 0 & 0.5 & 0 & 0 \\ 0 & 0 & 0.5 & 1 \end{bmatrix}.$$

It is clear that any probability vector

$$\mathbf{x} = \begin{bmatrix} a \\ 0 \\ 0 \\ 1-a \end{bmatrix} \tag{7.3}$$

is a steady-state vector. If one seeks to find systematically the steady-state vectors, one has to solve the homogeneous system $(P - I)\mathbf{x} = \mathbf{0}$ and choose the probability vectors in the null space $N(P-I)$. We will find that the solution consists exactly of the vectors in (7.3).

The increasing powers of P tend (by entry) to the matrix

$$Q = \begin{bmatrix} 1 & 2/3 & 1/3 & 0 \\ 0 & 0 & 0 & 0 \\ 0 & 0 & 0 & 0 \\ 0 & 1/3 & 2/3 & 1 \end{bmatrix}$$

(see EX 7.6.6 and EX 7.6.7*).*

- Given a system does it follow that whatever the initial state, the system will approach a steady-state vector? *No.* As is evident from Example 7.3, whatever the initial state vector $\mathbf{x}_0$ we start with, the system will always oscillate between $\mathbf{x}_0$ and $\mathbf{x}_1 = P\mathbf{x}_0$. In this case, the only steady-state vector is

$$\mathbf{e} = \begin{bmatrix} 0.5 \\ 0.5 \end{bmatrix}$$

and, unless $\mathbf{x}_0$ coincides with $\mathbf{e}$, the system will never converge to it.

We have seen that a Markov chain may have one or several steady-state vectors. However, having said that, it has also become clear that a Markov chain may not approach any steady-state vector.

One is naturally led to ponder:

- How to decide whether a Markov chain approaches a steady-state vector?

We end this section with a brief discussion of this question.

An $n \times n$ Markov matrix P is said to be **regular** if there exists a positive integer k such that all entries of P^k are positive, i.e., for all $i, j = 1, \dots, n$, $(P^k)_{ij} > 0$. A Markov chain whose transition matrix is regular is said to be a **regular Markov chain**

The proof of the next theorem can be found in [9]. The matrix limit in the next theorem is defined as the limit of each entry.

Theorem 7.1 *If P is an $n \times n$ regular Markov matrix, then*

$$P^k \to \begin{bmatrix} \mathbf{q} & \mathbf{q} & \dots & \mathbf{q} \end{bmatrix},$$

where $\mathbf{q}$ is an $n \times 1$ probability vector.

Under the circumstances of Theorem 7.1, notice that, given a probability vector $\mathbf{x}$, the limit of the sequence $(P^k\mathbf{x})$ is

$$\begin{bmatrix} \mathbf{q} & \mathbf{q} & \dots & \mathbf{q} \end{bmatrix} \mathbf{x} = (x_{11} + x_{21} + \dots + x_{n1})\mathbf{q} = \mathbf{q}. \tag{7.4}$$

Observe that the equality above does not depend on $\mathbf{x}$. Moreover, given the uniqueness of the limit, $\mathbf{q}$ is the unique steady-state vector of P. In fact, if $\mathbf{r}$ is a steady-state vector of P, then $P^k\mathbf{r} = \mathbf{r}$. However, by (7.4), $P^k\mathbf{r} \to \mathbf{q}$, yielding that $\mathbf{r} = \mathbf{q}$.

Summing up:

Proposition 7.3 *If P is a regular Markov matrix, then it has a unique steady-state vector $\mathbf{q}$ and, whatever the initial state vector $\mathbf{x}_0$ is, $P^k\mathbf{x}_0 \to \mathbf{q}$.*

Getting back to our initial example,

$$P = \begin{bmatrix} 0.6 & 0.3 \\ 0.4 & 0.7 \end{bmatrix}$$

is a regular Markov matrix. To obtain the steady-state vector we ought to solve the equation $(P - I)\mathbf{x} = \mathbf{0}$ or calculate $\lim P^k$.

We opt to solve the equation. A basis of $E(1)$ is $\{(3/4, 1)\}$ whose vector is not a probability vector. But, since $3/4 + 1 = 7/4$, the steady-state vector is $(3/7, 4/7)$ (*see* EX 7.6.8).

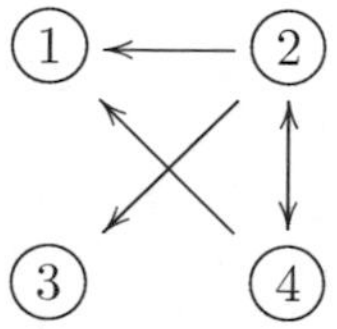

FIGURE 7.1: Links between pages in a 4-page web.

7.2.1 Google matrix and PageRank

It is reasonable that the importance of each internet page should be measured in terms of the relative frequency (overall the existing pages) it is visited by a random surfer. If one accounts for the links between pages and the probability of leaving a given page to another, one can model a web by means of a Markov matrix. If there were a steady-state vector for this matrix, then its entries would display the relative time spent visiting each page, allowing for a measure of the relevance of any given webpage.

We describe next the PageRank algorithm of webpage classification using the 4-page web in Figure 7.1. Suppose that the transition matrix corresponding to this example is

$$P = \begin{bmatrix} 1 & 1/3 & 0 & 1/2 \\ 0 & 0 & 0 & 1/2 \\ 0 & 1/3 & 1 & 0 \\ 0 & 1/3 & 0 & 0 \end{bmatrix}.$$

Observe that the non-zero entries in each column have the same value. This will be always the case for this type of transition matrices.

A surfer reaching pages 1 and 3 stays there, since no links to other pages exist. These are *dangling* pages. Hence, this matrix cannot have a unique steady-state vector, as, for any non-negative $a \leq 1$, the vector

$$\mathbf{x} = \begin{bmatrix} a \\ 0 \\ 1-a \\ 0 \end{bmatrix}$$

is a steady-state vector. This matrix P modelling our web is not a regular Markov matrix.

We see that, if we are to find a unique steady-state vector, an adjustment has to be made. The adjustment is that we assume that, when reaching this type of pages, the surfer stays there or goes to the remaining pages with equal probability. Hence, we get a new Markov matrix

$$P_1 = \begin{bmatrix} 1/4 & 1/3 & 1/4 & 1/2 \\ 1/4 & 0 & 1/4 & 1/2 \\ 1/4 & 1/3 & 1/4 & 0 \\ 1/4 & 1/3 & 1/4 & 0 \end{bmatrix}.$$

It is still the case that some zeros appear in the columns of the adjusted matrix P_1.

In order to ensure that we have a regular Markov matrix modelling the web, it is desirable to avoid zero entries. In particular, one case where this occurs is when two pages form a cycle. That is to say, when there are pages i, j such that there is a link from i to j and a link from j to i and no other links connect these pages to others.

The adjustment used is that the random surfer will move using the links from a given page with equal probability p and will pick any page of the web with probability $1 - p$. We get a transition matrix G, called the **Google matrix**,

$$\begin{aligned} G &= pP_1 + (1-p)Q \\ &= p \begin{bmatrix} 1/4 & 1/3 & 1/4 & 1/2 \\ 1/4 & 0 & 1/4 & 1/2 \\ 1/4 & 1/3 & 1/4 & 0 \\ 1/4 & 1/3 & 1/4 & 0 \end{bmatrix} + (1-p) \begin{bmatrix} 1/4 & 1/4 & 1/4 & 1/4 \\ 1/4 & 1/4 & 1/4 & 1/4 \\ 1/4 & 1/4 & 1/4 & 1/4 \\ 1/4 & 1/4 & 1/4 & 1/4 \end{bmatrix}. \end{aligned}$$

Rumour has it that Google uses $p = 0.85$. If we do the same, then

$$G = 0.85 \begin{bmatrix} 1/4 & 1/3 & 1/4 & 1/2 \\ 1/4 & 0 & 1/4 & 1/2 \\ 1/4 & 1/3 & 1/4 & 0 \\ 1/4 & 1/3 & 1/4 & 0 \end{bmatrix} + 0.15 \begin{bmatrix} 1/4 & 1/4 & 1/4 & 1/4 \\ 1/4 & 1/4 & 1/4 & 1/4 \\ 1/4 & 1/4 & 1/4 & 1/4 \\ 1/4 & 1/4 & 1/4 & 1/4 \end{bmatrix}.$$

The steady-state vector, i.e., the probability vector lying in the eigenspace $E(1)$ of this matrix is

$$\mathbf{q} \approx \begin{bmatrix} 0,314 \\ 0,245 \\ 0,220 \\ 0,220 \end{bmatrix}.$$

Hence, we have a webpage ranking: the most important page is 1 followed by page 2, and next we have pages 3 and 4 of equal importance.

7.3 Population Dynamics

In 1945, P.H. Leslie published the paper [11] on how matrices could be used to predict the evolution of populations. This **Leslie matrix model**, to be described below, only considers the female individuals of the population under study.

For example, suppose one has a population of human females divided into 5-year age groups: group g_1 corresponds to females whose age lies in $[0, 5[$, g_2 corresponds to ages in $[5, 10[$, etc. Obviously, the number n of groups depends on the maximum age in the female population under study and on the time spanned by each group.

Suppose also that we monitor this population every five years, that is, $t_0 = 0, t_1 = 5, t_2 = 10, \ldots, t_k = 5k, \ldots$.

The Leslie matrix model requires that the interval between consecutive observations have the same length as each age group.

Let

$$\mathbf{x}(0) = \begin{bmatrix} x_1(0) \\ x_2(0) \\ \vdots \\ x_n(0) \end{bmatrix}$$

be the **initial age distribution vector** displaying the number of females in each age group at time $t_0 = 0$ and let $\mathbf{x}(t_k)$ be the number of females in each group at time t_k, i.e., the **age distribution vector at time** t_k.

During a 5-year time span, it is expected to have deaths, births and aging in each age group. Hence, for $i = 1, \ldots, n$, let b_i denote the expected number of daughters born to a female in the age group i between the times t_k and t_{k+1}, and let s_i be the proportion of females in the group g_i at time t_k that are expected to be in the group g_{i+1} at time t_{k+1}.

Notice that, for each i, the number of daughters $b_i \geq 0$ and $0 < s_i \leq 1$, since we assume that the survival rate is never equal to zero, that is, not every female in the group dies between t_k and t_{k+1}.

It follows that

$$x_1(t_{k+1}) = x_1(t_k)b_1 + x_2(t_k)b_2 + \cdots + x_n(t_k)b_n \tag{7.5}$$

and, for $i = 2, \ldots, n$,

$$x_i(t_{k+1}) = s_{i-1}x_{i-1}(t_k) \tag{7.6}$$

In other words, if we consider the matrix

$$L = \begin{bmatrix} b_1 & b_2 & \ldots & \ldots & b_n \\ s_1 & 0 & 0 & 0 & 0 \\ 0 & s_2 & 0 & 0 & 0 \\ 0 & 0 & \ddots & 0 & 0 \\ 0 & 0 & 0 & s_{n-1} & 0 \end{bmatrix}, \tag{7.7}$$

we have

$$\mathbf{x}(t_{k+1}) = L\mathbf{x}(t_k)$$

and, in general, for $k = 0, 1, 2, \ldots,$

$$\mathbf{x}(t_k) = L^k \mathbf{x}(0), \tag{7.8}$$

where the matrix L in (7.7) is the **Leslie matrix**.

Example 7.5 *H. Bernadelli in 1941 [5] considered a population of beetles divided into three* 1*-year groups. The behaviour of this population is very simple: in the first year* $\frac{1}{2}$ *of the females survives and in the second year* $\frac{1}{3}$ *of the females survives. In the third year, the females give birth to an average of 6 females each. After reproduction, the females die.*

The Leslie matrix corresponding to this case is

$$L = \begin{bmatrix} 0 & 0 & 6 \\ \frac{1}{2} & 0 & 0 \\ 0 & \frac{1}{3} & 0 \end{bmatrix}.$$

Suppose one starts with 600 *females in each age group. Then, we have the following population distribution in years* $1, 2,$ *and* 3

$$\mathbf{x}(1) = \begin{bmatrix} 0 & 0 & 6 \\ \frac{1}{2} & 0 & 0 \\ 0 & \frac{1}{3} & 0 \end{bmatrix} \begin{bmatrix} 600 \\ 600 \\ 600 \end{bmatrix} = \begin{bmatrix} 3600 \\ 300 \\ 200 \end{bmatrix},$$

$$\mathbf{x}(2) = \begin{bmatrix} 0 & 0 & 6 \\ \frac{1}{2} & 0 & 0 \\ 0 & \frac{1}{3} & 0 \end{bmatrix}^2 \begin{bmatrix} 600 \\ 600 \\ 600 \end{bmatrix} = \begin{bmatrix} 0 & 2 & 0 \\ 0 & 0 & 3 \\ \frac{1}{6} & 0 & 0 \end{bmatrix} \begin{bmatrix} 600 \\ 600 \\ 600 \end{bmatrix} = \begin{bmatrix} 1200 \\ 1800 \\ 100 \end{bmatrix},$$

$$\mathbf{x}(3) = \begin{bmatrix} 0 & 0 & 6 \\ \frac{1}{2} & 0 & 0 \\ 0 & \frac{1}{3} & 0 \end{bmatrix}^3 \begin{bmatrix} 600 \\ 600 \\ 600 \end{bmatrix} = \begin{bmatrix} 1 & 0 & 0 \\ 0 & 1 & 0 \\ 0 & 0 & 1 \end{bmatrix} \begin{bmatrix} 600 \\ 600 \\ 600 \end{bmatrix} = \begin{bmatrix} 600 \\ 600 \\ 600 \end{bmatrix}.$$

Similarly, we will have in years $4, 5,$ *and* 6

$$\mathbf{x}(4) = \begin{bmatrix} 3600 \\ 300 \\ 200 \end{bmatrix}, \quad \mathbf{x}(5) = \begin{bmatrix} 1200 \\ 1800 \\ 100 \end{bmatrix}, \quad \mathbf{x}(6) = \begin{bmatrix} 600 \\ 600 \\ 600 \end{bmatrix}.$$

We have here the so-called ***population waves****: every* 3*-year period one observes the same population distribution. Notice that this behaviour does not depend on the particular initial age distribution vector* $x(0)$*. In fact, since* $L^3 = I$*, we will have always a 3-year cycle for this population.*

Example 7.6 *Consider now a female animal population whose life expectancy is 4 years. Let this population be divided into two age groups g_1, g_2 of 0 up to 2 years and from 2 to 4 years, respectively. Suppose that 50% of the females in group g_1 dies within each 2-year time span and each female is expected to give birth to 2 daughters whereas in g_2 each female gives birth to 4 daughters. Suppose that this animal population starts off with 30 individuals in g_1 and 10 in the age group g_2.*

We have now the Leslie matrix

$$L = \begin{bmatrix} 2 & 4 \\ \frac{1}{2} & 0 \end{bmatrix}$$

and the age distribution vectors

$$\mathbf{x}(2) = L\mathbf{x}_0 = L\begin{bmatrix} 30 \\ 10 \end{bmatrix} = \begin{bmatrix} 100 \\ 15 \end{bmatrix},$$

$$\mathbf{x}(4) = L^2\mathbf{x}_0 = \begin{bmatrix} 260 \\ 50 \end{bmatrix}, \qquad \mathbf{x}(6) = L^3\mathbf{x}_0 = \begin{bmatrix} 720 \\ 130 \end{bmatrix}, \qquad \mathbf{x}(8) = L^4\mathbf{x}_0 = \begin{bmatrix} 1960 \\ 360 \end{bmatrix}.$$

In this case we see a steady population growth.

In Bernadelli's example, it is clear that the population behaves in cycles. In Example 7.6, however, one may only infer that there might be a population growth in the long term. It would be desirable to have a clearer picture of the future.

As it happens, a crucial tool to answer our questions is the spectrum of the Leslie matrix. The characteristic polynomial of the Leslie matrix (7.7) is

$$p(\lambda) = (-1)^n\left(\lambda^n - b_1\lambda^{n-1} - b_2s_1\lambda^{n-2} - b_3s_1s_2\lambda^{n-3} - \cdots - b_ns_1s_2\dots s_{n-1}\right) \tag{7.9}$$

(see EX 7.6.12). Hence, for $\lambda \neq 0$,

$$p(\lambda) = (-1)^n\lambda^n\left(1 - \frac{b_1}{\lambda} - \frac{b_2s_1}{\lambda^2} - \frac{b_3s_1s_2}{\lambda^3} - \cdots - \frac{b_ns_1s_2\dots s_{n-1}}{\lambda^n}\right)$$

It follows that $\lambda \neq 0$ is a root of $p(\lambda)$ if and only if

$$\frac{b_1}{\lambda} + \frac{b_2s_1}{\lambda^2} + \frac{b_3s_1s_2}{\lambda^3} + \cdots + \frac{b_ns_1s_2\dots s_{n-1}}{\lambda^n} = 1. \tag{7.10}$$

It is a routine exercise in real function calculus to see that there exists a unique positive λ_1 satisfying (7.10). Hence, we have that $\lambda_1 > 0$ is a root of $p(\lambda)$. In fact, λ_1 is a simple root (see EX 7.6.12).

One can check directly that the vector

$$\mathbf{x}_1 = \begin{bmatrix} 1 & \frac{s_1}{\lambda_1} & \frac{s_1s_2}{\lambda_1^2} & \frac{s_1s_2s_3}{\lambda_1^3} & \cdots & \frac{s_1s_2s_3\dots s_{n-1}}{\lambda_1^{n-1}} \end{bmatrix}^T \tag{7.11}$$

is an eigenvector associated with the eigenvalue λ_1.

Summing up:

Theorem 7.2 *The Leslie matrix L in* (7.7)*, has a unique positive eigenvalue λ_1 with algebraic multiplicity* 1 *and such that there exists an eigenvector $\boldsymbol{x}_1$ in $E(\lambda_1)$ whose entries are all positive.*

The eigenvalue λ_1 of L is said to be the **dominant eigenvalue** if, for $\lambda \in \sigma(L)\backslash\{\lambda_1\}$, we have $|\lambda| < \lambda_1$.

It is not always the case that a Leslie matrix has a dominant eigenvalue. For example, the spectrum of the matrix in the Bernadelli's example is $\{1, \frac{-1\pm i\sqrt{3}}{2}\}$. This matrix has no dominant eigenvalue.

On the other hand, the Leslie matrix of Example 7.6, whose spectrum is $\{1 \pm \sqrt{3}\}$, has a dominant eigenvalue $\lambda_1 = 1 + \sqrt{3}$.

Criteria do exist to determine whether a Leslie matrix has a dominant eigenvalue but their analysis is beyond the scope of this book.

Our aim now is to show how the existence of a dominant eigenvalue helps to understand the long-term behaviour of the population. We assume in what follows that the Leslie matrix is diagonalisable, as is the case in the two previous examples.

Let the Leslie matrix (7.7) be diagonalisable, let $\lambda_1, \dots, \lambda_n$ be its eigenvalues, possibly repeated, and let S be a diagonalising matrix whose first column consists of the vector $\mathbf{x}_1$ in (7.11). The eigenvalue λ_1 is assumed to be dominant. It follows that

$$L = S \begin{bmatrix} \lambda_1 & 0 & \dots & 0 \\ 0 & \lambda_2 & 0 & 0 \\ & & \ddots & \\ 0 & 0 & & 0 \\ 0 & 0 & \dots & \lambda_n \end{bmatrix} S^{-1}.$$

Then, for time t_k, we have

$$\mathbf{x}(t_k) = L^k \mathbf{x}_0 = S \begin{bmatrix} \lambda_1^k & 0 & \dots & 0 \\ 0 & \lambda_2^k & 0 & 0 \\ & & \ddots & \\ 0 & 0 & & 0 \\ 0 & 0 & \dots & \lambda_n^k \end{bmatrix} S^{-1}\mathbf{x}(0)$$

$$= \lambda_1^k S \begin{bmatrix} 1 & 0 & \dots & 0 \\ 0 & (\frac{\lambda_2}{\lambda_1})^k & 0 & 0 \\ & & \ddots & \\ 0 & 0 & & 0 \\ 0 & 0 & \dots & (\frac{\lambda_n}{\lambda_1})^k \end{bmatrix} S^{-1}\mathbf{x}(0).$$

Since λ_1 is dominant and the first column of S is $\mathbf{x}_1$,

$$\lim_{k\to\infty}\left(\frac{1}{\lambda_1^k}\mathbf{x}(t_k)\right) = S\begin{bmatrix}1 & 0 & \dots & 0\\ 0 & 0 & 0 & 0\\ 0 & 0 & \ddots & 0\\ 0 & 0 & \dots & 0\end{bmatrix}S^{-1}\mathbf{x}(0) = \alpha\mathbf{x}_1, \tag{7.12}$$

for some scalar α. Hence, we have the approximation

$$\mathbf{x}(t_k) \approx \alpha\lambda_1^k\mathbf{x}_1. \tag{7.13}$$

Similarly, we also have

$$\mathbf{x}(t_{k-1}) \approx \alpha\lambda_1^{k-1}\mathbf{x}_1. \tag{7.14}$$

It follows that

$$\mathbf{x}(t_k) \approx \lambda_1\mathbf{x}(t_{k-1}), \tag{7.15}$$

that is, the proportion of females in consecutive age groups is (approximately) constant for a sufficiently large time.

We see that

(i) if $\lambda_1 < 1$, the population will decrease,

(ii) if $\lambda_1 = 1$, the population will stabilise,

(iii) if $\lambda_1 > 1$, the population will increase.

Example 7.6 (continued). *As seen before, the spectrum of this Leslie matrix is* $\{1 \pm \sqrt{3}\}$ *and* $\lambda_1 = 1 + \sqrt{3}$ *is dominant.*

Since the matrix is diagonalisable, we can apply the results above and can confidently say that the population is going to steadily increase (see EX 7.6.11*).*

7.4 Graphs

In this section, we give a short introduction to simple graphs, the main goal being to present, albeit briefly, a particular type of a symmetric matrix, the adjacency matrix, whose entries are either 0 or 1. We begin again with an example.

Suppose you have an archipelago of five islands some of which, possibly not all, are connected by bridges. Name the islands from 1 to 5, and consider Figure 7.2 where each line segment represents a bridge linking two islands.

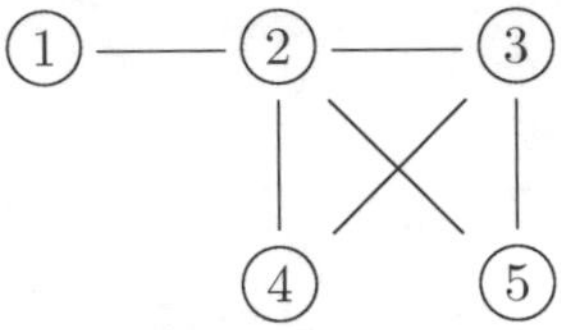

FIGURE 7.2: Graph of islands and bridges.

This is what is called a graph. In this example, each island is a **vertex** and each bridge is a line segment called an **edge** of the graph.

Formally, a **graph** is an ordered pair $G = (V, E)$ where

$$V = \{v_1, v_2, \ldots, v_n\}$$

is a non-empty (finite) set of vertices and $E \subseteq \{\{v_i, v_j\}\colon i, j = 1, \ldots, n\}$ is a subset consisting of edges, i.e., consisting of two-element subsets $\{v_i, v_j\}$ which lie in E whenever there is an edge connecting the vertices i, j.

In our example, we have $V = \{1, 2, 3, 4, 5\}$ and

$$E = \{\{1, 2\}, \{2, 3\}, \{2, 4\}, \{2, 5\}, \{3, 4\}, \{3, 5\}\}.$$

Hence, there is a total of six bridges linking the five islands.

You might ask yourself: How many ways are there of going, say, from 1 to 5? Or, what way is more efficient inasmuch as it has less bridges to cross?

We shall answer this questions by means of the adjacency matrix. The **adjacency matrix** $A = [a_{ij}]$ of the graph G is an $n \times n$ matrix whose columns and rows are associated with the vertices and such that $a_{ij} = 1$, if there is an edge between vertices v_i, v_j, and $a_{ij} = 0$, otherwise.

In other words, if v_i, v_j are **adjacent** vertices, i.e., there is an edge which they belong to, then $a_{ij} = 1 = a_{ji}$, otherwise $a_{ij} = 0 = a_{ji}$. An edge joining v_i, v_j is said to be **incident** to v_i and v_j.

The graph in Figure 7.2, has the adjacency matrix

$$A = \begin{bmatrix} 0 & 1 & 0 & 0 & 0 \\ 1 & 0 & 1 & 1 & 1 \\ 0 & 1 & 0 & 1 & 1 \\ 0 & 1 & 1 & 0 & 0 \\ 0 & 1 & 1 & 0 & 0 \end{bmatrix}.$$

Notice that A is a symmetric matrix and that the diagonal entries of A are all zero, since no *loop* is permitted from a vertex to itself. We are considering *simple graphs*: no loops and no more than one edge connecting two vertices.

The sum of the entries in a row i equals the number of edges connecting vertex i to other vertices. The same can be said for column i. This number is called the **degree** $\deg(v_i)$ of the vertex v_i. In our case, island 1 has degree $\deg(1) = 1$ whilst island 2 has degree $\deg(2) = 4$, for example.

A **walk** in a graph G is a sequence $(v_0, v_1, v_2, v_3, \dots, v_k)$ of vertices (possibly repeated) and edges

$$(\{v_0, v_1\}, \{v_1, v_2\}, \dots, \{v_{k-1}, v_k\})$$

which belong to G. The walk is said to start at v_0 and end at v_k. A **path** in a graph G is a walk whose sequence of vertices consists of distinct terms. Here it is allowed to have $v_0 = v_k$, in which case we have a **closed path**.

The **length** of the walk is the number $k \geq 0$ of edges. In what follows, for simplicity, we refer to walks and paths only specifying its sequence of vertices.

For example, the graph of Figure 7.2 has a path of length 2 consisting of the sequence of vertices $4, 2, 3$ and a walk $4, 2, 3, 2, 3$ of length 4. The endpoints of both path and walk are the same: the start is 4 and the end is 3. Another path is $3, 5, 2, 1$. This path has length 3.

In general, it is a more efficient way to go from an island to an other taking a path rather than a walk. The most efficient way is the shortest path connecting them.

The next theorem tells us how many walks there are connecting two vertices.

Theorem 7.3 *Let $A = [a_{ij}]$ be the $n \times n$ adjacency matrix of a graph G having n vertices. The number of walks of positive length k between vertices i, j is the entry-ij $a_{ij}^{(k)}$ of the matrix A^k.*

Proof *The theorem will be proved by induction. Let $i, j = 1, \dots, n$. If $k = 1$, by the definition of the adjacency matrix A, a_{ij} is either 0 or 1. This means that there is no edge or exactly one edge, respectively, connecting the vertices i and j. Consequently, the assertion holds when $k = 1$.*

Assume now that the statement is true for k. We will prove that it also holds for $k + 1$.

Let $l_{ir}^{(k)}$ be the number of k-length walks from v_i to v_r, where $r = 1, \dots, n$. The $k + 1$-length walks from v_i to v_j are all of k-length walks from v_i to v_r followed by all walks of length 1 from v_r to v_j. Hence, by the induction hypothesis, the number $l_{ij}^{(k+1)}$ of walks from v_i to v_j is

$$l_{ij}^{(k+1)} = \sum_{r=1}^{n} l_{ir}^{(k)} l_{rj}^{(1)} = \sum_{r=1}^{n} a_{ir}^{(k)} l_{rj}^{(1)}.$$

Since we have shown above that $l_{rj}^{(1)} = a_{rj}$, it follows that

$$l_{ij}^{(k+1)} = \sum_{r=1}^{n} a_{ir}^{(k)} a_{rj}^{(1)} = a_{ij}^{(k+1)},$$

as required.

In our example, we have

$$A^2 = \begin{bmatrix} 1 & 0 & 1 & 1 & 1 \\ 0 & 4 & 2 & 1 & 1 \\ 1 & 2 & 3 & 1 & 1 \\ 1 & 1 & 1 & 2 & 2 \\ 1 & 1 & 1 & 2 & 2 \end{bmatrix}, \qquad A^3 = \begin{bmatrix} 0 & 4 & 2 & 1 & 1 \\ 4 & 4 & 6 & 6 & 6 \\ 2 & 6 & 4 & 5 & 5 \\ 1 & 6 & 5 & 2 & 2 \\ 1 & 6 & 5 & 2 & 2 \end{bmatrix}.$$

Hence, if we consider islands 2 and 3, we have one walk of length 1 linking them, two walks of length 2 and 6 walks of length 3 (see EX 7.6.15).

Now that we can count the walks connecting two vertices, we shall see next that every walk contains a path.

Proposition 7.4 *Let $G = (V, E)$ be a graph and let $v, v' \in V$. If there exists a walk from v to v', then there exists a path from v to v'. Moreover, the length of the path is less than or equal to the length of the walk.*

Proof *Consider the set consisting of all the walks from v to v' (of any given length) and observe that there must be a minimum length walk in this set.*

If the walk has length 1, then the walk is itself a path. Suppose now that this minimum length walk has length $k \geq 2$. We want to show that this walk is a path. Suppose that, on the contrary, this walk is not a path. Consequently, there exist vertices v_r and v_{r+m}, with $m \neq 0$ and $v_r = v_{r+m}$. Then, if we remove from the walk all the vertices $v_{r+1}, \dots, v_{r+m}$ and the edges linking them, we obtain a strictly shorter length walk from v to v'. But this is a contradiction, since we assumed that our initial walk was the minimum length walk. The remaining assertion of the theorem is obvious.

For example, there is exactly one path between islands 1 and 2: this path has length 1. There are two paths between islands 1 and 3 of length 3 (see EX 7.6.15).

A subset of vertices of a graph G is said to be a **clique** if (i) the subset contains three vertices, at least; (ii) any two distinct vertices are adjacent; (iii) the subset is not contained in any strictly larger subset of vertices satisfying (ii).

For example, the island-bridge graph contains the cliques $\{2, 3, 4\}$ and $\{2, 3, 5\}$. Identifying these cliques was easy because our graph is small. It would be desirable, however, to have a systematic way of identifying the vertices lying in cliques. Why might this be of help? For example, if the edges of a graph identify the similarly politically inclined in a group of people, say, politicians voting a law, you might have an educated guess about whether the law will pass or not.

Corollary 7.1 *Let $A = [a_{ij}]$ be the $n \times n$ adjacency matrix of a graph $G = (V, E)$ having n vertices $v_1, v_2, \ldots, v_n$. A vertex v_i belongs to a clique if and only if the entry-ii of the matrix A^3 is non-zero, i.e., $a_{ii}^{(3)} \neq 0$.*

Proof *Suppose $a_{ii}^{(3)} \neq 0$. By Theorem 7.3, we know that $a_{ii}^{(3)}$ is the number of walks of length 3 from v_i to itself. Any walk of length 3 having the same vertex v_i as endpoints, needs to have another two distinct vertices, say, v_j, v_r. It follows that $\{v_i, v_j, v_r\}$ form a clique or are a subset of a clique.*

Conversely, suppose there is a clique containing v_i. By the definition of clique, there exist distinct vertices v_j, v_r such that $a_{ij}, a_{jr}, a_{ri} \neq 0$. It follows that $a_{ii}^{(3)} \geq a_{ij}a_{jr}a_{ri} \neq 0$.

We see that in our island-bridge graph, matrix A^3 indicates that the islands $2, 3, 4, 5$ belong to cliques, as expected, while this is not the case for island 1.

7.5 Differential Equations

The aim of this section is to introduce the exponential of a square matrix. This will be motivated by and then applied to the solution of differential equations.

Let

$$x'(t) = cx(t) \tag{7.16}$$

be a linear differential equation over the reals. It is known that its set of solutions consists of all the functions $x(t) \colon \mathbb{R} \to \mathbb{R}$ such that

$$x(t) = \alpha e^{ct},$$

where α is a(ny) real constant. Suppose that we increase the 'complexity' of the problem: we have now the system of (first-order) linear differential equations

$$\begin{cases} x_1'(t) &= -x_1(t) + x_2(t) \\ x_2'(t) &= 5x_1(t) + 3x_2(t) \end{cases} \tag{7.17}$$

in the unknowns $x_1(t), x_2(t) \colon \mathbb{R} \to \mathbb{R}$.

This system can be re-written in matrix form as

$$\begin{bmatrix} x_1'(t) \\ x_2'(t) \end{bmatrix} = \begin{bmatrix} -1 & 1 \\ 5 & 3 \end{bmatrix} \begin{bmatrix} x_1(t) \\ x_2(t) \end{bmatrix} \tag{7.18}$$

or, more compactly,

$$\mathbf{x}' = A\mathbf{x}, \tag{7.19}$$

where A is the matrix above and the derivative of a vector (or matrix) is to be calculated by entry. Here the dependence on t is omitted to simplify the notation.

Can we still mimic in some way the solution of (7.16) to solve the system (7.17)? In other words, is it possible to give a meaning to the exponential e^A of a matrix A such that a solution of (7.19) can be constructed somehow using the exponential?

If one were to define the **exponential of a matrix** A by generalising what is known for the exponential function over $\mathbb{R}$, one would naturally be led to write

$$e^A = I + A + \frac{1}{2!}A^2 + \frac{1}{3!}A^3 + \cdots + \frac{1}{n!}A^n + \cdots = \sum_{n=0}^{\infty} \frac{1}{n!}A^n. \tag{7.20}$$

Although, each term of the formal power series is meaningful, does the series *converge*? And what does *convergence* even mean?

Let us begin by calculating the matrix powers for our example. The matrix A is diagonalisable and

$$A = \begin{bmatrix} -1 & 1 \\ 1 & 5 \end{bmatrix} \begin{bmatrix} -2 & 0 \\ 0 & 4 \end{bmatrix} \begin{bmatrix} -1 & 1 \\ 1 & 5 \end{bmatrix}^{-1}$$

(see EX 7.6.18). It follows that

$$A^n = \begin{bmatrix} -1 & 1 \\ 1 & 5 \end{bmatrix} \begin{bmatrix} (-2)^n & 0 \\ 0 & 4^n \end{bmatrix} \begin{bmatrix} -1 & 1 \\ 1 & 5 \end{bmatrix}^{-1}$$

and

$$\sum_{n=0}^{k} \frac{1}{n!}A^n = \begin{bmatrix} -1 & 1 \\ 1 & 5 \end{bmatrix} \begin{bmatrix} \sum_{n=0}^{k} \frac{1}{n!}(-2)^n & 0 \\ 0 & \sum_{n=0}^{k} \frac{1}{n!}4^n \end{bmatrix} \begin{bmatrix} -1 & 1 \\ 1 & 5 \end{bmatrix}^{-1}.$$

When $k \to \infty$, we have that this matrix product tends by entry to

$$\underbrace{\begin{bmatrix} -1 & 1 \\ 1 & 5 \end{bmatrix}}_{S} \begin{bmatrix} \sum_{n=0}^{\infty} \frac{1}{n!}(-2)^n & 0 \\ 0 & \sum_{n=0}^{\infty} \frac{1}{n!}4^n \end{bmatrix} \underbrace{\begin{bmatrix} -1 & 1 \\ 1 & 5 \end{bmatrix}^{-1}}_{S^{-1}} = S \begin{bmatrix} e^{-2} & 0 \\ 0 & e^4 \end{bmatrix} S^{-1}.$$

In general, let A be a $n \times n$ diagonalisable matrix with (possibly repeated) eigenvalues $\{\lambda_1, \lambda_2, \ldots, \lambda_n\}$. Then

$$A = S \operatorname{diag}(\lambda_1, \lambda_2, \ldots, \lambda_n) S^{-1},$$

where $\operatorname{diag}(\lambda_1, \lambda_2, \ldots, \lambda_n)$ is a diagonal matrix whose diagonal consists of the eigenvalues of A repeated as many times as the corresponding algebraic multiplicities, and S is a diagonalising matrix (see Theorem 4.3 and the 'How to diagonalise a matrix' box).

A reasoning similar to that above yields that the **exponential e^{A}** of the matrix A is

$$e^{A} = S \operatorname{diag}(e^{\lambda_1}, e^{\lambda_2}, \ldots, e^{\lambda_n}) S^{-1}.$$

It follows that the exponential of tA, where $t \in \mathbb{R}$, is

$$\begin{aligned} e^{tA} &= I + tA + \frac{1}{2!}(tA)^2 + \frac{1}{3!}(tA)^3 + \cdots + \frac{1}{n!}(tA)^n + \ldots \\ &= \sum_{n=0}^{\infty} \frac{1}{n!}(tA)^n = S\Big(\sum_{n=0}^{\infty} \frac{1}{n!}(tD)^n\Big) S^{-1}. \end{aligned}$$

Hence

$$e^{tA} = S \operatorname{diag}(e^{\lambda_1 t}, e^{\lambda_2 t}, \ldots, e^{\lambda_n t}) S^{-1}, \tag{7.21}$$

i.e.,

$$e^{tA} = S e^{tD} S^{-1}. \tag{7.22}$$

Notice that

$$(e^{tA})' = Ae^{tA}$$

(see EX 7.6.19). Hence, given a(ny) $n \times 1$ vector $\mathbf{c}$, we have

$$(e^{tA}\mathbf{c})' = (e^{tA})'\mathbf{c} = Ae^{tA}\mathbf{c}$$

from which follows that,

$$\mathbf{x}(t) = e^{tA}\mathbf{c} \tag{7.23}$$

is a solution of the system of linear differential equations

$$\mathbf{x}' = A\mathbf{x}.$$

It is possible to show that, in fact, any solution of this system is of the form (7.23). Moreover, since

$$\mathbf{x}(0) = e^{\mathbf{0}}\mathbf{c} = \mathbf{c},$$

we have that fixing the initial conditions at $t = 0$ determines a unique solution. (Observe that $e^{\mathbf{0}} = I$.) It follows that this unique solution is

$$\mathbf{x}(t) = e^{tA}\mathbf{x}(0). \tag{7.24}$$

Example 7.7 *Consider the system of linear differential equations*

$$\begin{cases} x' &= 2x + z \\ y' &= -y \\ z' &= x + 2z \end{cases}$$

which we want to solve with the initial conditions $x(0) = -1, y(0) = 1, z(0) = 2$. We begin by writing the system in matrix form

$$\begin{bmatrix} x' \\ y' \\ z' \end{bmatrix} = \underbrace{\begin{bmatrix} 2 & 0 & 1 \\ 0 & -1 & 0 \\ 1 & 0 & 2 \end{bmatrix}}_{A} \begin{bmatrix} x \\ y \\ z \end{bmatrix}.$$

This 3×3 *matrix is diagonalisable:*

$$A = \begin{bmatrix} 0 & 1 & 1 \\ 1 & 0 & 0 \\ 0 & -1 & 1 \end{bmatrix} \begin{bmatrix} -1 & 0 & 0 \\ 0 & 1 & 0 \\ 0 & 0 & 3 \end{bmatrix} \begin{bmatrix} 0 & 1 & 1 \\ 1 & 0 & 0 \\ 0 & -1 & 1 \end{bmatrix}^{-1}.$$

By (7.22),

$$e^{tA} = \begin{bmatrix} 0 & 1 & 1 \\ 1 & 0 & 0 \\ 0 & -1 & 1 \end{bmatrix} \begin{bmatrix} e^{-t} & 0 & 0 \\ 0 & e^{t} & 0 \\ 0 & 0 & e^{3t} \end{bmatrix} \begin{bmatrix} 0 & 1 & 1 \\ 1 & 0 & 0 \\ 0 & -1 & 1 \end{bmatrix}^{-1}$$

$$e^{tA} = \left[e^{-t} \begin{bmatrix} 0 \\ 1 \\ 0 \end{bmatrix} \quad e^{t} \begin{bmatrix} 1 \\ 0 \\ -1 \end{bmatrix} \quad e^{3t} \begin{bmatrix} 1 \\ 0 \\ 1 \end{bmatrix} \right] \begin{bmatrix} 0 & 1 & 0 \\ 1/2 & 0 & -1/2 \\ 1/2 & 0 & 1/2 \end{bmatrix}. \tag{7.25}$$

Hence

$$e^{tA} = \frac{1}{2} \begin{bmatrix} e^{t} + e^{3t} & 0 & -e^{t} + e^{3t} \\ 0 & 2e^{-t} & 0 \\ -e^{t} + e^{3t} & 0 & e^{t} + e^{3t} \end{bmatrix}.$$

Using the initial conditions,

$$\begin{bmatrix} x \\ y \\ z \end{bmatrix} = \frac{1}{2} \begin{bmatrix} e^{t} + e^{3t} & 0 & -e^{t} + e^{3t} \\ 0 & 2e^{-t} & 0 \\ -e^{t} + e^{3t} & 0 & e^{t} + e^{3t} \end{bmatrix} \begin{bmatrix} -1 \\ 1 \\ 2 \end{bmatrix} = \frac{1}{2} \begin{bmatrix} -3e^{t} + e^{3t} \\ 2e^{-t} \\ 3e^{t} + e^{3t} \end{bmatrix},$$

i.e.,

$$x(t) = -\tfrac{3}{2}e^{t} + \tfrac{1}{2}e^{3t}, \quad y(t) = e^{-t}, \quad z(t) = \tfrac{3}{2}e^{t} + \tfrac{1}{2}e^{3t}.$$

In (7.25), the right hand side of the equality consists of a product of two matrices. Notice that the columns of the first matrix are formed by eigenvectors of A multiplied by certain 'weights'. Each column shows a pairing between the eigenvector and the corresponding eigenvalue inasmuch as the weight is a exponential in whose exponent appears this eigenvalue. This behaviour can be seen in general.

Let A be an $n \times n$ diadonalisable matrix such that $A = SDS^{-1}$ and consider the system (7.19). By (7.22) and (7.24), we have

$$\mathbf{x}(t) = e^{tA}\mathbf{x}(0) = Se^{tD}S^{-1}\mathbf{x}(0)$$

and, therefore,

$$\mathbf{x}(t) = \left[e^{\lambda_1 t}\mathbf{x_1} \mid e^{\lambda_2 t}\mathbf{x_2} \mid \dots \mid e^{\lambda_n t}\mathbf{x_n} \right] \underbrace{\begin{bmatrix} \alpha_1 \\ \alpha_2 \\ \vdots \\ \alpha_n \end{bmatrix}}_{S^{-1}\mathbf{x}(0)},$$

where $\mathbf{x_1}, \mathbf{x_2}, \ldots, \mathbf{x_n}$ are the eigenvectors used to diagonalise this matrix. In other words, the solution of the initial value problem is a linear combination of the column vectors

$$e^{\lambda_1 t}\mathbf{x_1}, \quad e^{\lambda_2 t}\mathbf{x_2}, \quad \ldots, \quad e^{\lambda_n t}\mathbf{x_n}.$$

We can apply this to solve the system of Example 7.7 in an alternative way. We know now that the solution has the form

$$\begin{bmatrix} x(t) \\ y(t) \\ z(t) \end{bmatrix} = \alpha_1 e^{-t} \begin{bmatrix} 0 \\ 1 \\ 0 \end{bmatrix} + \alpha_2 e^{t} \begin{bmatrix} 1 \\ 0 \\ -1 \end{bmatrix} + \alpha_3 e^{3t} \begin{bmatrix} 1 \\ 0 \\ 1 \end{bmatrix}.$$

Using the initial conditions at $t = 0$,

$$\begin{bmatrix} -1 \\ 1 \\ 2 \end{bmatrix} = \alpha_1 \begin{bmatrix} 0 \\ 1 \\ 0 \end{bmatrix} + \alpha_2 \begin{bmatrix} 1 \\ 0 \\ -1 \end{bmatrix} + \alpha_3 \begin{bmatrix} 1 \\ 0 \\ 1 \end{bmatrix},$$

we have $\alpha_1 = 1, \alpha_2 = -\frac{3}{2}, \alpha_3 = \frac{1}{2}$. Hence,

$$x(t) = -\tfrac{3}{2}e^{t} + \tfrac{1}{2}e^{3t}, \quad y(t) = e^{-t}, \quad z(t) = \tfrac{3}{2}e^{t} + \tfrac{1}{2}e^{3t}.$$

Summing up: the solution is a linear combination of n weighted eigenvectors of A – any n linearly independent eigenvectors will do. The specific linear combination is obtained by forcing the solution to satisfy the initial conditions.

7.6 Exercises

EX 7.6.1. Show that the spectrum of a projection matrix is a non-empty subset of $\{0, 1\}$. Find the corresponding eigenspaces.

EX 7.6.2. Find the projection matrix onto the the subspace W of $\mathbb{R}^4$ such that

$$W = \{(x, y, z, w) \in \mathbb{R}^4 \colon x + y - z = 0, z = w\}.$$

EX 7.6.3. Find the straight line that best fits the points $(2, 4), (3, 5), (7, 9)$.

EX 7.6.4. Show that the product of two Markov matrices is a Markov matrix.

EX 7.6.5. Prove Proposition 7.2. Hint: Show that, given an $n \times n$ stochastic matrix, if $A\mathbf{x} = \mathbf{x}$ then

$$\sum_{i=1}^{n} \sum_{j=1}^{n} |(Ax)_{ij}| \leq \sum_{i=1}^{n} |x_{i1}|$$

and that the equality holds if and only if the entries in $\mathbf{x}$ have the same sign.

EX 7.6.6. For the Markov matrix of Example 7.4, calculate P^k with $k = 0, 1, 2, \ldots, 10$.

EX 7.6.7. Calculate the steady-state vectors of the Markov matrices P, Q in Example 7.4. Does this contradict Proposition 7.3? Why?

EX 7.6.8. Calculate the steady-state vector of

$$P = \begin{bmatrix} 0.6 & 0.3 \\ 0.4 & 0.7 \end{bmatrix}$$

and compare with (7.2).

EX 7.6.9. Let

$$P = \begin{bmatrix} 0 & 0 & 0 & 0 & 0 \\ 1/2 & 1 & 0 & 1/2 & 1/2 \\ 1/2 & 0 & 1 & 0 & 0 \\ 0 & 0 & 0 & 0 & 1/2 \\ 0 & 0 & 0 & 1/2 & 0 \end{bmatrix}$$

be the transition matrix of a 5-page web. Draw this web, find its Google matrix and rank its pages.

EX 7.6.10. Consider a female animal population whose life expectancy is 10 years. Let this population be divided in two age groups g_1, g_2 of 0 up to 5 years and from 5 to 10 years, respectively. Suppose that 20% of the females in group g_1 dies within each time span of 5 years and each female is expected to give birth to 2 daughters whereas in g_2 each female gives birth to 4 daughters.

Suppose that this animal population starts off with 30 individuals in g_1 and 20 in the age group g_2.

Find the Leslie matrix and the age distribution vectors $\mathbf{x}(5), \mathbf{x}(10)$. How many females are in g_1 and g_2 after 15 years?

EX 7.6.11. Consider a female animal population whose life expectancy is 4 years. Let this population be divided in two age groups g_1, g_2 of 0 up to 2 years and from 2 to 4 years, respectively. Suppose that 50% of the females in group g_1 dies within each 2-year time span and each female is expected to give birth to 2 daughters whereas in g_2 each female gives birth to 4 daughters. Suppose that this animal population starts off with 30 individuals in g_1 and 10 in the age group g_2.

Find the Leslie matrix and the age distribution vectors from $\mathbf{x}(2)$ to $\mathbf{x}(8)$. Determine the dominant eigenvalue λ_1 and calculate the same distribution vectors with the approximation formula. How is the population expected to behave?

EX 7.6.12. Show that the equality (7.9) holds and that there exists only one $\lambda_1 > 0$ satisfying (7.10). Hint: show that the real function $f(\lambda)$ on the left hand side of (7.10) is decreasing, that $\lim_{\lambda\to 0^+} f(\lambda) = +\infty$ and that $\lim_{\lambda\to+\infty} f(\lambda) = 0$.

EX 7.6.13. Show that λ_1 in the previous exercise is a simple root of $p(\lambda)$. Hint: recall that a root a of a polynomial $q(t)$ is simple if and only if $q'(a) \neq 0$.

Show that λ_1 is a simple root of $p(\lambda)$. Hint: recall that a root a of a polynomial $q(t)$ is simple if and only if $q'(a) \neq 0$.

EX 7.6.14. Suppose that

$$L = \begin{bmatrix} 0 & 1 & 1 \\ \frac{1}{2} & 0 & 0 \\ 0 & \frac{1}{5} & 0 \end{bmatrix}.$$

is the Leslie matrix of some female population. How many age groups are there? What is the approximate proportion of the number of females in two consecutive age groups for large enough time? Is the population going to increase eventually? Why?

EX 7.6.15. For the island-bridge example of §7.4, find the walks of length 2 and 3 connecting islands 1 and 3. Which of them are paths?

EX 7.6.16. Draw the simple graph whose adjacency matrix is

$$\begin{bmatrix} 0 & 1 & 0 & 0 & 1 & 0 \\ 1 & 0 & 1 & 1 & 0 & 1 \\ 0 & 1 & 0 & 0 & 1 & 0 \\ 0 & 1 & 0 & 0 & 1 & 0 \\ 1 & 0 & 1 & 1 & 0 & 1 \\ 0 & 1 & 0 & 0 & 1 & 0 \end{bmatrix}$$

and find the degree of its vertices.

EX 7.6.17. Consider the graph

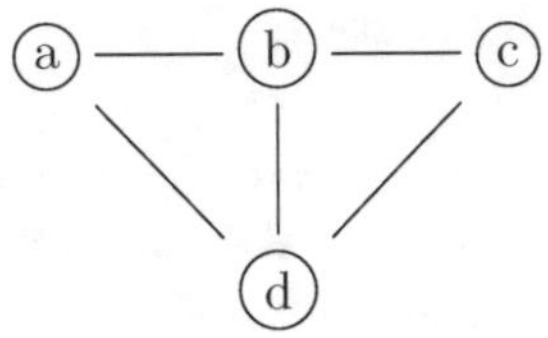

Find the adjacency matrix, the vertices lying in cliques and those cliques having b. Find the paths whose start point is b and endpoint is d.

EX 7.6.18. Calculate e^A with

$$A = \begin{bmatrix} -1 & 1 \\ 5 & 3 \end{bmatrix}.$$

EX 7.6.19. Let A be a $n \times n$ diagonalisable matrix. Show that

$$(e^{tA})' = Ae^{tA}.$$

EX 7.6.20. Solve the system of differential equations

$$\begin{cases} x' = 3x + 5y \\ y' = x - y \end{cases}$$

with $x(0) = 1, y(0) = 1$.

EX 7.6.21. Solve the initial value problem $\mathbf{x}' = A\mathbf{x}$ with

$$A = \begin{bmatrix} 2 & -2 & 2 \\ 0 & 0 & 1 \\ 0 & 0 & 1 \end{bmatrix}, \qquad \mathbf{x}(0) = \begin{bmatrix} 1 \\ 0 \\ 0 \end{bmatrix}.$$

7.7 At a Glance

Special matrices were introduced in this chapter in connection with applications of Linear Algebra:

- The projection matrix (least squares solutions/minimisation of error) – a matrix corresponding to an orthogonal projection onto a subspace of a real vector space for a fixed basis. It can be calculated through a formula using a matrix whose columns are the mentioned basis.

- The stochastic matrix (Markov chains) – a real square matrix whose entries are non-negative and such that the sum of the entries in each column equals 1. The spectrum of a stochastic matrix contains 1. The probability eigenvectors associated with the eigenvalue 1 might indicate that the system modelled by the matrix reaches equilibrium eventually.

- The Google matrix (PageRank) – a particular type of a stochastic matrix, it has always a (probability) eigenvector associated with the eigenvalue 1 whose entries rank the webpages the matrix models.

- The Leslie matrix (evolution of populations) – a real square matrix

$$L = \begin{bmatrix} b_1 & b_2 & \dots & \dots & b_n \\ s_1 & 0 & 0 & 0 & 0 \\ 0 & s_2 & 0 & 0 & 0 \\ 0 & 0 & \ddots & 0 & 0 \\ 0 & 0 & 0 & s_{n-1} & 0 \end{bmatrix}, \tag{7.26}$$

  where $b_1, b_2, \dots b_n \geq 0$ and $0 < s_1, s_2, \dots, s_{n-1} \leq 1$. Under some circumstances, L has a dominant eigenvalue whose value indicates the evolution of a population in the long run.

- The adjacency matrix (graphs) – a real square matrix whose entries are either 0 or 1 which models the edges linking the vertices of a simple graph. It allows for determining the number of walks between two vertices and identifying vertices lying in cliques in the graph, for example.

- The exponential of a matrix (differential equations) – the matrix e^A is obtained using the spectrum of A. It can be used to solve differential equations.

The author wishes to thank J. Teixeira and M.J. Borges for Example 7.7, EX 7.6.20, and EX 7.6.21.

Chapter 8

Appendix

8.1	Uniqueness of Reduced Row Echelon Form	285
8.2	Uniqueness of Determinant	286
8.3	Direct Sum of Subspaces	287

8.1 Uniqueness of Reduced Row Echelon Form

We prove here the uniqueness of the reduced row echelon form of a matrix. This proof is essentially that in [16].

Proposition 8.1 *Let A be a $k \times n$ matrix over $\mathbb{K}$ and let R, R' be reduced row echelon matrices obtained from A through elementary row operations. Then $R = R'$.*

Proof *This result will be proved by induction on the the number of columns n of the matrix. The result is clear enough for $n = 1$. Let A be an $k \times n$ matrix with $n > 1$, and suppose now that the result holds for all number of columns less than or equal to $n - 1$.*

If R, R' are reduced row echelon forms of A, then, by the induction hypothesis, their first $n-1$ columns must be equal. Indeed, if we remove the last column of all three matrices, thus obtaining $k \times n-1$ matrices, say, $A_{n-1}, R_{n-1}, R'_{n-1}$, respectively, then R_{n-1}, R'_{n-1} are reduced row echelon forms of A_{n-1}. Hence, $R_{n-1} = R'_{n-1}$.

It follows that R, R' may differ only in the nth column. Suppose, then, that they differ in row i, that is, $r_{in} \neq r'_{in}$.

Let $\mathbf{x}$ be a vector such that $A\mathbf{x} = \mathbf{0}$. Then, we have also that $R\mathbf{x} = \mathbf{0} = R'\mathbf{x}$. Consequently, $(R - R')\mathbf{x} = \mathbf{0}$ from which follows that $x_n = 0$, since $r_{i,n} \neq r'_{in}$, as assumed above. This forces both columns n of R and R' to have a pivot (equal to 1), since otherwise x_n would be an independent variable. However, since $R_{n-1} - R'_{n-1} = 0$, we have finally that these pivots must be located in the same row. This yields a contradiction, since it forces $R = R'$ which we assumed to be different.

DOI: 10.1201/9781351243452-8

8.2 Uniqueness of Determinant

Proposition 8.2 *There exists a unique function $f\colon M_n(\mathbb{K}) \to \mathbb{K}$ such that, for all $A \in M_n(\mathbb{K})$,*

(Ax1) $f(I) = 1$;

(Ax2) $f(P_{ij}A) = -f(A) \quad$ *(with $i \neq j$, $i, j = 1, \ldots, n$);*

(Ax3) Given $\alpha \in \mathbb{K}$ and $i \in \{1, \ldots, n\}$,

$$f\left(\begin{bmatrix} \vdots \\ \alpha \mathbf{l}_i \\ \vdots \end{bmatrix}\right) = \alpha f\left(\begin{bmatrix} \vdots \\ \mathbf{l}_i \\ \vdots \end{bmatrix}\right)$$

$$f\left(\begin{bmatrix} \mathbf{l}_1 \\ \vdots \\ \mathbf{l}_{i-1} \\ \mathbf{l}_i + \mathbf{l}'_i \\ \mathbf{l}_{i+1} \\ \vdots \\ \mathbf{l}_n \end{bmatrix}\right) = f\left(\begin{bmatrix} \mathbf{l}_1 \\ \vdots \\ \mathbf{l}_{i-1} \\ \mathbf{l}_i \\ \mathbf{l}_{i+1} \\ \vdots \\ \mathbf{l}_n \end{bmatrix}\right) + f\left(\begin{bmatrix} \mathbf{l}_1 \\ \vdots \\ \mathbf{l}_{i-1} \\ \mathbf{l}'_i \\ \mathbf{l}_{i+1} \\ \vdots \\ \mathbf{l}_n \end{bmatrix}\right),$$

where $\mathbf{l}_i, \mathbf{l}'_i$ are matrix rows.

Proof *This proof is inspired by that of [2, Theorem 3.2]*
Let $g\colon M_n(\mathbb{K}) \to \mathbb{K}$ be a function defined, for all $A \in M_n(\mathbb{K})$, by

$$g(A) = f(A) - \det A,$$

where $\det A$ *is the function given by the Leibniz's formula (see §2.2,* (2.4)*).*

Suppose firstly that A is invertible and that we reduce A to the identity I, the reduced row echelon form of A, using elementary operations. Then, by (Ax2), (Ax3), and Proposition 2.1 (iii),

$$g(A) = c(f(I) - \det I),$$

where $c \neq 0$ is a scalar which accounts for the changes in the determinant brought along by the elementary operations. Since

$$f(I) = 1 = \det I,$$

if follows that $g(A) = 0$, i.e., the functions f and $\det$ *coincide on the invertible matrices.*

If A is not invertible, then its reduced row echelon form has a zero row. Hence, by Proposition 2.1 (i),

$$f(A) = 0 = \det A.$$

Notice that Proposition 2.1 must hold also for f, since we used only the Axioms (Ax1) – (Ax3) in its proof.

The proof is complete.

8.3 Direct Sum of Subspaces

In Chapter 3, we defined the direct sum of two subspaces. Here we generalise this to a finite number of subspaces.

Let V be a vector space over $\mathbb{K}$ and let $S_1, S_2, \dots, S_k$ be subspaces of V. Define the sum $\sum_{i=1}^k S_i$ of these subspaces by

$$\sum_{i=1}^{k} S_i = \{\boldsymbol{x}_1 + \dots \boldsymbol{x}_2 + \cdots + \boldsymbol{x}_k \colon \boldsymbol{x}_i \in S_i,\ i = 1, 2, \dots, k\}.$$

The set $\sum_{i=1}^k S_i$ is itself also a subspace of V. This sum is said to be a **direct sum** $S_1 \oplus S_2 \oplus \cdots \oplus S_k$ if, for all $i = 1, 2, \dots, k$,

$$S_i \cap \sum_{l \in \{1,2,\dots,k\} \setminus i} S_l = \{0\}. \tag{8.1}$$

Notice that this definition is exactly that we presented in Chapter 3 when referring to the direct sum of two subspaces.

Proposition 8.3 *Let V be a vector space over $\mathbb{K}$ with $\dim V = n$, and let $S_1, S_2, \dots, S_k$ be subspaces of V. Then the following are equivalent.*

(i) $V = S_1 \oplus S_2 \oplus \cdots \oplus S_k$.

(ii) $\sum_{i=1}^k \dim S_i = n$ *and, for all* $i = 1, 2, \dots, k$,

$$S_i \cap \sum_{l \in \{1,2,\dots,k\} \setminus i} S_l = \{0\}. \tag{8.2}$$

(iii) $V = \sum_{i=1}^k S_i$ *and, for all $x \in V$, the decomposition*

$$\boldsymbol{x} = \boldsymbol{x}_1 + \dots \boldsymbol{x}_2 + \cdots + \boldsymbol{x}_k$$

is unique, with $\boldsymbol{x}_i \in S_i$, for all $i = 1, \dots, k$.

Proof *We begin by showing that (i) implies (ii). Observe that, by* (8.1) *the union of the bases of all subspaces is a linearly independent set. Indeed, if it were not, then some non-zero vector spanned by one of the bases would be a linear combination of the vectors in the bases of the remaining spaces,* *contradicting* (8.1)*. Since* $V = \sum_{i=1}^{k} S_i$*, it follows that the union of these bases spans a subspace of dimension* n *and, hence,* $\sum_{i=1}^{k} \dim S_i = n$*.*

To see that (ii) $\Rightarrow$ *(iii), consider that, given some* $\boldsymbol{x}$ *in* V*, we have*

$$\boldsymbol{x}_1 + \ldots \boldsymbol{x}_2 + \cdots + \boldsymbol{x}_k = \boldsymbol{x} = \boldsymbol{z}_1 + \ldots \boldsymbol{z}_2 + \cdots + \boldsymbol{z}_k,$$

with $\boldsymbol{x}_i, \boldsymbol{z}_i \in S_i$*, for all* $i = 1, \ldots, k$*. Then,*

$$(\boldsymbol{x}_1 - \boldsymbol{z}_1) + (\boldsymbol{x}_2 - \boldsymbol{z}_2) + \cdots + (\boldsymbol{x}_k - \boldsymbol{z}_k) = \mathbf{0}.$$

But by (8.2)*, each of the summands must coincide with* $\mathbf{0}$*, yielding the uniqueness of the decomposition. Consequently, the union of the bases of all subspaces is a linearly independent set from which follows the remaining assertion.*

Suppose now (iii) holds. To show that this implies (i), if suffices to show that (8.1) *holds. Suppose, on the contrary, that, for some* i*, there existed a non-zero* $\boldsymbol{x} \in S_i \cap \sum_{l \in \{1,2,\ldots,k\} \setminus i} S_l$*. But then*

$$\mathbf{0} = \boldsymbol{x} - \sum_{l \in \{1,2,\ldots,k\} \setminus i} \boldsymbol{x}_l,$$

yielding two different ways of decomposing $\mathbf{0}$*, which cannot be.*

Chapter 9

Solutions

9.1 Solutions to Chapter 1 289
9.2 Solutions to Chapter 2 294
9.3 Solutions to Chapter 3 294
9.4 Solutions to Chapter 4 299
9.5 Solutions to Chapter 5 300
9.6 Solutions to Chapter 6 301
9.7 Solutions to Chapter 7 303

9.1 Solutions to Chapter 1

EX 1.5.1 (a) and (c) are linear equations, (b) and (d) are not linear equations.
EX 1.5.2 The solution sets are:

(a) $\{(0,0,0)\}$

(b) $\{(-\frac{1}{3}x_3, -\frac{2}{3}x_3 - x_4, x_3, x_4) : x_3, x_4 \in \mathbb{R}\}$

(c) $\{(w,x,y,z) : x = -w \wedge y = w \wedge z = 0 \quad (w \in \mathbb{C})\}$

EX 1.5.3 The systems a) and c) have non-trivial solutions. System c) might have only the trivial solution or non-trivial solutions, depending on its coefficients a_{ij}.
EX 1.5.5

(a) $\begin{cases} x = 1 \\ y = 2 \\ z = 3 \end{cases}$

(b) $\begin{cases} x = 1 \\ y = 2 \end{cases}$

(c) $\begin{cases} 3x + y = 0 \\ y + 3z = 0 \end{cases}$

DOI: 10.1201/9781351243452-9

(d) $2x - y - z - w = 0$

EX 1.5.4 The solution sets are:

(a) The solution set is $\mathcal{S} = \{(3, 1, 2)\}$.

(b) $\mathcal{S} = \{(-\frac{1}{7} - \frac{3}{7}x_3, \frac{1}{7} - \frac{4}{7}x_3, x_3) \colon x_3 \in \mathbb{R}\}$.

(c) $\mathcal{S} = \emptyset$.

(d) $\mathcal{S} = \{(-6 - 2v - 3y, v, -2 - y, 3 + y, y) \,:\, v, y \in \mathbb{R}\}$.

EX 1.5.6

(a) Yes (it is a row echelon matrix); Yes (it is in reduced row echelon form); rank 3.

(b) Yes; Yes; 2.

(c) Yes; Yes; 2.

(d) Yes; Yes; 2.

(e) No; No; 2.

(f) Yes; No; 2.

(g) No; No; 2.

(h) Yes; Yes; 0.

(j) No; No; 1.

(k) No; No; 3.

(l) No; No; 2.

(m) Yes; No; 2.

EX 1.5.7

$$\begin{bmatrix} 1 & 0 & 0 & 1 & 1 \\ 0 & 1 & 0 & 2 & -1 \\ 0 & 0 & 1 & -1 & 0 \end{bmatrix},$$

rank (A)=3.

EX 1.5.8 $\operatorname{rank}(A_\alpha) = 2$ when $\alpha = -1, 0, 1$ and $\operatorname{rank}(A_\alpha) = 3$ otherwise.

If $\alpha = 0$, then the systems are consistent and have one independent variable.

If $\alpha = \pm 1$ and $\beta = 0$, then the systems are consistent and have one independent variable.

If $\alpha = \pm 1$ and $\beta \neq 0$, then the systems are inconsistent.

If $\alpha \in \mathbb{R} \backslash \{-1, 0, 1\}$, then the systems are consistent and have no independent variable.

EX 1.5.9 The system is always consistent, since it is homogeneous.

If $\alpha \neq 2$ and $\alpha \neq -2$, then the system has only the trivial solution $(0,0,0,0)$.

If $\alpha = 2$, then the system has two free variables and the solution set is

$$\mathcal{S} = \{(2z, -z-w, z, w) : z, w \in \mathbb{R}\}.$$

If $\alpha = -2$, , then the system has one free variable and the solution set is

$$\mathcal{S} = \{(2z, -z, z, 0) : z \in \mathbb{R}\}.$$

EX 1.5.10

$$B + C = \begin{bmatrix} 7 & \sqrt{3} & 0 \\ 2 & -5 & 1 \\ \pi & 2 & 9 \end{bmatrix}, \qquad 2A = \begin{bmatrix} 2 & -4 \\ 8 & 2 \\ 2\sqrt{2} & 6 \end{bmatrix},$$

$$AB = \begin{bmatrix} 1+4\sqrt{3} & -2+\sqrt{3} \\ -2+\sqrt{2} & -2 \\ \pi+8-\sqrt{2} & -2\pi-1 \end{bmatrix}, \qquad CB = \begin{bmatrix} 6 & 6\sqrt{3} & 0 \\ -8 & 4 & -4 \\ 10\pi & 20 & -10 \end{bmatrix},$$

$\operatorname{tr} B = -1, \operatorname{tr} C = 12$. The remaining operations are not possible.

EX 1.5.11

a) $(AB)_{23} = -10$ and the column 2 of AB is $\begin{bmatrix} 1 \\ 8 \\ 2 \end{bmatrix}$.

b)

$$(A-B)^T = \begin{bmatrix} 2 & 1 & -2 \\ 0 & -6 & -2 \\ 2 & 0 & 2 \end{bmatrix}, \qquad \operatorname{tr}((A-B)^T) = -2.$$

EX 1.5.12 The correct assertion is D.

EX 1.5.13

$$A = \begin{bmatrix} 0 & -1 & -2 \\ 1 & 0 & -1 \\ 2 & 1 & 0 \end{bmatrix}.$$

EX 1.5.14

$$A = \begin{bmatrix} 2 & 4 & 8 \\ 4 & 8 & 16 \\ 8 & 16 & 32 \end{bmatrix}.$$

EX 1.5.15 If n is even, i.e., $n = 2k$, with $k \in \mathbb{N}_0$, then

$$A^n = A^{2k} = (-1)^k \begin{bmatrix} i & 0 \\ 0 & i \end{bmatrix}.$$

If n is odd, i.e.,$n = 2k+1$, with $k \in \mathbb{N}_0$, then

$$A^n = A^{2k+1} = (-1)^k A.$$

EX 1.5.16

$$A = \begin{bmatrix} -1 & 2 & 1 \\ 0 & 1 & 1 \\ 1 & -2 & -1 \end{bmatrix}.$$

EX 1.5.17

(a) $\begin{bmatrix} -7 & 2 \\ 4 & -1 \end{bmatrix}$

(b) $-\frac{1}{39}\begin{bmatrix} 5 & -4 \\ -6 & -3 \end{bmatrix}$

(c) Not invertible.

(d) $\begin{bmatrix} \frac{3}{2} & -1 & -\frac{1}{2} \\ -\frac{11}{10} & 1 & \frac{7}{10} \\ -\frac{6}{5} & 1 & \frac{2}{5} \end{bmatrix}$

(e) Not invertible.

(f) $\begin{bmatrix} \frac{7}{2} & -1 & 0 \\ 0 & 1 & -1 \\ -3 & 0 & 1 \end{bmatrix}$

(g) $\begin{bmatrix} 1 & -\frac{1}{3} & 0 & 0 \\ 0 & \frac{1}{3} & -\frac{1}{5} & 0 \\ 0 & 0 & \frac{1}{5} & -\frac{1}{7} \\ 0 & 0 & 0 & \frac{1}{7} \end{bmatrix}$

(h) Not invertible.

EX 1.5.18

$$A^3 = \begin{bmatrix} 1 & 6 \\ 0 & 1 \end{bmatrix}, \quad A^{-3} = \begin{bmatrix} 1 & -6 \\ 0 & 1 \end{bmatrix}, \quad A^2 - 2A + I = (A - I)^2 = \begin{bmatrix} 0 & 0 \\ 0 & 0 \end{bmatrix},$$

$$X = \begin{bmatrix} 3/2 & -3/2 \\ 1/2 & -1/2 \end{bmatrix}.$$

EX 1.5.19

(a) Elementary operation: $L_2 - 5L_1$ Elementary matrix: $E_{21}(3) = \left[\begin{smallmatrix} 1 & 0 \\ -5 & 1 \end{smallmatrix}\right]$

(b) Elementary operation: $-\frac{1}{3}L_3$ Elementary matrix: $D_3(-\frac{1}{3}) = \left[\begin{smallmatrix} 1 & 0 & 0 \\ 0 & 1 & 0 \\ 0 & 0 & -\frac{1}{3} \end{smallmatrix}\right]$

(c) Elementary operation: $L_2 \leftrightarrow L_4$ Elementary matrix: $P_{24} = \begin{bmatrix} 1&0&0&0\\0&0&0&1\\0&0&1&0\\0&1&0&0 \end{bmatrix}$

(d) Elementary operation: $L_3 + \frac{1}{2}L_2$ Elementary matrix: $E_{32}(\frac{1}{2}) = \begin{bmatrix} 1&0&0&0\\0&1&1&0\\0&\frac{1}{2}&1&0\\0&0&0&1 \end{bmatrix}$

EX 1.5.20

$$E_1 = E_{24}(-1) = \begin{bmatrix} 1&0&0&0\\0&1&0&-1\\0&0&1&0\\0&0&0&1 \end{bmatrix} \qquad E_2 = D_2(-5) = \begin{bmatrix} 1&0&0&0\\0&-5&0&0\\0&0&1&0\\0&0&0&1 \end{bmatrix}$$

$$A^{-1} = D_2(-5)^{-1}E_{24}(-1)^{-1} = \underbrace{\begin{bmatrix} 1&0&0&0\\0&-1/5&0&0\\0&0&1&0\\0&0&0&1 \end{bmatrix}}_{D_2(-1/5)} \underbrace{\begin{bmatrix} 1&0&0&0\\0&1&0&1\\0&0&1&0\\0&0&0&1 \end{bmatrix}}_{E_{24}(1)}$$

$$= \begin{bmatrix} 1&0&0&0\\0&-1/5&0&-1/5\\0&0&1&0\\0&0&0&1 \end{bmatrix}$$

EX 1.5.21 D).

EX 1.5.22 Let $A = [a_{ij}]$ be a matrix such that, for all $n \times n$ matrices B we have $AB = BA$. It follows immediately that A is an $n \times n$ matrix.

Let $B = E_{ii}$, where E_{ii} the matrix having all entries equal to zero except for the entry-ii which is equal to 1. Then,

$$L_i^A = E_{ii}A = AE_{ii} = C_i^A,$$

where L_i^A is a matrix whose row i is the row i of A and whose remaining rows are zero rows, and C_i^A is a matrix with zero columns except for the ith column which is that of A. Hence, since the only possibly non-zero entry in common in both matrices is a_{ii}, all the remaining entries in both matrices are equal to zero. Letting i vary one gets that A is a diagonal matrix.

Suppose now that $B = P_{ij}$. Then

$$P_{ij}A = AP_{ij},$$

from which follows that, for all $i, j = 1, \ldots, n$, $a_{ii} = a_{jj}$. Recall that the operation $P_{ij}A$ interchanges the rows i and j of A whilst the operation AP_{ij} swaps the columns i and j of A. Hence, $A = \alpha I$.

EX 1.5.23

$$A=\begin{bmatrix}1&0&0\\-5&1&0\\-3&4&1\end{bmatrix}\begin{bmatrix}4&0&0\\0&5&0\\0&0&1\end{bmatrix}\begin{bmatrix}1&4&1\\0&1&-5\\0&0&1\end{bmatrix}=\begin{bmatrix}1&0&0\\-5&1&0\\-3&4&1\end{bmatrix}\begin{bmatrix}4&16&4\\0&5&-25\\0&0&1\end{bmatrix},$$

$\boldsymbol{x}=(1,-1,0)$.

9.2 Solutions to Chapter 2

EX 2.4.1 $\det A=-912$.
EX 2.4.2 $\det(-2A^{-3})=-8/27$, $\det\left(\left(AB^T\right)^2\right)=81$, $\det(E_{32}(-2)D_3(2)P_{34}A)=-6$.
EX 2.4.3 C)
EX 2.4.4 $\det B=-10$.
EX 2.4.5 104.
EX 2.4.6 A).
EX 2.4.7

$$\begin{bmatrix}1&1&1\\0&1&0\\0&0&1\end{bmatrix}.$$

EX 2.4.8 $C_{14}=60$; $(A^{-1})_{41}=i\frac{15}{480}$.
EX 2.4.9 $x_1=-28/11, x_2=-34/11, x_3=-30/11$.
EX 2.4.10 (a) $\alpha=-1-i$; (b) $\boldsymbol{x}=(1,1,0)$.
EX 2.4.11 (a) $a\in\mathbb{R}\backslash\{0,3\}$; (b) $\frac{1}{a-3}$.

9.3 Solutions to Chapter 3

EX 3.7.1 (a), (b), and (d).
EX 3.7.2

(a) $(-3,-\frac{7}{3},-5)=-2\boldsymbol{u}+\boldsymbol{v}-2\boldsymbol{w}$

(b) $(2,\frac{11}{3},2)=4\boldsymbol{u}-5\boldsymbol{v}+1\boldsymbol{w}$

(c) $(0,0,0)=0\boldsymbol{u}+0\boldsymbol{v}+0\boldsymbol{w}$

(d) $(\frac{7}{3},\frac{8}{3},3)=0\boldsymbol{u}-2\boldsymbol{v}+3\boldsymbol{w}$

EX 3.7.3 (a), (b), and (d).
EX 3.7.4 (a), (c), and (e).
EX 3.7.5 (b) and (f).
EX 3.7.6 We show firstly that $\mathbb{P}_n$, together with the usual addition of polynomials and multiplication of a polynomial by a scalar, is a real vector space. Clearly, axioms (i) and (ii) are verified since the addition is commutative and associative. On the other hand, the zero polynomial $p(t) = 0$ is the additive identity, showing that axiom (iii) is satisfied.

Let $p(t) = a_0+a_1t+a_2t^2+\cdots+a_nt^n$ be a generic polynomial. The additive inverse of $p(t)$ is $p(t) = -a_0 - a_1t - a_2t^2 - \cdots - a_nt^n$ which settles (iv).

(v) Let α be a real number and let $q(t) = b_0 + b_1t + b_2t^2 + \cdots + b_nt^n$ be a polynomial. Then

$$\begin{aligned}\alpha(p+q)(t) &= \alpha(a_0 + a_1t + a_2t^2 + \cdots + a_nt^n + b_0 + b_1t + b_2t^2 + \cdots + b_nt^n)\\ &= \alpha(a_0 + a_1t + a_2t^2 + \cdots + a_nt^n) + \alpha(b_0 + b_1t + b_2t^2 + \cdots + b_nt^n)\\ &= \alpha p(t) + \alpha q(t).\end{aligned}$$

Axioms (vi) and (vii) are shown to hold similarly and it is obvious that (viii) also holds since $1p(t) = p(t)$. Hence $\mathbb{P}_n$ is a real vector space.

We can show similarly that $\mathbb{P}$ is a real vector space. Propositions 1.4 and 1.5 show that $M_{n,k}(\mathbb{K})$ is a vector space over $\mathbb{K}$.

The addition of continuous real functions on $[a, b]$ is commutative, associative and has the zero function on $[a, b]$ as the additive identity. The additive inverse of a function $f(t)$ is $-f(t)$ and $1f(t) = f(t)$.

If f, g are functions on $[a, b]$ and $\alpha, \beta \in \mathbb{R}$, then

$$\alpha(f+g)(t) = \alpha(f(t) + g(t)) = \alpha f(t) + \alpha g(t),$$

$$(\alpha\beta)f(t) = \alpha(\beta f(t))$$

and

$$(\alpha + \beta)f(t) = \alpha f(t) + \beta f(t).$$

Hence $C([a, b])$ is a real vector space.
EX 3.7.7 (a) and (c).
EX 3.7.8 These operations satisfy all axioms of a vector space except for $1\boldsymbol{u} = \boldsymbol{u}$, where $u \in \mathbb{C}^2$. Hence $\mathbb{C}^2$ is not a vector space for these operations.
EX 3.7.9

(a) It is linearly independent and a basis is $\{(1, -1, 0), (0, 0, 2)\}$

(b) It is linearly dependent and a basis is $\{(2, 4, 12), (-1, -1, -1)\}$

(c) It is linearly dependent and a basis is $\{(1, 2, 3, 4), (0, 1, 1, 0)\}$

(d) It is linearly dependent and a basis is $\{(1 + i, 2i, 4 - i)\}$

(e) It is linearly dependent and a basis is $\{(1, 2, 6, 0), (3, 4, 1, 0), (4, 3, 1, 0)\}$

EX 3.7.10

(a) A possible basis is $\mathcal{B} = ((2,1,0),(-3,0,1))$ and $\dim W = 2$.

(b) $(\boldsymbol{v})_{\mathcal{B}} = (-1,2)$.

EX 3.7.11

(a) A possible basis is $\mathcal{B} = \big((i,0,0,i)\big)$ and $\dim W = 1$.

(b) $\mathbf{v}_{\mathcal{B}} = (-i)$.

EX 3.7.12

(a) These vectors do not form a basis for W because they are linearly dependent. For example, $\boldsymbol{w} = 2\boldsymbol{u} + \boldsymbol{v}$.

(b) $\mathcal{B}_W = \{\boldsymbol{u}, \boldsymbol{v}\}$, $\dim W = 2$.

(c) Vector equation:

$$(x,y,z,w) = t(1,0,0,1) + s(2,2,0,1), \qquad t,s \in \mathbb{R}.$$

Parametric equations:

$$\begin{cases} x = t + 2s \\ y = 2s \\ z = 0 \\ w = t + s \qquad\qquad t,s \in \mathbb{R} \end{cases}.$$

Cartesian equations:

$$W = \{(x,y,z,w) \in \mathbb{R}^4 \colon z = 0 \wedge -2x + y + 2w = 0\}.$$

EX 3.7.13 Add, for example, the vector $(1,0,0)$.
EX 3.7.14 For example, the vectors $(0,0,1,0)$, $(0,0,0,1)$ can be adjoined to S to form a basis of $\mathbb{R}^4$. A way to see this is observing that the subspace spanned by S is defined by the cartesian equations

$$z + 2w = 0 \quad \wedge \quad y = 0,$$

which are not satisfied by $(0,0,1,0)$, $(0,0,0,1)$.
EX 3.7.15

(a)

$$\mathcal{B}_{N(A)} = \{(-2,3,1,0)\},$$

$$\mathcal{B}_{L(A)} = \{(1,1,-1,1),(0,-1,3,0),(0,0,0,1)\},$$

$$\mathcal{B}_{C(A)} = \{(1,0,-1,0),(1,-1,0,-1),(1,0,0,1)\}.$$

(b) $4 = \dim N(A) + \dim L(A) = 1 + 3$.

(c)

$$\mathbf{x} \in N(MA^T) \iff MA^T\mathbf{x} = \mathbf{0}.$$

Hence

$$\boldsymbol{x} \in N(MA^T) \iff M^{-1}MA^T\mathbf{x} = \mathbf{0} \iff A^T\mathbf{x} = \mathbf{0} \iff \mathbf{x} \in N(A^T).$$

We have

$$4 = \dim N(A^T) + \dim L(A^T) = \dim N(A^T) + \dim C(A) = \dim N(A^T) + 3$$

from which follows that $N(MA^T) = \dim N(A^T) = 1$. Consequently, $\dim(MA^T) = 4 - 1 = 3$.

EX 3.7.16

$\mathcal{B}_{N(A)} = \{(0, 0, 1)\}$, $\dim N(A) = 1$.

$\mathcal{B}_{L(A)} = \{(1, 1, 0), (-3, 1, 0)\}$, $\dim L(A) = 2$.

$\mathcal{B}_{C(A)} = \{(1, -3, 0, 1), (1, 1, 0, -1)\}$, $\dim C(A) = 2$.

EX 3.7.17

(a) No because $(2, 1, -1) \notin C(A)$.

(b) The solution set is

$$\{(x, y, z) \in \mathbb{R}^3 \colon (x, y, z) = t(1, 1, 1) + s(1, -1, -1) + (1, 2, 3)\ \forall t \in \mathbb{R}\}.$$

(b) $\dim N(B^TB) \geq 2$.

EX 3.7.18

(a) $\mathcal{B}_{U \cap V} = \{(0, 0, 1, 1)\}$.

(b) $\dim(\mathbb{R}^4 + (U \cap V)) = 4$.

EX 3.7.19

$\mathcal{B}_{U+W} = \{(0, 0, 3 - i, 0), (1 - 2i, 0, 0, 1 - 2i), (0, 1, 2, 0)\}$, $\dim(U + W) = 3$.

$\mathcal{B}_{U \cap W} = (1 - 2i, 0, 0, 1 - 2i)$, $\dim(U \cap W) = 1$.

Since $\dim U = 2 = \dim W$, the formula holds.

EX 3.7.20 $5 + 9t + 3t^2 + 5t^3 = 3\boldsymbol{p_1} - 4\boldsymbol{p_2} + \boldsymbol{p_3}$.
EX 3.7.21 (a) Linearly independent, (b) linearly dependent, (c) linearly independent.
EX 3.7.22 $(3 - 2t)_{\mathcal{B}} = (8, 1)$ and $(3 - 2t)_{\mathcal{P}_1} = (3, -2)$.
EX 3.7.23 A possible basis is

$$\mathcal{B}_S = (1 + t - t^3, t + t^2 - t^3, 2 - 2t)$$

and $(3 - 2t^3 + t^2)_{\mathcal{B}_S} = (1, 1, 1)$.
EX 3.7.24 $M = -2A + B - 2C$.
EX 3.7.25 The only matrix which does not lie in the space spanned by the given matrices is that of (d).
EX 3.7.26 $A_{\mathcal{B}} = (2, 1, -1, 1 + i)$.
EX 3.7.27 A possible basis is

$$\mathcal{B} = \left(\begin{bmatrix} -1 & 3 \\ 0 & 0 \end{bmatrix}, \begin{bmatrix} 0 & 1 \\ 1 & 0 \end{bmatrix}, \begin{bmatrix} 2 & -2 \\ 0 & 1 \end{bmatrix}\right).$$

EX 3.7.28

(a) $M_{\mathcal{B}\leftarrow\mathcal{E}_2} = \begin{bmatrix} -1 & -1 \\ 0 & 1 \end{bmatrix}$ $\quad (2, 2)_{\mathcal{B}} = (-4, 2)$

(b) $M_{\mathcal{E}_2\leftarrow\mathcal{B}'} = \begin{bmatrix} 1 & -2 \\ 2 & 1 \end{bmatrix}$

(c) $M_{\mathcal{B}\leftarrow\mathcal{B}'} = \begin{bmatrix} -3 & 1 \\ 2 & 1 \end{bmatrix}$

EX 3.7.29

1. $M_{\mathcal{B}\leftarrow\mathcal{E}_3} = \begin{bmatrix} -1 & 2 & 1 \\ 0 & 0 & 2 \\ 1 & 0 & -1 \end{bmatrix}$

2. $\boldsymbol{v} = (\boldsymbol{v})_{\mathcal{E}_3} = (12, 12, -4)$

3. $M_{\mathcal{B}'\leftarrow\mathcal{B}} = \begin{bmatrix} 0 & -1 & 0 \\ 0 & 0 & 1 \\ 1 & 0 & 0 \end{bmatrix}$

4. $M_{\mathcal{B}'\leftarrow\mathcal{E}_3} = M_{\mathcal{B}'\leftarrow\mathcal{B}} M_{\mathcal{B}\leftarrow\mathcal{E}_3} = \begin{bmatrix} 0 & 0 & -2 \\ 1 & 0 & -1 \\ -1 & 2 & 1 \end{bmatrix}$

EX 3.7.30 a) $(2, -6)$, b) $\mathcal{B} = (\frac{5}{16} + \frac{1}{16}t, -\frac{1}{16} + \frac{3}{16}t)$.
EX 3.7.31 A possible solution is

$$\mathcal{B}_S = \left(\begin{bmatrix} -1 & 0 \\ 0 & 1 \end{bmatrix}, \begin{bmatrix} 0 & 1 \\ 0 & 0 \end{bmatrix}\right), \quad \mathcal{B} = \left(\begin{bmatrix} -1 & 0 \\ 0 & 1 \end{bmatrix}, \begin{bmatrix} 0 & 1 \\ 0 & 0 \end{bmatrix}, \begin{bmatrix} 1 & 0 \\ 0 & 0 \end{bmatrix}\right).$$

$$M_{\mathcal{B}_1\leftarrow\mathcal{B}} = \begin{bmatrix} \frac{1}{2} & 0 & 0 \\ 0 & 1 & 0 \\ 0 & 0 & 1 \end{bmatrix}$$

9.4 Solutions to Chapter 4

EX 4.5.1

a) Yes (eigenvalue -3)

b) No.

c) No.

d) Yes (eigenvalue 1)

e) Yes (eigenvalue 0)

EX 4.5.2 It is: eigenvector $(1, -1, 0)$.
EX 4.5.3
a) $p(\lambda) = (-3i - \lambda)(i - \lambda)$; eigenvalues are $-3i, i$; bases $\{(1, 2)\}$ and $\{(0, 1)\}$.
b) $p(\lambda) = (4 - \lambda)^3$; basis $\{(1, 1, -3)\}$.
EX 4.5.4 a) No; b) yes; c) yes, d) no, and e), f), g) yes.
EX 4.5.5

$$A = SDS^{-1}, \quad D = \begin{bmatrix} 1 & 0 & 0 \\ 0 & -1 & 0 \\ 0 & 0 & -1 \end{bmatrix}, \qquad S = \begin{bmatrix} -2 & 1 & 1 \\ 1 & -1 & 0 \\ -1 & 0 & -1 \end{bmatrix}; \qquad A^{21} = A.$$

EX 4.5.8 D).
EX 4.5.10 Hint: Let $E_{i,j}$ be the matrix whose entries are all zero except for the entry-ij which is equal to 1. Begin by showing that

$$J_n(\lambda) - \lambda I = \sum_{i=1}^{n-1} E_{i,i+1}$$

and

$$(J_n(\lambda) - \lambda I)^2 = \sum_{i=1}^{n-2} E_{i,i+1} E_{i+1,i+2} = \sum_{i=1}^{n-2} E_{i,i+2}.$$

Generalise for $(J_n(\lambda) - \lambda I)^p$, with $1 < p < n$, and show that

$$(J_n(\lambda) - \lambda I)^{n-1} = E_{1,n}.$$

Then obtain the result.
EX 4.5.11 (i) No, no; (ii) no, yes; (iii) no, yes; (iv) no, yes.
EX 4.5.12 Up to a permutation of blocks:

$$\begin{bmatrix} -2 & 0 & 0 \\ 0 & -2 & 0 \\ 0 & 0 & 1 \end{bmatrix}, \qquad \begin{bmatrix} -2 & 1 & 0 \\ 0 & -2 & 0 \\ 0 & 0 & 1 \end{bmatrix}.$$

EX 4.5.13 a) $J = \begin{bmatrix} 0 & 1 \\ 0 & 0 \end{bmatrix}, \quad S = \begin{bmatrix} -1 & -\frac{1}{2} \\ 1 & 0 \end{bmatrix}$;

$b)J = \begin{bmatrix} 3 & 1 & 0 \\ 0 & 3 & 0 \\ 0 & 0 & 1 \end{bmatrix}, \quad S = \begin{bmatrix} 1 & 0 & 0 \\ 0 & 1 & -1 \\ 0 & 0 & 1 \end{bmatrix}$; c) $J = \begin{bmatrix} 1 & 0 & 0 & 0 \\ 0 & 3 & 1 & 0 \\ 0 & 0 & 3 & 1 \\ 0 & 0 & 0 & 3 \end{bmatrix}, \quad S = \begin{bmatrix} \frac{3}{2} & 1 & 0 & 0 \\ 1 & 0 & -1 & -1 \\ 0 & 0 & -2 & -1 \\ 0 & 0 & 0 & -2 \end{bmatrix}$

9.5 Solutions to Chapter 5

EX 5.8.1 a) Yes, b) yes, c) no, d) no, and e), f), g) yes.
EX 5.8.2

$$a) \begin{bmatrix} 2 & -1 \\ 1 & 0 \\ -1 & 1 \end{bmatrix}, \qquad b) \begin{bmatrix} 0 & -i & 0 \\ 0 & -1 & 5-3i \\ -3 & 0 & 0 \end{bmatrix}.$$

a) $N(T) = \{(0,0)\}, I(T) = \text{span}\{(2,1,-1),(-1,0,1)\}$, not an isomorphism; b) $N(T) = \{(0,0,0)\}, I(T) = \mathbb{C}^3$, it is an isomorphism.
EX 5.8.3 $N(TS) = \{(x,y) \in \mathbb{R}^2 \colon x = -y\} = I(TS)$. This transformation is not an isomorphism.

EX 5.8.4 T is invertible and $[T^{-1}]_{\mathcal{E}_2,\mathcal{E}_2} = \begin{bmatrix} \frac{3}{10} & \frac{-1}{10} \\ \frac{1}{10} & \frac{3}{10} \end{bmatrix}$.

EX 5.8.5 $[T]_{\mathcal{E}_2,B} = \begin{bmatrix} 2 & 1 \\ 1 & 1 \end{bmatrix}$.

EX 5.8.6 Considering the basis $B = \{(1,2)\}$,

$$[T]_{\mathcal{E}_2,B} = \begin{bmatrix} -9 & 0 \\ -6 & 0 \end{bmatrix}.$$

$N(T)$ is the straight line $x = 0$, and $I(T)$ is the straight line $y = \frac{2}{3}x$. T is neither injective nor surjective.
EX 5.8.7

$$[T]_{\mathcal{E}_2,\mathcal{E}_2} = \begin{bmatrix} 1 & 0 & 0 & 0 \\ 0 & 0 & 1 & 0 \\ 0 & 1 & 0 & 0 \\ 0 & 0 & 0 & 1 \end{bmatrix},$$

$$[T]_{\mathcal{B},\mathcal{B}} = M_{\mathcal{B}\leftarrow\mathcal{E}_2}[T]_{\mathcal{E}_2,\mathcal{E}_2}M_{\mathcal{E}_2\leftarrow\mathcal{B}} = \begin{bmatrix} 1 & 0 & 0 & 0 \\ -1 & -1 & 1 & 0 \\ 0 & 0 & 1 & 0 \\ 0 & 0 & 0 & 1 \end{bmatrix}.$$

EX 5.8.8 $N(T) = \{-a + 2at \colon a \in \mathbb{R}\}$,

$$I(T) = \left\{ \begin{bmatrix} a & a \\ a & -a \end{bmatrix} : a \in \mathbb{R} \right\}.$$

EX 5.8.9

$$[T]_{\mathcal{B},\mathcal{B}} = M_{\mathcal{B}\leftarrow\mathcal{E}_2}[T]_{\mathcal{P}_2,\mathcal{P}_2}M_{\mathcal{P}_2\leftarrow\mathcal{B}} = \begin{bmatrix} 0 & -2 & 0 \\ 0 & 0 & 0 \\ 1 & 3 & 0 \end{bmatrix}.$$

EX 5.8.10 a) True, b) true, c) true, d) false, e) true, f) true, g) false.
EX 5.8.12 $\sigma(T) = \{0,1\}$, $E(0)$ is the straight line $x = 0 = y$, $E(1)$ is the

plane $z = 0$. The invariant subspaces are $\{(0,0,0)\}$, the eigenspaces and any straight line contained in the plane $z = 0$.

EX 5.8.13 a) $\sigma(T) = \emptyset$; $\sigma(S) = \{-1\}, E(-1) = \mathbb{R}^2$. b) $\sigma(A) = \{\pm i\}, B_E(i) = \{(i,1)\}, B_E(-i) = \{(-i,1)\}$.

EX 5.8.14 a)

$$[T_1]_{B_s,B_s} = \begin{bmatrix} 1 & 0 & 0 & 0 \\ 0 & \frac{1}{2} & \frac{1}{2} & 0 \\ 0 & \frac{1}{2} & \frac{1}{2} & 0 \\ 0 & 0 & 0 & 1 \end{bmatrix}, \qquad [T_2]_{B_s,B_s} = \begin{bmatrix} 1 & 0 & 0 & 0 \\ 0 & \frac{1}{2} & -\frac{1}{2} & 0 \\ 0 & -\frac{1}{2} & \frac{1}{2} & 0 \\ 0 & 0 & 0 & 1 \end{bmatrix};$$

b) $\sigma(T_1) = \{0,1\} = \sigma(T_2)$;

$$E_{T_1}(1) = \text{span}\left\{ \begin{bmatrix} 1 & 0 \\ 0 & 0 \end{bmatrix}, \begin{bmatrix} 0 & 1 \\ 1 & 0 \end{bmatrix}, \begin{bmatrix} 0 & 0 \\ 0 & 1 \end{bmatrix} \right\} \text{(symmetric matrices)},$$

$$E_{T_1}(0) = \text{span}\left\{ \begin{bmatrix} 0 & -1 \\ 1 & 0 \end{bmatrix} \right\} \text{(skew-symmetric matrices)},$$

$$E_{T_2}(1) = \text{span}\left\{ \begin{bmatrix} 0 & -1 \\ 1 & 0 \end{bmatrix} \right\} \text{(skew-symmetric matrices)},$$

$$E_{T_2}(0) = \text{span}\left\{ \begin{bmatrix} 1 & 0 \\ 0 & 0 \end{bmatrix}, \begin{bmatrix} 0 & 1 \\ 1 & 0 \end{bmatrix}, \begin{bmatrix} 0 & 0 \\ 0 & 1 \end{bmatrix} \right\} \text{(symmetric matrices)}.$$

9.6 Solutions to Chapter 6

EX 6.7.1

(a) $\sqrt{37}$

(b) $\sqrt{21} + \sqrt{14}$

(c) $3\sqrt{21}$

(d) $(\frac{2}{\sqrt{14}}, -\frac{1}{\sqrt{14}}, \frac{3}{\sqrt{14}})$

(e) 1

(f) $\arccos \frac{1}{\sqrt{14}\sqrt{21}}$

(g) $\sqrt{33}$

EX 6.7.2 $\frac{1}{\sqrt{3}}(1,1,1), -\frac{1}{\sqrt{3}}(1,1,1)$.
EX 6.7.4 $i(-\frac{3}{5}, -\frac{6}{5})$.
EX 6.7.5 $\{\frac{1}{4}(-2,2,-2,2), \frac{1}{\sqrt{2}}(0,1,1,0), \frac{1}{\sqrt{12}}(1,-1,1,3)\}$.
EX 6.7.6 Basis: $\{(0,1,2),(1,2,5)\}$; equation $x+2y-z=0$.
EX 6.7.7 $\mathrm{d}(\boldsymbol{u}, S^{\perp}) = \sqrt{\frac{8}{3}}$; $\begin{bmatrix} \frac{2}{3} & \frac{1}{3} & \frac{1}{3} & 0 \\ \frac{1}{3} & \frac{1}{6} & \frac{1}{6} & 0 \\ \frac{1}{3} & \frac{1}{6} & \frac{1}{6} & 0 \\ 0 & 0 & 0 & 0 \end{bmatrix}$; $\boldsymbol{u}_1 = \frac{2}{3}(2,1,1,0), \boldsymbol{u}_2 = \frac{1}{3}(-1,1,1,3)$.
EX 6.7.8 (a) $\langle (\frac{4}{3},1,-1),(0,-1,1)\rangle = 0$; (b) $\|(\frac{4}{3},1,-1)\| = 2/\sqrt{3}$ and $\|(0,-1,1)\| = 2$; (c) $u_1 = (7/3,1,-1)$ and $u_2 = (-4/3,-2,0)$.
EX 6.7.9 $(1,0,1),(1,0,1)$.
EX 6.7.10 (a) $\dim(W^{\perp}) = 1$; (b) $\mathrm{d}(\boldsymbol{p}, W^{\perp}) = \sqrt{2}$; (c)$(1/\sqrt{3}, 1/\sqrt{3}, 1/\sqrt{3}, 0)$.
EX 6.7.11

(a) No because $(2,1,-1) \notin C(A)$.

(b) The solution set is

$$\{(x,y,z) \in \mathbb{R}^3 \colon (x,y,z) = t(1,1,1) + s(1,-1,-1) + (1,2,3)\ \forall t,s \in \mathbb{R}\}.$$

EX 6.7.12 (a) Parametric equations: $x = 1, y = -t, z = -1-3t, w = 0\, t \in \mathbb{R}$; cartesian equations: x=1, 3y-z=1, w=0. (b) $x+z=0$, $w=0$. (c) Parametric equations: $x = 1, y = -t, z = -1-3t, w = 0\, t \in \mathbb{R}$; cartesian equations: $x = 1-\alpha, y = \alpha+\beta, x = 1-\alpha, w = 0\, \alpha,\beta \in \mathbb{R}$. (d) $x+2y-z=2$, w=0; $(1,2,-1,0)$. (e) $\frac{1}{\sqrt{6}}$.
EX 6.7.13 (b) $\dim U = 1, \dim U^{\perp} = 3$; (c) orthonormal bases of U and $U^{\perp}$ are, respectively, $\left\{\begin{bmatrix} 0 & \frac{1}{\sqrt{2}} \\ -\frac{1}{\sqrt{2}} & 0 \end{bmatrix}\right\}$ and $\left\{\begin{bmatrix} 0 & \frac{1}{\sqrt{2}} \\ \frac{1}{\sqrt{2}} & 0 \end{bmatrix}, \begin{bmatrix} 1 & 0 \\ 0 & 0 \end{bmatrix}, \begin{bmatrix} 0 & 0 \\ 0 & 1 \end{bmatrix}\right\}$; (d) $\mathrm{proj}_U \begin{bmatrix} 3 & 3 \\ -3 & 6 \end{bmatrix} = \begin{bmatrix} 0 & 3 \\ -3 & 0 \end{bmatrix}$, $\mathrm{proj}_{U^{\perp}} = \begin{bmatrix} 3 & 0 \\ 0 & 6 \end{bmatrix}$; (e) $\begin{bmatrix} 0 & 3 \\ -3 & 0 \end{bmatrix}$; (d) $\sqrt{45}$.
EX 6.7.14 $\{(-2a-b) + at + bt^2 \,:\, a,b \in \mathbb{R}\}$.
EX 6.7.15

$$\begin{bmatrix} 1 & 1 & 1 \\ 1 & 1 & 1 \\ 1 & 1 & 1 \end{bmatrix} = \begin{bmatrix} -\frac{1}{\sqrt{2}} & -\frac{1}{\sqrt{6}} & \frac{1}{\sqrt{3}} \\ 0 & \frac{2}{\sqrt{6}} & \frac{1}{\sqrt{3}} \\ \frac{1}{\sqrt{2}} & \frac{-1}{\sqrt{6}} & \frac{1}{\sqrt{3}} \end{bmatrix} \begin{bmatrix} 0 & 0 & 0 \\ 0 & 0 & 0 \\ 0 & 0 & 3 \end{bmatrix} \begin{bmatrix} -\frac{1}{\sqrt{2}} & 0 & \frac{1}{\sqrt{2}} \\ -\frac{1}{\sqrt{6}} & \frac{2}{\sqrt{6}} & -\frac{1}{\sqrt{6}} \\ \frac{1}{\sqrt{3}} & \frac{1}{\sqrt{3}} & \frac{1}{\sqrt{3}} \end{bmatrix}.$$

.

EX 6.7.17 $A = UD\overline{U}^T$,

$$D = \begin{bmatrix} 7 & 0 & 0 \\ 0 & 1 & 0 \\ 0 & 0 & -1 \end{bmatrix}, \qquad U = \begin{bmatrix} 1 & 0 & 0 \\ 0 & \frac{1}{\sqrt{3}} & \frac{1+i}{\sqrt{3}} \\ 0 & \frac{1-i}{\sqrt{3}} & -\frac{1}{\sqrt{3}} \end{bmatrix};$$

$$A = 7\begin{bmatrix} 1 & 0 & 0 \\ 0 & 0 & 0 \\ 0 & 0 & 0 \end{bmatrix} + \begin{bmatrix} 0 & 0 & 0 \\ 0 & \frac{1}{3} & \frac{1+i}{3} \\ 0 & \frac{1-i}{3} & \frac{2}{3} \end{bmatrix} - \begin{bmatrix} 0 & 0 & 0 \\ 0 & \frac{2}{3} & \frac{-1-i}{3} \\ 0 & \frac{-1+i}{3} & \frac{1}{3} \end{bmatrix}$$

EX 6.7.19

$$\begin{bmatrix} 3 & 2 & 2 \\ 2 & 3 & -2 \end{bmatrix} = \begin{bmatrix} \frac{1}{\sqrt{2}} & \frac{1}{\sqrt{2}} \\ \frac{1}{\sqrt{2}} & -\frac{1}{\sqrt{2}} \end{bmatrix} \begin{bmatrix} 5 & 0 & 0 \\ 0 & 3 & 0 \end{bmatrix} \begin{bmatrix} \frac{1}{\sqrt{2}} & \frac{1}{\sqrt{2}} & 0 \\ \frac{1}{3\sqrt{2}} & -\frac{1}{3\sqrt{2}} & \frac{4}{3\sqrt{2}} \\ \frac{2}{3} & -\frac{2}{3} & -\frac{1}{3} \end{bmatrix}.$$

9.7 Solutions to Chapter 7

EX 7.6.1 For a projection matrix $P \neq 0, I$, the eigenspaces are $E(1) = C(A)$, $E(0) = N(A)$. Here $P = A(A^TA)^{-1}A^T$.

EX 7.6.2

$$\frac{1}{5}\begin{bmatrix} 3 & -2 & 1 & 1 \\ -2 & 3 & 1 & 1 \\ 1 & 1 & 2 & 2 \\ 1 & 1 & 2 & 2 \end{bmatrix}$$

EX 7.6.3 $y = -22/7 + 12/7x$.

EX 7.6.4 Let P, Q be two $n \times n$ Markov matrices. Obviously P, Q has positive entries only. It remains to show that the sum of the entries in any column equals 1. Fixing column j,

$$\sum_{i=1}^{n}(PQ)_{ij} = \sum_{i=1}^{n}\sum_{k=1}^{n} p_{ik}q_{kj} = \sum_{k=1}^{n}\sum_{i=1}^{n} p_{ik}q_{kj} = \sum_{k=1}^{n} q_{kj}\sum_{i=1}^{n} p_{ik} = \sum_{k=1}^{n} q_{kj} = 1.$$

EX 7.6.7 The steady-state vectors of both matrices are $(a, 0, 0, 1-a)$, $a \in \mathbb{R}$. There is no contradiction as these matrices are not regular Markov matrices and, therefore, are not forced to have unique steady-state vectors.

EX 7.6.8 The steady-state vector is $(3/7, 4/7)$.

EX 7.6.9

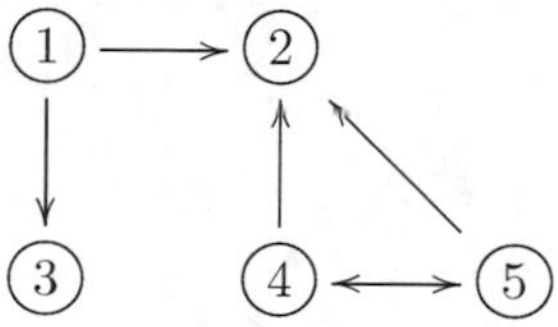

$$G = 0.85\begin{bmatrix} 0 & 1/5 & 1/5 & 0 & 0 \\ 1/2 & 1/5 & 1/5 & 1/2 & 1/2 \\ 1/2 & 1/5 & 1/5 & 0 & 0 \\ 0 & 1/5 & 1/5 & 0 & 1/2 \\ 0 & 1/5 & 1/5 & 1/2 & 0 \end{bmatrix} + 0.15\begin{bmatrix} 1/5 & 1/5 & 1/5 & 1/5 & 1/5 \\ 1/5 & 1/5 & 1/5 & 1/5 & 1/5 \\ 1/5 & 1/5 & 1/5 & 1/5 & 1/5 \\ 1/5 & 1/5 & 1/5 & 1/5 & 1/5 \\ 1/5 & 1/5 & 1/5 & 1/5 & 1/5 \end{bmatrix};$$

steady-state vector

$$\mathbf{q} \approx \begin{bmatrix} 0,11 \\ 0,33 \\ 0,16 \\ 0,20 \\ 0,20 \end{bmatrix};$$

ranking of pages: 2,4–5,3,1.

EX 7.6.10 The Leslie matrix for this population is

$$L = \begin{bmatrix} 2 & 4 \\ 0.8 & 0 \end{bmatrix}.$$

After $5, 10$ and 15 years, one has the age distribution vectors

$$\mathbf{x}(5) = \begin{bmatrix} 2 & 4 \\ 0.8 & 0 \end{bmatrix} \begin{bmatrix} 30 \\ 20 \end{bmatrix} = \begin{bmatrix} 140 \\ 24 \end{bmatrix},$$

$$\mathbf{x}(10) = \begin{bmatrix} 2 & 4 \\ 0.8 & 0 \end{bmatrix}^2 \begin{bmatrix} 30 \\ 20 \end{bmatrix} = \begin{bmatrix} 2 & 4 \\ 0.8 & 0 \end{bmatrix} \begin{bmatrix} 140 \\ 24 \end{bmatrix} = \begin{bmatrix} 376 \\ 112 \end{bmatrix},$$

$$\mathbf{x}(15) = \begin{bmatrix} 2 & 4 \\ 0.8 & 0 \end{bmatrix}^3 \begin{bmatrix} 30 \\ 20 \end{bmatrix} = \begin{bmatrix} 2 & 4 \\ 0.8 & 0 \end{bmatrix} \begin{bmatrix} 376 \\ 112 \end{bmatrix} = \begin{bmatrix} 1200 \\ 113100.8 \end{bmatrix}.$$

After 15 years, there are 1200 females aged below 5 years and approximately $113, 101$ aged between 5 and 10 years.

EX 7.6.11 We have now the Leslie matrix

$$L = \begin{bmatrix} 2 & 4 \\ \frac{1}{2} & 0 \end{bmatrix}$$

and the age distribution vectors

$$\mathbf{x}(2) = L\mathbf{x}_0 = L \begin{bmatrix} 30 \\ 10 \end{bmatrix} = \begin{bmatrix} 100 \\ 15 \end{bmatrix},$$

$$\mathbf{x}(4) = L^2\mathbf{x}(0) = \begin{bmatrix} 260 \\ 50 \end{bmatrix}, \ \mathbf{x}(6) = L^3\mathbf{x}(0) = \begin{bmatrix} 720 \\ 130 \end{bmatrix}, \ \mathbf{x}(8) = L^4\mathbf{x}(0) = \begin{bmatrix} 1960 \\ 360 \end{bmatrix}.$$

The dominant eigenvalue is $\lambda_1 = 1 + \sqrt{3}$ and, therefore, the population will increase. Using the approximate formula, we have

$$\mathbf{x}(2) = (1 + \sqrt{3})\mathbf{x}(0) = \begin{bmatrix} 81.9 \\ 27.3 \end{bmatrix},$$

$$\mathbf{x}(4) = (1 + \sqrt{3})^2\mathbf{x}(0) = \begin{bmatrix} 223.9 \\ 74.6 \end{bmatrix}, \qquad \mathbf{x}(6) = (1 + \sqrt{3})^3\mathbf{x}(0) = \begin{bmatrix} 611,8 \\ 203.9 \end{bmatrix},$$

$$\mathbf{x}(8) = (1 + \sqrt{3})^4\mathbf{x}(0) = \begin{bmatrix} 1672,3 \\ 557,1 \end{bmatrix}.$$

EX 7.6.14 We have three age groups. The dominant eigenvalue $\lambda_1 \approx 0,79$ is the approximate proportion of females in two consecutive age groups in the long run. The population decreases, since $\lambda_1 < 1$.

EX 7.6.15 Islands 1 and 3: walk of length 2 is $\{1,2\},\{2,3\}$; walks of length 3 are $(\{1,2\},\{2,4\},\{4,3\})$, $(\{1,2\},\{2,5\},\{5,3\})$; all walks are paths.

EX 7.6.16

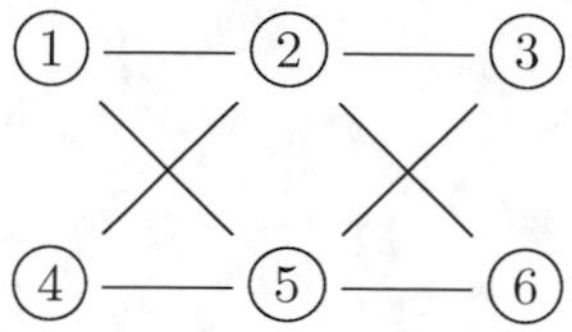

$\deg(1) = 2 = \deg(3) = \deg(4) = \deg(6), \deg(2) = 4 = \deg(5)$.

EX 7.6.17

$$A = \begin{bmatrix} 0 & 1 & 0 & 1 \\ 1 & 0 & 1 & 1 \\ 0 & 1 & 0 & 1 \\ 1 & 1 & 1 & 0 \end{bmatrix}, \qquad A^3 = \begin{bmatrix} 2 & 5 & 2 & 5 \\ 5 & 4 & 5 & 5 \\ 2 & 5 & 2 & 5 \\ 5 & 5 & 5 & 4 \end{bmatrix},$$

All vertices lie in cliques. Cliques having b: $\{a,b,d\}$ and $\{b,c,d\}$. Paths : $(\{b,a\},\{a,d\})$; $(\{b,d\})$; $(\{b,c\},\{c,d\})$.

EX 7.6.18

$$e^A = \begin{bmatrix} -1 & 1 \\ 1 & 5 \end{bmatrix} \begin{bmatrix} -2 & 0 \\ 0 & 4 \end{bmatrix} \begin{bmatrix} -5/6 & 1/6 \\ 1/6 & 1/6 \end{bmatrix} = \begin{bmatrix} -1 & 1 \\ 5 & 3 \end{bmatrix}.$$

EX 7.6.19

$$\begin{aligned} (e^{tA})' &= S\big(\operatorname{diag}(e^{\lambda_1 t}, e^{\lambda_2 t}, \dots, e^{\lambda_n t})\big)' S^{-1} \\ &= S \operatorname{diag}(\lambda_1 e^{\lambda_1 t}, \lambda_2 e^{\lambda_2 t}, \dots, \lambda_n e^{\lambda_n t}) S^{-1} \\ &= S \operatorname{diag}(\lambda_1, \lambda_2, \dots, \lambda_n) \operatorname{diag}(e^{\lambda_1 t}, e^{\lambda_2 t}, \dots, e^{\lambda_n t}) S^{-1} \\ &= S \operatorname{diag}(\lambda_1, \lambda_2, \dots, \lambda_n) S^{-1} S \operatorname{diag}(\lambda_1 e^{\lambda_1 t}, \lambda_2 e^{\lambda_2 t}, \dots, \lambda_n e^{\lambda_n t}) S^{-1} \\ &= A e^{tA}. \end{aligned}$$

EX 7.6.20 $x(t) = -\frac{2}{3}e^{-t} + \frac{5}{3}e^{4t}, \quad y(t) = \frac{2}{3}e^{-t} + \frac{1}{3}e^{4t}$.

EX 7.6.21

$$\mathbf{x}(t) = \begin{bmatrix} e^{2t} \\ 0 \\ 0 \end{bmatrix}.$$

Bibliography

[1] H. Anton and C. Rorres. *Elementary Linear Algebra.* Wiley, 11th edition, 2014.

[2] T.M. Apostol. *Calculus vol. II.* Xerox, 2nd edition, 1969.

[3] M. Artin. *Algebra.* Prentice-Hall, 1991.

[4] S. Axler. *Linear Algebra Done Right.* Springer Verlag, 3rd edition, 2015.

[5] H. Bernadelli. Population waves. *Journal of the Burma Research Society 31 (Part 1)*, pages 1–18, 1941.

[6] P.R. Halmos. *Finite-Dimensional Vector Spaces.* Springer Verlag, 2000.

[7] R.A. Horn and C.R. Johnson. *Matrix Analysis.* Cambridge University Press, 1985.

[8] K. Jänich. *Linear Algebra.* Springer Verlag, 1994.

[9] J. Kemeny and J. Snell. *Finite Markov Chains.* Springer-Verlag, 1976.

[10] D.C. Lay, S.R. Lay, and J.J. McDonald. *Linear Algebra and Its Applications.* Pearson, 5th edition, 2014.

[11] P.H. Leslie. On the use of matrices in certain population mathematics. *Biometrika 33*, pages 183–212, 1945.

[12] C.D. Meyer. *Matrix Analysis and Applied Linear Algebra.* SIAM, 2000.

[13] A.P. Santana and J.F. Queiró. *Introdução à Álgebra Linear.* Gradiva, 4th edition, 2018.

[14] E. Sernesi. *Linear Algebra, A Geometric Approach.* Chapman & Hall, 1993.

[15] P.R. Strang. *Introduction to Linear Algebra.* Wellesley-Cambridge Press, 5th edition, 2016.

[16] T. Yuster. The reduced row echelon form of a matrix is unique: a simple proof. *Mathematics Magazine 57 (2)*, pages 93–94, 1984.

Index

angle, 218

basis
orthogonal basis, 228
orthonormal basis, 228
standard, 99
best approximation, 235, 252
bilinear function, 206
positive definite, 206
symmetric, 206

Cauchy–Schwarz inequality, 210, 216
Cayley–Hamilton Theorem, 152
characteristic equation, 132
characteristic polynomial, 132

determinant, 56
function, 56
Sarrus' rule, 78
distance, 207, 215
distance to a subspace, 235
distance to an affine subspace, 252

eigenspace, 132
eigenvalue, 131
dominant, 270
eigenvector, 131
elementary row operations, 3
Euclidean space, 207, 216

Gauss–Jordan elimination (GJE), 12
Gaussian elimination (GE), 7
generalised eigenspace, 157
generalised eigenvector, 156
order, 156
graph, 272
adjacent vertices, 272
clique, 274
closed path, 273
degree of a vertex, 273
edge, 272
incident edge, 272
path, 273
vertex, 272
walk, 273

hermitian matrix
positive definite, 217
hyperplane, 249

inner product, 205, 215
complex inner product space, 215
real inner product space, 206

Jordan block, 164
degree, 164
Jordan canonical form, 164
Jordan chain, 156
length, 156

least squares, 258
error vector, 258
least squares solution, 258
Leslie matrix model, 266
population waves, 268
linear equation, 5
unknowns, 5
variables, 5
linear space, 82
linear subspace, 85
linear transformation, 176
characteristic polynomial, 200
codomain, 176
composition, 191

diagonalisation, 200
domain, 176
eigenspace, 199
eigenvalue, 199
eigenvector, 199
image, 185
invariant subspace, 202
kernel , 185
null space , 185
nullity, 186
rank, 186
spectrum, 199
linearly dependent set, 91
linearly independent set, 90

Markov chain, 261
regular, 264
matrix, 1
orthogonal, 238
size, 1
transpose, 28
unitary, 243
addition, 17
additive inverse, 18
adjacency, 272
adjugate, 73
anti-diagonal of a matrix, 30
anti-symmetric, 30
canonical row echelon form, 11
change of basis, 120
cofactor, 70
column, 1
column space, 108
column vector, 3
complex matrix, 2
diagonal, 32
diagonal of a matrix, 30
diagonalisable, 141
diagonalising, 141
elementary, 37
entry, 1
exponential, 276, 277
Google, 266
Gram matrix, 213, 217
hermitian, 137
idempotent, 233
identity, 32
inverse, 32, 33
invertible, 33
Leslie matrix, 268
linear combination of columns, 21
lower triangular, 43
Markov, 261
matrix multiplication, 22
minor, 70
multiplication by a column vector, 19
multiplication by a scalar, 18
nilpotent, 150
non-singular, 33
null space, 107
nullity, 108
of cofactors, 73
order, 3
orthogonally diagonalisable, 239
permutation matrix, 66
pivot, 3
positive definite, 213
power of a matrix, 28, 36
probability, 261
rank, 14
real matrix, 2
rectangular, 3
reduced row echelon form, 11, 13
regular, 264
row, 1
row echelon form, 3
row echelon matrix, 3
row space, 108
row vector, 3
similar, 139
singular, 33
spectrum, 131
stochastic, 261
strictly upper triangular, 153
symmetric, 30
trace, 31
transition, 261
unitarily diagonalisable, 243

upper triangular, 43
zero column, 3
zero row, 3
matrix equation, 25
multiplicity
algebraic, 134
geometric, 134

norm, 207, 215

orthogonal projection
on a vector, 230
on a subspace, 230
orthogonal set, 226

parallelogram law, 211, 216
permutation, 66
even, 67
odd, 67
sign, 67
plane, 249
point, 249
Pythagorean theorem, 219

Rank-nullity Theorem, 112
reflection, 176

scalars, 1
Schur's Theorem, 147
sesquilinear function, 215
positive definite, 215
singular value, 245
singular value decomposition, 246
singular values, 247
spectral decomposition, 241, 244
straight line, 249
subspace, 85
A-invariant, 153
affine subspace, 249
cartesian equations, 97
direct sum, 119
finite direct sum, 287
finite sum, 287
parametric equations, 97
sum of subspaces, 117
vector equation, 97
system of linear equations, 5
augmented matrix, 6
back substitution, 8
coefficient matrix, 6
column vector of independent terms, 6
consistent, 5
dependent variables, 10
equivalent systems, 5
forward substitution, 47
free variables, 10
general solution, 5
homogeneous, 5
inconsistent, 5
independent variables, 10
non-homogeneous, 5
particular solution, 10
solution set, 5
trivial solution, 5

theorem
Rank-nullity Theorem, 186, 190
translation, 176

vector, 82
coordinate, 96
linear combination, 87
linearly dependent, 91
linearly independent, 90
orthogonal, 219
probability, 262
span, 87
state, 261
steady-state, 262
vector orthogonal to a set, 220
vector space, 82
addition, 82
additive identity, 82
additive inverse, 82
basis, 93
complex, 82
dimension, 102
finite dimensional, 102
infinite dimensional, 102
ordered basis, 95

point, 82
real, 82
scalar multiplication, 82
spanning set, 89
standard basis, 93, 94

Printed in the United States
by Baker & Taylor Publisher Services